BEYOND SETTLER TIME

*Mark Rifkin*

# BEYOND SETTLER TIME

TEMPORAL SOVEREIGNTY AND INDIGENOUS SELF-DETERMINATION

Duke University Press • Durham and London • 2017

© 2017 DUKE UNIVERSITY PRESS
*All rights reserved*

Designed by Courtney Leigh Baker
Typeset in Garamond Premier Pro and Scala Sans by Westchester Publishing Services

Library of Congress Cataloging-in-Publication Data
Names: Rifkin, Mark, [date] author.
Title: Beyond settler time : temporal sovereignty and indigenous self-determination / Mark Rifkin.
Description: Durham : Duke University Press, 2017. | Includes bibliographical references and index.
Identifiers: LCCN 2016038104 (print)
LCCN 2016038932 (ebook)
ISBN 9780822362852 (hardcover)
ISBN 9780822362975 (pbk.)
ISBN 9780822373421 (ebook)
Subjects: LCSH: Indians of North America— Colonization. | Indians, Treatment of— United States— History. | Time perception. | Geographical perception. | United States— History— Colonial period, ca. 1600–1775.
Classification: LCC E77 .R54 2017 (print) | LCC E77 (ebook) | DDC 970.004/97—dc23 LC record available at https://lccn.loc.gov/2016038104

COVER ART: Andrea Carlson, Blanket Sphincter (detail). Courtesy of the artist and Bockley Gallery.

CONTENTS

Preface • vii
Acknowledgments • xv

ONE. Indigenous Orientations • 1

TWO. The Silence of Ely S. Parker • 49

THREE. The Duration of the Land • 95

FOUR. Ghost Dancing at Century's End • 129

CODA. Deferring Juridical Time • 179

Notes • 193
Bibliography • 241
Index • 269

PREFACE

Native peoples occupy a double bind within dominant settler reckonings of time.¹ Either they are consigned to the past, or they are inserted into a present defined on non-native terms. From this perspective, Native people(s) do not so much exist within the flow of time as erupt from it as an anomaly, one usually understood as emanating from a bygone era. In *Everything You Know about Indians Is Wrong,* Paul Chaat Smith offers a particularly pointed commentary on non-natives' "absolute refusal to deal with [Indians] as just plain folks living in the present and not the past" (18), further noting, "Silence about our own complicated histories supports the colonizers' idea that the only real Indians are full-blooded, from a reservation, speak their language, and practice the religion of their ancestors" (26). Smith suggests that a fuller, less blinkered and amnesiac, version of history that can attend to the complexities of Native lives would counter the stereotypical circulation of images that position Indians as anachronisms. However, he also observes, "History promises to explain why things are and how they came to be this way, and it teases us by suggesting that if only we possessed the secret knowledge, the hidden insight, . . . we could perhaps master the present," adding that "no history is complete without knowing the history of the history" (53). Can a more capacious narrative of history provide a remedy to the appearance of Indians as temporal aberrations? Is "history" itself neutral with respect to the process of dislodging indigeneity from the flow of time? Is "the present"?

Arguing for the importance of a "history of the history" indicates the need to move beyond a broadened version of the same.² While insisting that Natives and non-natives "have a common history" after 1492 (74), Smith also emphasizes that Indigenous people(s) "see things differently. We come from a different place," one specifically shaped by "the land question," which "just won't go away" (85). In these formulations, he captures rather precisely the problem with which I began, namely, the need to assert Indigenous being-in-time but the danger of doing so in ways that take the temporal frames generated in and by settler governance as themselves given—engaging with "complicated histories" whose distinction from those of non-natives is shaped by the ongoing dynamics of "the land question." As Anna Lee Walters asks, "When the real Indians succumbed a century ago, were their unborn grandchildren expected to yield their birthright also? Was the future laid to rest with the ancestors? Were Indians of two hundred years ago more Indian than those a century after them?"[3] If Native peoples are portrayed as always in the process of vanishing and as ceasing to be truly Indigenous if their practices deviate from a (stereotypical) model implicitly pegged to a particular moment in the past, usually the eighteenth or nineteenth century, then the answer seems to be, in Johannes Fabian's well-worn articulation, to insist on their *coevalness*.[4] Consequently, a good deal of scholarship has insisted that Indigenous persons and peoples inhabit the same time as settlers, engage with historical developments and change as a result, are moving toward the future like all other populations and peoples, and can adapt their modes of social life to current circumstances without ceasing to be authentic. However, an emphasis on coevalness tends to bracket the ways that the idea of a shared present is not a neutral designation but is, instead, defined by settler institutions, interests, and imperatives. To the extent that "the land question" means that the impression of the singularity of the space of the nation-state operates as an ongoing colonial imposition that denies Indigenous peoples' histories, sovereignties, and self-determination, why would the concept of inherently shared time be more liberatory or less conducive to settler superintendence? If, in Smith's terms quoted above, Natives "see things differently" due to Indigenous relations to place and peoplehood, would that not affect the meaning, conceptualization, and experience of time?

*Beyond Settler Time: Temporal Sovereignty and Indigenous Self-Determination* demonstrates the need for not just a more expansive or inclusive version of "history" or the "present" but an examination of the principles, procedures, inclinations, and orientations that constitute settler time as a particular way of narrating, conceptualizing, and experiencing temporality. I argue that asserting the shared modernity or presentness of Natives and non-natives implicitly casts Indigenous peoples as inhabiting the current moment and moving toward the future in ways that treat dominant non-native geographies, intellectual and political categories, periodizations, and conceptions of causality as given—as the background against which to register and assess Native being-in-time. In this way I seek to raise questions about the meaning and implications of the pursuit of forms of temporal recognition. Conversely, the book explores the texture of Indigenous temporalities, seeking to theorize and engage the presence of Native experiences of becoming that shift in relation to new circumstances while remaining irreducible to non-native spatial and temporal formations. Examining a range of kinds of sources, including film, government reports, fiction, histories, and autobiography, I explore the potential for conceptualizing and tracing modes of Native time that exceed the terms of non-native mappings and histories. The project is organized as a series of meditations on particular kinds of temporal tensions—ways that Indigenous forms of time push against the imperatives of settler sovereignty.

The book takes inspiration from the role of relativity within physics in challenging the commonsensical conception of time as neutral, universal, and inherently shared. Within post-Einsteinian notions of time, there is no such thing as an absolute time that applies everywhere at once. Instead, the experience and calculation of time are contingent. Simultaneity depends on one's inertial frame of reference, such that two observers who are moving with respect to each other will not agree on when an event occurs or on other aspects of time's passage. If in physics a *frame of reference* refers to relative motion, we also can think about that concept in more socially resonant ways. Such collective frames comprise the effects on one's perception and material experience of patterns of individual and collective memory, the legacies of historical events and dynamics, consistent or recursive forms of inhabitance, and the length and character of the timescales in which current events are situated. Together, these elements of temporal experience provide a background that orients quotidian experiences of time and change, giving shape, direction, and meaning to them. As in the account offered by relativity, there is no inherently privileged or mutual "now" (or sense of time's passage more broadly) shared by disparate frames of reference. Through Indian law and policy, Native peoples have been subjected to profound reorganizations of prior geographies and modes of inhabitance, forms of governance, networks of exchange, tempos of ordinary life, and dynamics of individual maturation in an attempt to reorder Indigenous temporalities, to remake them in ways that fit non-native timescapes of expansion and dispossession. Employing notions of temporal multiplicity opens the potential for conceptualizing Native continuity and change in ways that do not take non-native frames of reference as the self-evident basis for approaching Indigenous forms of persistence, adaptation, and innovation. This book aims, then, to pluralize temporality so as to open possibilities for engaging with Indigenous self-articulations, forms of collective life, and modes of self-determination beyond their incorporation or translation into settler frames of reference. In this way it seeks to open conceptual room for addressing Native collective articulations and experiences of time that exceed non-native accounts—for engaging expressions of temporal sovereignty.

The book focuses on particular kinds of temporal knottings. While proceeding from the mid-nineteenth century through to the late twentieth century, it does not offer a history as such, and each chapter reaches across periods. In place of arguing for temporal recognition, being seen as equally "modern" or part of a shared "present," the chapters gesture toward temporal sovereignty—the need to address the role of time (as narrative, as experience, as immanent materiality of continuity and change) in struggles over Indigenous landedness, governance, and everyday socialities. Chapters 2, 3, and 4 each take up a particular issue that poses problems in seeking to think in Native time, and each one seeks to trace varied impositions of settler time and the ways they foreclose Indigenous temporal frames of reference. After an initial chapter that lays out the project's theoretical and methodological commitments, the chapters move from analysis of the limits of dominant accounts of national time to consideration of everyday negotiations with the temporality imposed by Indian policy, and the final chapter discusses Indigenous people's (and peoples') means of envisioning futurity through connections across time and with nonhuman entities.

The second chapter, "The Silence of Ely S. Parker," addresses the representation of the Civil War in the movie *Lincoln* (2012) as an occasion for considering the marginalization of Native peoples and processes of settler occupation in conventional narratives of national history. The Civil War usually functions as the fundamental demarcation line in periodizing U.S. history, and it often is taken as marking a crucial change in the character of the political union and the nation as a whole. The movie affectively invests in the war—through the defeat of the Confederacy and the end of the institution of chattel slavery—as redeeming the principles of equality and freedom promised in the Revolution. This way of narrating the Civil War and emancipation might be understood as performing a temporality of exception, in which the war functions as a caesura in the evolution of the national union. *Lincoln* suggests how the Civil War's almost ubiquitous role in envisioning national history occludes engagement with histories of Native presence and dispossession—an elision that can be registered in the mute figure of Ely S. Parker, who appears in the film alongside Ulysses S. Grant while never being named. Focusing on the continuities of Indian policy and its aggressions across the period of the war, the chapter addresses the Dakota War of 1862 and Parker's role in Indian affairs. I attend to how figures of exception are mobilized within official and popular discourses in ways that efface Native experiences of time, casting Indigenous resistance to displacement as an inexplicable eruption rather than a response to accreting forms of state-sanctioned invasion. The focus on the Civil War as a signal event within national time gains meaning in the context of the presumption of the necessary persistence of the settler-state, normalizing settler sovereignty as a condition for narrating and experiencing U.S. history. In contrast, the Dakota War and Parker's career highlight alternative renderings of persistence in which the national union continually reemerges through its violent imposition on existing peoples, territorialities, sovereignties, and temporalities. Turning to the writings of Charles Alexander Eastman, the chapter addresses how the violence of the Indian Wars and of the treaty system becomes part of the self-understanding of the next generation. Eastman's texts offer an account of nineteenth-century history and its legacies in which the coordinates, trajectories, and implications center on the continuing possibilities for Native life and peoplehood (including as national subjects) amid ongoing occupation, in ways quite disjunct from conventional fixation on the supposed epochal shift brought by the Civil War.

The third chapter, "The Duration of the Land," considers the difficulty of negotiating between an allotment-imposed framework and extant Osage modes of becoming. Allotment sought to inculcate particular kinds of temporal consciousness and practice, in an attempt to "civilize" Natives into normative non-native life cycles in ways that would reaffirm the coherence and dominance of U.S. jurisdiction. Under this policy the federal government worked to reorganize everyday Native activity at all levels, from homemaking to work, education, and land use, aiming to reorder the social landscape of Indigenous territories. However, even while subject to these forms of compulsion, Native people continued to experience such changes from the perspective of their own temporal formations, shaped by their ongoing occupancy in their homelands. In the novel *Sundown*, John Joseph Mathews offers an account of the everyday affects generated by inhabiting allotment's field of force and its temporal inscriptions while also having a frame of reference shaped by Osage forms of sociospatiality. Critiquing the violence of allotment, the novel traces how it pressures Osages to conform to a vision of futurity defined by the state's extension of authority over Native peoples and lands. Reciprocally, Mathews explores how Osage histories (including the timescale of inhabitance in, and rhythms of relation to, that place) influence ordinary perception in ways that exceed the imaginings of Indian policy, while also indicating how such duration remains open to change on its own terms (including the emergence of the I'n-Lon-Schka and the Peyote religion). These experiences of time provide a background for the characters' sensations in ways that make them irreducible to a "now" shared with non-natives. Moreover, the text often marks the lived incommensurability of these temporal formations through figures of queerness. Mathews's repeated invocation of the term *queer* alludes to the linkage within sexological and popular discourses of people of color with perversity due to their supposedly less advanced forms of family formation and polymorphous desire. *Sundown* plays on this set of associations to suggest how the main character's inability to fit in, including his supposed failure to be properly heterofamilially directed, might open onto a larger set of questions about how the imposition of settler governance becomes naturalized by presenting its rearrangement of ordinary life as merely expressive of the normal temporality of procreation. Conversely, such associations illustrate how Indigenous modes of history and placemaking are dismissed by being coded as an endemic, racially transmitted incapacity for civilization. In narrating the main character's sensation of disorientation with respect to events unfolding around him, the novel suggests that his feeling of queerness within the social formations created by allotment indicates less an Indian inability to adapt—to give up the deviant fixation on the past—than continuing and evolving Osage experiences of time that emerge out of enduring connections to their homeland.

The final chapter, "Ghost Dancing at Century's End," turns to the question of futurity, specifically the role of prophecy in Indigenous temporal formations. The most well-known example of this phenomenon is the Ghost Dance, which usually refers to the late nineteenth-century movement engendered by the visions of a Northern Paiute man named Wovoka. Inasmuch as the Ghost Dance has been cast as a response to the deprivations caused by Indian policy, culminating in the Wounded Knee massacre (1890), it circulates as the sign of the end of an era, in which the closing of the frontier indicates the becoming past of Native sovereignties that are not directly superintended (or overridden entirely) by settler claims and governance. For this very reason, the memory of the Ghost Dance serves as a powerful entry point for considering the work of prophecy, in its challenge to settler narratives of the historical inevitability of Indian subordination and disappearance. Novels at the end of the twentieth century take up the Ghost Dance and its continuing influence, highlighting the capacity of prophecy to disorient non-native conceptions of realness with respect to time and Native peoples' place in it. Sherman Alexie's *Indian Killer* and Leslie Marmon Silko's *Gardens in the Dunes* mark and refuse the ways that Native histories continually are translated as tales of loss (of authenticity, of proper bloodedness, of connection to "tradition") within dominant settler conceptions of time. More than highlighting the vitality of Indigenous presentness, these novels offer accounts of the Ghost Dance that unfold the power of vision and spirit in connecting quotidian experiences of time to ongoing formations of being and becoming by Indigenous peoples in their homelands (the scope of which is broadly framed). Rather than being a remarkable occurrence, prophecy in the texts emerges in response to everyday forms of relationship and struggle. Its occurrence indicates less a rupture in time than the ways other-than-chronological forms of experience remain immanent within daily life, and the texts suggest how such ordinary sensations reciprocally give rise to prophecy and are intensified by it, with commonplace events and dynamics creating the conditions for action by entities that likely would be characterized by non-natives as supernatural. Rejecting reproductively inflected narratives of inheritance or declension, Alexie's and Silko's texts elaborate the intimacy of modes of prophetic reach across time, emphasizing the possibilities for self-determination and Indigenous duration that arise in being out of sync with settler time.

ACKNOWLEDGMENTS

This book began while I was watching the movie *Lincoln*. I found myself wondering about the mute appearance of Ely S. Parker and what it meant for thinking about how Native peoples fit within the story of the Civil War as an epochal break in U.S. history. Those musings became the basis for the earlier essay version of chapter 2, which then became the kernel for the project as a whole. I began wondering how dominant ways of narrating national history displaced, effaced, and foreclosed engagements not only with Native histories but with the continuities of settler colonial policy and its violences. For this initial thought experiment, I owe a debt to Beth Piatote and Jason Cooke. In a comment on a draft of my previous book *Settler Common Sense* (2014), Beth asked me to consider more fully the character and significance of Native temporalities, and Jason's dissertation on the regionally specific representation of Native peoples in the removal era as subjects/objects of settler-oriented histories suggested the importance of conceptualizing non-native experiences of history as themselves key forces in shaping Indian policy and interactions with Native peoples. To both of them, I am deeply grateful for explicitly or implicitly pushing me toward more substantive reflection on the relation between time, indigeneity, and settler colonialism. Fairly quickly, my initial interest in periodization turned phenomenological. I began thinking less about the narration of time than about differences in how it is lived, and for this particular way of turning toward affect, I would like to thank Zach Laminack, whose work on feeling prompted my own phenomenological reflections.

I am deeply appreciative of those who have offered feedback on this project over the years. Thanks to Mishuana Goeman, Jean Dennison, Lisa Tatonetti, Coll Thrush, Joe Genetin-Pilawa, and Beth Freeman for their incredibly generous readings and immensely helpful comments. Parts of the project were presented at the University of California, Berkeley; the University of Pennsylvania; the University of North Carolina at Charlotte; the Dartmouth American Studies Institute; Wake Forest University; and the Native American and Indigenous Studies Association. Thanks so much to everyone who invited me and to all the people who attended for their wonderful questions. Parts of the project were published previously as articles, and while they have been significantly revised, I would like to thank the presses for their consent to have the pieces included here. Parts of chapter 2 were published as "The Silence of Ely S. Parker: The Emancipation Sublime and the Limits of Settler Memory" in *NAIS* 1, no. 2 (2014), and parts of chapter 3 were published as "The Duration of the Land: The Queerness of Spacetime in *Sundown*" in *Studies in American Indian Literatures* 27, no. 1 (2015). I also would like to thank the journals' anonymous readers for their comments. I have benefited immeasurably from ongoing conversations with a range of scholars, including Neel Ahuja, Joanne Barker, Nancy Bentley, Kevin Bruyneel, Jodi Byrd, Jessica Cattelino, David Chang, Eric Cheyfitz, Pete Coviello, Jennifer Denetdale, Jean Dennison, Emilio del Valle Escalante, Beth Freeman, Mishuana Goeman, Alyosha Goldstein, Sandy Grande, Lisa Kahaleole Hall, Shona Jackson, Malinda Maynor Lowery, Scott Morgensen, Dana Nelson, Robert Nichols, Josie Saldaña-Portillo, Bethany Schneider, Audra Simpson, Kyla Wazana Tompkins, and Priscilla Wald.

The University of North Carolina at Greensboro has been my home base for the past nine years, and my colleagues there continue to give me space and support in just the right balance. In particular, I would to thank Risa Applegarth, Danielle Bouchard, Sarah Cervenak, Michelle Dowd, Asa Eger, Jen Feather, Tara Green, Sarah Hamrick, Ellen Haskel, Karen Kilcup, Derek Krueger, Isabell Moore, Christian Moraru, Noelle Morrisette, Gene Rogers, Scott Romine, María Sánchez, Amy Vines, and Karen Weyler.

Thanks so much to my editor, Courtney Berger. I've been thrilled to have the chance to work with you.

Beyond the walls of my study lies a whole other world of life. For reminding me of that fact, I'm deeply grateful to Sheila and Alex Avelin, Zivia Avelin, Jon Dichter, Kevin and Justin Dichter, Mike Hardin, Lisa Dilorio Smith, Tiffany Eatman Allen and Will Allen, Alicia and Bobby Murray, Debbie and Andy Johnson, Craig Bruns, Keith Brand, Kent Latimer, JJ McArdle, and Ali Cohen. My debts to Erika Lin know no bounds (and defy easy description). My parents, Neal and Sharon Rifkin, and sister, Gail Dichter, remain the absolute constants in my life, and having them there—marking the continuities and changes across my past, present, and future—makes all the difference.

Finally, the flow of my life changed after meeting Rich Murray. The passing of time feels different now, and I'm deeply grateful to dwell with him in the movement of that tide.

ONE. **INDIGENOUS ORIENTATIONS**

For things to be simultaneous, they must be situated within a single frame of reference, in the sense that there is not an absolute time against which all events can be measured.[1] With respect to the contemporaneity of non-natives and Indigenous peoples, the frame for thinking their synchronicity usually is provided by settler discourses, structures, and perceptions. More than offering invidious portraits of Indians as backward and disappearing, non-native accounts, governmental and popular, treat the space of the United States as a given in which to set the unfolding of events, and in this way the political union functions as something of an atemporal container for the occurrences, movements, conjunctures, periodicities, and pulsations of history, providing the background against which the movement of time can be registered. Native activists and intellectuals have argued against the idea of inclusion within the United States, understanding that gesture as an erasure of the specificity of Indigenous geopolitical claims, rights to self-determination, autochthonous existence as polities distinct from the settler state, and, perhaps most pointedly, the ways the colonial violence of settler rule has worked through forced incorporation of Indigenous peoples into the "domestic" space of the nation. Yet the insistence that Native people(s) occupy a singular present with non-natives and that the notion of being-in-time or the potential for change remain contingent on belonging to that shared, unified "now" (which includes a shared "then" of the past) seems to eerily resemble the representation of Indigenous populations and territories as necessarily part of the United States. Asserting Indigenous people's and peoples' presence in the present, as opposed to casting them as anachronisms, does not necessarily redress the violence perpetrated through the organization of history around the coordinates of settler occupation—the treatment of non-native temporalities as the baseline for marking Native being-in-time.

Rather than approaching time as an abstract, homogeneous measure of universal movement along a singular axis, we can think of it as plural, less as a temporality than *temporalities*. From this perspective, there is no singular unfolding of time, but, instead, varied temporal formations that have their own rhythms—patterns of consistency and transformation that emerge immanently out of the multifaceted and shifting sets of relationships that constitute those formations and out of the interactions among those formations. As V. F. Cordova observes, "time is an abstraction derived from the fact that there is motion and change in the world."[2] U.S. settler colonialism produces its own temporal formation, with its own particular ways of apprehending time, and the state's policies, mappings, and imperatives generate the frame of reference (such as plotting events with respect to their place in national history and seeing change in terms of forms of American progress). More than just affecting ideologies or discourses of time, that network of institutionalized authority over "domestic" territory also powerfully shapes the possibilities for interaction, development, and regularity within it. Such imposition can be understood as the denial of Indigenous *temporal sovereignty*, in the sense that one vision or way of experiencing time is cast as the only temporal formation—as the baseline for the unfolding of time itself. However, such compulsory interpellation of Natives into U.S. life is never fully accomplished nor fully able to displace Indigenous temporal orientations.

To speak of temporal *orientation* suggests the ways that time can be regarded less as a container that holds events than as potentially divergent processes of becoming. Being temporally oriented suggests that one's experiences, sensations, and possibilities for action are shaped by the existing inclinations, itineraries, and networks in which one is immersed, turning toward some things and away from others. More than a question of relations in space, orientation involves reiterated and nonconscious tendencies, suggesting ways of inhabiting time that shape how the past moves toward the present and future. In *Queer Phenomenology* Sara Ahmed asks, "What does it mean to be oriented? How do we begin to know or to feel where we are, or even where we are going, by lining ourselves up with the features of the grounds we inhabit, the sky that surrounds us, or the imaginary lines that cut through maps?"; she observes, "The direction we take excludes things for us, before we even get there," adding, "Depending on which way one turns, different worlds might even come into view. If such turns are repeated over time, then bodies acquire the very shape of such direction."[3] Being oriented, having a feeling of place and self in relation to other places and selves as well as a feeling of where one is going and the pace at which one is heading there, entails moving in particular directions in line with extant patterns.

This persistent (and largely unwilled) regeneration of continuity not only happens "in time" but is the substance, feel, and force of time unfolding.[4] If one's perception of the world might be quite different depending on where one turns, we might understand the paths traced out by one's orientations—following those particular paths in those specific ways—as giving rise to a kind of temporality, qualitatively distinguishable from other experiences of time. We further might understand collective modes of orientation as a temporal formation that has its own frame of reference and processes of becoming.

Native peoples remain oriented in relation to collective experiences of peoplehood, to particular territories (whether or not such places are legally recognized as reservations or given official trust status), to the ongoing histories of their inhabitance in those spaces, and to histories of displacement from them. Such orientations open up "different worlds" than those at play in dominant settler orderings, articulations, and reckonings of time.[5] Developing such notions of temporal orientation and multiplicity opens the potential for conceptualizing Native continuity and change in ways that move beyond the modern/traditional binary; that do not take non-native frameworks as the self-evident basis for approaching Indigenous forms of persistence, adaptation, and innovation; and that enable consideration of temporal sovereignty, how sensations and articulations of time take part in Indigenous peoples' operation as polities and their pursuit of self-determination. As Deborah Miranda observes with respect to the history of her people (the Esselen), "Story, like culture, is constantly moving. It is a river where no gallon of water is the same gallon it was one second ago. Yet it is still the same river. . . . Even if the whole is in constant change. In fact, *because* of that constant change."[6] What does it mean to consider Native temporalities as having their own flow—as coherent yet changing, affected by other flows but not the same as them? In this way *Beyond Settler Time* explores how Native peoples' varied experiences of duration can remain nonidentical with respect to the dynamics of settler temporal formations, indicating ways of being-in-time that are not reducible to participation in a singular, given time—a unitary flow—largely contoured by non-native patterns and priorities.

Rather than marking an absolute distinction between Natives and nonnatives, suggesting that there are unbreachable barriers that generate utterly incommensurable and hermetically sealed Indian and white forms of experience, I am suggesting the presence of discrepant temporalities that can be understood as affecting each other, as all open to change, and yet as not equivalent or mergeable into a neutral, common frame—call it time, modernity, history, or the present. The aim is not to search for an authentic Indigenous conception of time as against degrading forms of settler influence. The effects of non-native expropriation, superintendence, and exploitation can be understood as intimate parts of Native experiences of time and becoming, as contributing to Indigenous orientations but in ways that exceed the paths of development envisioned by such interventions and invasions. Instead of juxtaposing the past and the present in order to preserve the former from the ravages of the latter, I am suggesting the importance of attending to Native conceptualizations, articulations, and impressions of time that do not easily fit within a framework explicitly or implicitly oriented around settler needs, claims, and norms—a pluralization of time that facilitates Indigenous peoples' expressions of self-determination.

In this way I seek to offer ways of marking Native peoples' translation into an account of time already oriented around settlement. My focus on the force, effects, and limits of temporal inclusion is not in the interest of authenticating certain ways of becoming as truly Native, or of invalidating others as a version of false consciousness.[7] Instead, my aim lies in trying to open greater conceptual and discursive room for addressing time in ways that avoid the following: falling back into the dichotomization of tradition and modernization, mandating that Native modes of being-in-time be understood as inherently occupying an experience of the present shared with non-natives, implicitly distinguishing between beliefs about time and its supposed universal facts, and insisting on the adoption of settler modes of time as the real in order to engage with Euro-American historicism(s).[8] There is an inherently speculative quality to what I'm doing. The position I am taking up is negative dialectical and offered in solidarity.[9] Dale Turner observes, "The project of unpacking and laying bare the meaning and effects of colonialism will open up the physical and intellectual space for Aboriginal voices."[10] *Beyond Settler Time* is such an effort "of unpacking and laying bare" from the perspective of a non-native, highlighting the violence of extant forms of temporal recognition (and their de facto modes of translation). The critical question, then, is, Does this critical orientation open useful intellectual, imaginative, and/or affective potentials? The materials I work with in this study are intended to be generative for exploring the interpretive possibilities of this mode of analysis—investigating what intellectual and political possibilities are opened through this way of approaching the question of time. In this sense I'm offering less an explanation than a hermeneutic, one that emerges out of a careful and ongoing engagement with Native texts and Native scholars and that hopefully can contribute to the pursuit of Indigenous self-determination by proposing additional conceptual tools for marking the force, effects, and endurance of settlement.[11] In this vein my insistence on the potential distinction between Native and non-native experiences of time may be understood as aiming to facilitate possibilities for temporal sovereignty.

*Modernity/ies, Temporal Recognition, and the Limits of "Now"*

What does it mean to be recognized as existing in time? The representation of Native peoples as either having disappeared or being remnants on the verge of vanishing constitutes one of the principal means of effacing Indigenous sovereignties. Such a portrayal of Indigenous temporal stasis or absence erases extant forms of occupancy, governance, and opposition to settler encroachments. Moreover, it generates a prism through which any evidence of such survival will be interpreted as either vestigial (and thus on the way to imminent extinction) or hopelessly contaminated (as having lost—or quickly losing—the qualities understood as defining something, someone, or some space as properly "Indian" in the first place). These kinds of elisions and anachronizations can be understood as a profound denial of Native being. They perform a routine and almost ubiquitous excision of Indigenous persons and peoples from the flux of contemporary life, such that they cannot be understood as participants in current events, as stakeholders in decision making, and as political and more broadly social agents with whom non-natives must engage. This making of Indians into ghostly remainders enacts what Kevin Bruyneel has referred to as "colonial time," in which "temporal boundaries" are constructed between "an 'advancing' people and a 'static' people, locating the latter out of time," and, within this dynamic, "increasingly . . . tribal sovereignty [appears] as a political expression that is out of (another) time, and therefore a threat to contemporary American political life and political space."[12] The temporal trick whereby Indians are edited out of the current moment—or cast as inherently anachronistic—emerges out of the refusal to accept the (geo)political implications of persistent Indigenous becoming, the ways that the presentness of Native peoples challenges settler claims to possession now and for the future. As Jean O'Brien observes in *Firsting and Lasting*, her study of nineteenth-century town histories in New England, "non-Indians refused to regard culture change as normative for Indian peoples"; "Indians, then, can never be modern."[13] However, is acknowledgment of Native timeliness the same as according Indigenous peoples status as modern? In what ways is conceptualizing Native being-in-time as the inhabiting of *modernity* (or a shared present with non-natives) equivalent to a bid for inclusion within settler modes of recognition? How might an implicit imperative to become temporally intelligible to non-natives limit possibilities for envisioning other Indigenous experiences of time and expressions of temporal sovereignty?

The pursuit of recognition by the settler state often results in a translation of Indigenous histories, modes of collectivity, and relations to place into forms that better fit extant legal and administrative frames. Official non-native discourses themselves employ temporal narratives that produce limited visions of Native collective selfhood. In *Native Acts*, Joanne Barker argues, "Native traditions have been fixed in an authentic past and then used as the measure of a cultural-as-racial authenticity in the present," adding that "through the discourses of recognition, U.S. national narrations represent recognition as an expected outcome of Native cultural authenticity." She later observes, "If Native peoples are to secure the recognition and protection of their legal status and rights as defined therein, they must be able to demonstrate their *aboriginality*—as pursuit, as essence, as a truth that transcends," and this standard "makes it impossible for Native peoples to narrate the historical and social complexities of cultural exchange, change, and transformation—to claim cultures and identities that are conflicted, messy, uneven, modern, technological, mixed."[14] To be authentic means to preserve forms of tradition that emanate from the past in pristine ways; that performance of stasis is the condition of possibility for being accorded status as proper Indians. Such enactments of aboriginality explicitly and implicitly serve as the basis for (grudging, partial, and circumscribed) governmental acknowledgment of Native sovereignty. From this perspective, being recognized as Indian means staging a version of pastness that disavows the "complexities" of Native life, including "the historical realities of accident, succession, alienation, passion, personal conflict, dissension, and disparity."[15] Miranda wonders, "Those who will not change do not survive; but who are we, when we have survived?" and if as part of that process of survival, as she says, "my tribe must reinvent ourselves—rather than try to copy what isn't there in the first place"—that very process of reinvention in relation to changing circumstances can become the basis for declaring that a people has ceased to exist as such.[16]

If Indigenous peoples are called on to embody an older and purer version of themselves (and understood as actually descended from groups identified as Native only when they do so), the alternative to such time warping seems to lie in a turn to history. Yet what is the relation of history—the narration of the connections among the past, present, and future—to settler institutionalities and imperatives?[17] Viewing Natives as being *historical*, in the sense of acknowledging Native existence in and change over time, includes addressing the effects of settler colonialism on Native lifeways, choices, and modes of collective self-expression and organization. That awareness of how Native people(s) are affected by the passage of time—or, more precisely, of the operation of Native processes of becoming that are animated by the multifaceted and shifting social, political, and environmental networks in which they are enmeshed—often is portrayed as participation in a singular history alongside non-natives. In *Indians in Unexpected Places*, Philip Deloria sets out "to consider the kinds of frames that have been placed around a shared past," and later, noting the similarity of patterns of life and material culture among Indians and rural whites in the early twentieth century, he observes that "otherwise critical Indian agents . . . , when pressed, sometimes confessed that Indians and non-Indians were experiencing the world together."[18] This insistence on synchrony, interaction, and co-implication in unfolding events works against the denial of Native persistence as well as the attempt to freeze Indigenous persons and peoples into a simulacrum of pastness, a fantasized construction of Indian realness cast as immanently tied to a bygone era.

However, in countering anachronization, this approach generates a different set of temporal difficulties. Deloria argues "that some Indian people—more than we've been led to believe—leapt quickly into modernity," adding, "They leapt, I think, because it became painfully clear that they were not distinct from the history that was even then being made. Whether they liked it or not, other people were building a world around, on top of, and through Native American people. That world took as its material base the accumulation of capital ripped from indigenous lands, resources, and labor over the course of centuries."[19] The sharedness of Native and non-native coexistence and influence on each other appears here as mutual participation in modernity, but given that some Native people "leapt" into it, modernity also indicates something that exists separately from the temporal experience of Natives prior to that point. The shift from that earlier experience of time to modernity is explained through Native subjection to enduring kinds of expropriation and exploitation of their homelands, communities, and bodies. The resulting "history," then, clearly involves Indigenous people(s) but arrives as a "painful" and violent disruption whose propulsive force arises from the "other people" who "were building a world" around and on top of them, primarily against their will (or at least without their meaningful consent). Characterizing such an unfolding as "shared" seems to emphasize the facticity and importance of Native presence while still putting it in relation to settler-driven change, noting the important and varied role of Natives within a story that is still ultimately oriented around non-native transformations. What would an account of Indigenous experiences of time on their own terms look like, one that also suggests the profound effects of the forms of invasion, seizure, and occupation to which Deloria points? Put another way, can the complex choices made by the people Deloria addresses be understood in ways other than as a "leap" into some other kind of time called modernity, and what possibilities for envisioning temporality (and its relation to self-determination) might such a shift enable?

Characterizing Native and non-native experiences, trajectories, and orientations as all occurring within a singular shared modernity (or history) engages in important intellectual work: insisting on Native survival and significance, refusing the idea that people can be assessed against frozen images of tradition, and highlighting the role of settler colonialism in shaping non-native lives (in ways spectacular and quotidian). However, positing such temporal sharedness implicitly affirms a kind of recognition that merges Native people(s) into a conception of the present whose contours emerge from the ongoing assault on Indigenous sovereignties. What precisely does (entry into) modernity entail? Deloria insists, "Lives lived around liberating travel and cosmopolitan sophistication mattered. So did engagement with technology—not just cars, but sewing machines, merry-go-rounds, telephones, and film cameras. All these things pointed to the ways in which Indian people created modernity in dialogue with others." Movement outside the boundaries of reservations (including transoceanic journeys), alterations in material culture, and participation in new industries and forms of exchange are collated here as participation in *modernity*, which then can be characterized as something that Natives cocreated. Deloria further suggests, "The members of this Native cohort [in the late nineteenth and early twentieth centuries] were not the first Indian people to engage American popular culture, to be sure. They were the first to do so, however, in its particularly modern form, at the very moment when those forms were developing."[20] The notion of the modern here suggests a certain way of inhabiting and experiencing time, one that is not reducible to *engaging* with forms of non-native cultural production and commerce per se or to adopting, appropriating, or adapting particular once-alien practices, patterns, objects, or beliefs.

Instead, being modern, or inhabiting modernity as a shared experience, involves a qualitative shift from something that came before. Deloria observes that "the final moment, of conquest, pacification, and incorporation of Indian people, then, might also be seen to represent one of the many critical instants in which the United States became aware of its own modernity."[21] The emergence and recognition of modernity as a specific sort of temporal experience appears intimately connected to the decimation of Native peoples, but more than simply providing a period marker (with *modern* serving as a name for what comes in the wake of allotment, for example), the use of modernity as a means of describing and understanding forms of presentness in which both Natives and non-natives were enmeshed (a "world" inhabited "together") seems to be shaped by forms of settler extension and extraction that are taken as fundamentally altering the conditions of being-in-time for Native peoples. Co-participation in chronologically unfolding time, then, is not distinct from immersion in the modern as a particular epoch defined by specific kinds of shifts, including in the experience of time itself.[22] O'Brien suggests that "non-Indians actively produced their own modernity by denying modernity to Indians," raising questions about the capacity of modernity—or any way of marking and relating time oriented around settler interests—to express Native modes of temporality.[23] If "Indian people created modernity in dialogue with others," as Deloria says, such contribution seems to involve *leaping* into a world whose condition of possibility lies in "conquest."

To be clear, my questions are not about whether to emphasize the extent of ongoing settler violence, to highlight its intensive continuing effects on Native peoples, or to explore the significance of settlement for all aspects of non-native life. Rather, my inquiries tend toward the following: how does conceptualizing time as itself a mutual experience for those initiating and subjected to such violence make temporality into an extension of settler colonialism? What happens to the possibilities for conceptualizing Indigenous sovereignty and self-determination when they, a priori, are understood as occurring within a singular temporal formation oriented by settler coordinates?[24] How might we see, in Veena Das's terms, "the signature of the state" at play in these ways of marking time?[25] The notion of a shared past and present depends here on joint participation within a period whose character is defined by non-native actions and frameworks. How does settler time—notions, narratives, and experiences of temporality that de facto normalize non-native presence, influence, and occupation—come to serve as the background for articulating and recognizing Native being-in-time? How might such temporal incorporation be understood as part of the dynamics of settler colonialism, constraining and effacing other ways of apprehending Indigenous temporality and processes of becoming? Describing the famous image of the Chiricahua Apache warrior and leader Geronimo riding in a car, Deloria argues, "A powerful and important cultural vitality coheres around the figure of Geronimo in an automobile. It insists on the autonomy of Native individuals, cultures, and societies, and it demands recognition that perhaps your modernity is not distinct from—or better than—mine."[26] While Deloria indicates that Natives may experience modernity in ways that sustain their "autonomy," what remains unclear here is what makes the act of riding in an automobile an index of participation in modernity in the first place. What other kinds of temporal orientations—other sorts of "vitality," ways of being-in-time, and relations to Chiricahua pasts and futures—might that act have, without it being construed as part of an encompassing synchronous formation called *modernity*?

Such a de facto unification can implicitly establish Euramerican frameworks as the standard against which to assess Native deviations. In the introduction to his collection *Alternative Modernities*, Dilip Parameshwar Gaonkar illustrates this dynamic. He states, "Born in and of the West some centuries ago under relatively specific sociohistorical conditions, modernity is now everywhere," and its qualities entail the creation and extension of capitalism, the self-reflexive rejection of tradition as such in potentially freeing ways, and a struggle around forms of mechanized standardization. "Non-Western people," then, have "hybrid modernities," which have arrived as "modernity has travelled from the West to the rest of the world," and such hybridities enable one to "think with a difference" about modernity writ large.[27] The global movement of a formation called modernity whose qualities and emergence are tied to conditions in "the West" can somehow encompass presents elsewhere in ways that bear little relation to the rhythms, trajectories, and momentum of time that preceded the modern in the spaces in which it arrives. Or such dynamics appear as forms of *hybridity* or *difference*, epiphenomenal variations in the face of a presumed temporal linkage within modernity. Saying that modernity arose out of confrontations between *the West* and *the not-West* (however these terms are mapped) does not obviate the problems with presuming participation in a common temporal formation, in which the dominant coordinates of Euro-American sociality and governance still provide the basis through which to register processes of becoming. This problematic arises in a number of prominent efforts to try to map the violences of what is envisioned as the contemporary world system. For example, Walter Mignolo emphasizes the centrality of the conquest of the Americas to the emergence of the modern world, and while he continually foregrounds the ability of Europe and the United States to extend and impose the terms of their "local histories" on a planetary scale, these various histories remain part of an encompassing formation—a modernity defined by "coloniality" and the production of "colonial difference"—in which "border epistemologies emerg[e] from the wounds of colonial histories, memories, and experiences."[28] Similarly, Sylvia Wynter speaks of "our present single world order and single world history," and she suggests that the forms of global connection that have proliferated in the wake of the Columbian encounter can enable a new revolution in human global consciousness that transcends the

European-derived racial hierarchy, which itself has thwarted a conception of collective human identity and history as such.²⁹ Addressing the dynamics of forced incorporation into Euro-American social systems, José Rabasa observes that "nonmodern subjects might actually learn the ways of European thinking without necessarily abandoning their capacity to dwell in their own worlds," later adding, "Revisionist histories produce narratives that assess the contributions of non-Western peoples to modernity without giving much thought to the paradoxical integration of the onetime nonhistorical peoples into what ends up constituting by definition a universal single history."³⁰

To the extent that the existence of the U.S. nation-state and its jurisdictional authority over Native peoples provides a constant for forms of temporal reckoning, including a "shared" role by Native people in national history, it serves as the background through which understandings and experiences of the unfolding of time gain orientation. Ahmed notes, "We can think . . . of the background not simply in terms of what is around what we face . . . , but as produced by acts of relegation: some things are relegated to the background in order *to sustain* a certain direction." More than referencing that which has been consigned to the role of set piece as opposed to active agent, the background indicates what is held constant in order to perceive movement, including the passage of time. It serves less as an inert setting than as the condition of possibility for registering action, change, survival. Ahmed further suggests that "the figure 'figures' insofar as the background both is and is not in view" and that "a background is what explains the conditions of emergence or an arrival of something as the thing that it appears to be in the present."³¹ Absent a background, nothing can figure in or as the foreground and be available for attention, perception, or acknowledgment. If the coherence of the settler state and its presumptive absorption of Native peoples serve as the implicit structuring frame through which to approach and understand temporality on lands claimed by the United States, both the sharedness and the direction of unfolding events will be experienced as consonant with that geopolitical imaginary. All those subject to the state's jurisdiction in domestic space will appear as occupying a common time. Exploring what constitutes the background for marking and experiencing time, then, draws attention not only to the milieu, at whatever scale, that serves as the context for thinking and feeling time's unfolding but also to the taken-for-granted processes through which temporal dynamics are figured, including the following: the timeframe thought relevant to address events (especially the editing out of generations, if not centuries or millennia, of inhabitance in a given place or region); the kinds of causal explanations offered, as well as conceptions of who or what constitutes an actor in the making of history (including discounting the presence and effectivity of ancestors, nonhuman entities, aspects of the landscape, collective stories, and ceremonies); the coordinates one uses for conceptualizing relevance (like measuring forms of continuity and change against phenomena that ostensibly illustrate national subjects, in Deloria's terms quoted earlier, "experiencing the world together"—for example, events taken as of national significance, like the Civil War); and assumptions about what can be held constant (such as national jurisdiction) and, thus, how to conceptualize the potentials for change.

While not denying numerous and ongoing forms of interaction among Natives and non-natives and the profound influence of settler governance on the shape of those relationships, I want to trouble the idea that asserting the existence of a singular present into which Indigenous peoples are always already incorporated serves as a means of breaking the hold of colonial influence by recognizing Native agency and contemporaneity. The positing of inherently mutual participation in the unfolding of time—itself imagined de facto as a line reaching from the past toward the future—contributes to the adoption of a standard model of development in which non-Euro-American conceptions and experiences of time appear as deviations that are transitioning toward a dominant framework. As Dipesh Chakrabarty notes in his critique of Euro-American historicism, "This transition is also a process of translation of diverse life-worlds and conceptual horizons about being human into the categories of Enlightenment thought," a process in which "the overriding (if often implicit) themes are those of development, modernization, and capitalism."[32] In other words, when Euro-American temporal formations provide the background for conceptualizing time itself, "diverse life-worlds" are implicitly translated into the normative frame of those formations, limiting possibilities for (Indigenous) self-determination by presuming the necessity of transitioning to particular forms of self-organization, narration, and governance.

In *X-Marks* Scott Richard Lyons offers a powerful account of Native being-in-time that illustrates this problem. For him, the *x-mark*, which literally refers to a treaty signature, "symbolize[s] Native assent to things (concepts, policies, technologies, ideas) that, while not necessarily traditional in origin, can sometimes turn out all right and occasionally even good." Repudiating cultural purity as a goal (and referring to those who seek it, and who police others in its name, as "culture cops"), he embraces what he characterizes as contamination and hybridity, seeking to move toward a reckoning with the complexities and diversities of contemporary Native identity, practice, collectivity, and self-articulation. However, his refusal of stasis and unanimity is accompanied by an epochal account of the onset of the modern that performs the kind of transition narrative to which Chakrabarty refers. Lyons argues, "X-marks are made in a different kind of Indian time that must be characterized in some potentially problematic ways. First, I distinguish between traditional and modern time, clocking the supplanting of the former by the latter at around 1492, or really when the treaties were made," further noting that "the original x-marks were pledges to adopt new ways of living that, looking backwards, seem more accurately described as modern." The modern constitutes a new "kind" of time, one that appears to alter the dynamics of change itself. Certain "ways of living" count as modern and, as such, are inherently disjunct from what transpired previously—in "traditional" time. Lyons later notes that "indigenous people have the right to move in modern time" and that "our ancestors promised that their descendants would be part of the modern world."[33] To live "in modern time" is to be on the other side of the break, in a time and "world" shared with everyone else, and in this way *modern* functions less as simply descriptive (later in chronological time) than as normative, a right to inclusion in a certain kind of shared time. Being in the present, changing over time, being in a universally common time, and having specifically modern "ways of living" become fused with each other, and the processes and legacies of settler coercion provide the background that orients this unity.[34]

The price for Indigenous peoples of such forms of temporal recognition is being enfolded into frames not of their making that can normalize non-native presence, privilege, and power. In *Liberalism and Empire* Uday Singh Mehta says of English notions of universal personhood that, in theory, also could apply to non-European populations, "Behind the capacities ascribed to all human beings exists a thicker set of social credentials that constitute the real bases of political inclusion," adding, "They draw on and encourage conceptions of human beings that are far from abstract and universal, and in which the anthropological minimum is buried under a thick set of social inscriptions and signals."[35] One could say similar things with respect to notions of participation in a shared modernity: that behind the apparent extension to all lies "a thicker set" of assumptions about what it means to be modern and to participate in this formation, including treating specific (geo)political formations ("social inscriptions and signals") as the background against which to register—in Ahmed's terms, to *figure*—meaningful being-in-time.[36] Although the formulation of Indigenous being-in-time as inclusion in the present (or in a mutually made past or prospective future) may operate as a way of challenging racializing forms of anachronization, it threatens to elide other ways of envisioning the multivectored dynamics of Native peoples' continuity and change that exceed a frame that centers on coparticipation with non-natives. As part of her critique of the ways white histories deny modernness to Indians, O'Brien suggests they "narrate Indian degeneracy, whereas for non-Indians... change is inextricable from the progress narrative that signals their difference and superiority." Such accounts cannot acknowledge the possibility of Native survival into the future via adaptations to altered conditions: "their field of vision narrowed to the local, and they refused to understand the persistence of Indian kinship and mobility on the landscape, not to mention their ongoing measured separateness as political entities."[37] Various forms of persistence and separateness cannot be grasped within this framework, since Native peoples can signify only as remnants. However, does incorporating Indigenous peoples into, in Deloria's words quoted earlier, "a world [built] around, on top of, and through Native American people"—a temporality in which the ongoing existence of the settler state provides the de facto background—engage such continuities and autonomies any better? As Glen Coulthard suggests, "Instead of ushering in an era of peaceful coexistence grounded on the ideal of *reciprocity* or *mutual* recognition, the politics of recognition in its contemporary liberal form promises to reproduce the very configurations of... power that Indigenous peoples' demands for recognition have historically sought to transcend." To what extent does the notion of shared time, of temporal recognition, engender possibilities for Indigenous self-determination, and to what extent does it reproduce the normalization of, in Coulthard's terms, the "inherited background field" of "colonial relation"?[38]

What can figure in this context, and what remains unrecognized? In *Mohawk Interruptus* Audra Simpson argues, "There is a political alternative to 'recognition,' the much sought-after and presumed 'good' of multicultural politics. This alternative is 'refusal.'" Such "refusal" entails a rejection of being translated as "different" within the dynamics of settler governance, being seen as possessing a "culture that is defined by others and will be accorded a protected space of legal recognition *if* your group evidences that 'difference' in terms that are sufficient to the settlers' legal eye," with Simpson further insisting that such transposition of indigeneity into multicultural *difference* "is politically untenable and thus normatively should be refused."[39] Rather than suggesting that Native peoples (the Mohawks of Kanawahke in particular) live outside the orbit of settler imposition, she presents them as "operating in the teeth of Empire, in the face of state aggression," and as "exist[ing] without recognition, in states of strangulation." Thus, the idea of refusing recognition is less about being unimplicated in the choices, affects, policies, imaginaries, and brutalities of non-natives than about insisting that Indigenous peoples have an existence not a priori tethered to settler norms and frames.[40] As Miranda says of Tom Miranda (her paternal grandfather), "His stories about his parents and their parents before them remind me with painful but enlightening clarity how it is that California Indians lost so much culture, language, land, identity—and yet still have an identity and community, albeit often fragmented and/or reinvented."[41] Despite not being acknowledged by the U.S. federal government, Miranda's people, the Esselen, maintain "an identity and community," passed on through stories that themselves enact forms of continuity while also indicating the collective effects of settler violence that influence the Esselens' experiences of time.[42] To be officially recognized would entail manifesting "culture, language, land, identity" in ways that could testify to documentable forms of continuity that could be correlated to dominant historical reckonings, figuring Indigenous collectivity against the background of modes of settler time. Moreover, Esselen responses to fragmentation and processes of reinvention have their own rhythms that are not necessarily commensurable with, in Mehta's terms, the "thicker set of social credentials" that constitute participation in the modern or the present.[43]

The issue I seek to raise is not whether Native peoples can choose to engage in practices that could be characterized as modern, or whether they could characterize their own experiences of time as modern, but what the stakes are of treating such participation or experience as necessarily indicating entry into a singular temporal formation that itself marks the sole possibility for moving toward the future. Might such practices (including treaties, centralized modes of government, particular forms of infrastructure, kinds of commodification and exchange, etc.) also gain meaning and be envisioned as choosable within the context of existing and evolving Indigenous experiences of time, change, and continuity—Native lifeworlds—rather than as a shift to some other sense of time called modern? Might such practices be understood as helping to (re)orient existing Indigenous social trajectories but in ways that do not necessarily create a temporal break (or, in Chakrabarty's terms, "transition") from what came before?[44] How can we think about the effectivity of the kinds of stories Miranda cites and the ways resulting forms of "identity and community" remain oriented by the persistence of peoplehood while also giving rise to forms of reinvention? Lyons offers another account of Indigenous experiences of time somewhat different from what I addressed earlier. He says, "It is also the case that since modernity's onset in Native America—a process that happened by way of conquest and colonization—there has [sic] always been a great number of different, interlocking 'epochs' or *durées* at any given moment: multiple modes of production, diversities of belief, contending memories, and competing future visions—in other words, different times unfolding in common space," adding, "If the expression 'Indian time' means anything, it should signify this history of temporal multiplicity."[45] The variability of Native responses to conquest, choices made when faced with its imperatives, and social practices and visions while living under it can be understood as *temporal multiplicity*. While this phrase might mean the copresence of various stages of being and becoming modern existing side by side, it also opens the possibility for considering the copresence of varied ways of living time, the coexistence of temporal formations that cannot be assessed against a presumptively modern present—a singular background for a necessarily shared history.

What possibilities are there for temporal multiplicity under the conditions of settler dominance? In seeming to grant temporal equality or recognition, the sense of shared time can efface collective forms of becoming and ways of being-in-time that arise out of Indigenous histories, territorialities, and ordinary experiences of peoplehood. In *Translating Time* Bliss Cua Lim notes, "The rhetoric of anachronism is consistently employed by proponents of homogeneous time whenever a stubborn heterogeneity is encountered. One comes to expect that wherever anachronism is shouted, conflicting, coexisting times are being hastily denounced."[46] In this sense the rejoinder to the anachronization of Native peoples may be to argue not that they occupy the homogeneous time of the present but that Indigenous temporalities may conflict with, or simply be heterogeneous to, settler time. To be clear, though, I do not seek to cast the modern as somehow inherently anti-Indigenous or a sign of a loss of Indigenous authenticity, nor do I want to police the boundaries of indigeneity or of Indigenous temporal self-understanding. Instead, I seek to explore the following: What possibilities does the pursuit of temporal recognition bracket or defer? What ways of engaging Indigenous historicity and futurity—and of contesting settler epistemological privilege—does such recognition forgo, and what might be the value of conceptual alternatives? Another way of posing this question might be, what possibilities are opened by the effort to think Indigenous temporal sovereignty (in terms of both the relative autonomy of Native experiences and articulations of time and the violence of imposing settler temporal frameworks through which Native experiences of time are assaulted, denied, and reordered)?

## *Temporal Formations, Frames of Reference, and the (Im)Possibilities of Translation*

In the absence of recourse to a sense of time as simply marching forward in universal synchrony, with everyone occupying a singular now, there must be a way of thinking the plurality of time.[47] Rather than a successive series of presents, each becoming past in turn, being-in-time can be understood as fundamentally oriented. More than simply existing as a unit unto itself, the present bears within itself an impetus born from what's been and directed toward particular goals, ends, horizons. Neither inalterable nor ephemeral, these inclinations contour and animate processes of becoming that have their own trajectories.[48] Each trajectory might be thought of as having specific tendencies and itineraries that exceed the notion of the present as a series of slices of time. Without a notion of supervening, encompassing, and singular time in which all events unfold, there are only disparate temporal formations emerging in their own ways that have shifting effects on each other as they come into contact.

Such an accounting of temporality/ies starts from a different set of presumptions than those of post-Enlightenment historicist time. As Cordova argues, "There can be no universals in the face of an infinity of complexity. There are no absolutes. The complexity is infinite because part of that complexity is change, motion. Whatever is, is in motion, and change is inevitable in the world."[49] Similarly, Russell West-Pavlov describes time as "the pulsating drive of the unceasing transformation of being itself," adding that "there is no 'time' outside of the multiple ongoing processes of material becoming."[50] Without a homogenizing conception of contemporaneity and succession, in which the universal movement of time itself functions as a causative principle, change, motion, and relation immanently arise through extant and emergent dynamics, as they shift and develop through their own internal processes and in connection with each other. West-Pavlov further notes, "Temporality is not the environment of these processes, or the measuring stick to calibrate them, but rather, the processes themselves," and these processes give rise to "multiple temporalities which are immanent to the very processes of material being itself in all its manifestations."[51] Noting the existence of multiple temporalities that cannot be unified into a singular time, then, means acknowledging the diversity of processes of becoming and the variety of potential interrelations among those processes.[52] Attending to such multiplicity, though, is not the same as offering a broad typology as the basis for distinguishing settler and Indigenous experiences of time—such as linear versus circular or having a sensibility based on time versus one based on space. These blanket descriptions tend to freeze the terms (settlers and Indigenous peoples, as well as time and space) into a static opposition that denies internal forms of difference as well as meaningful relation.[53] Thinking in terms of a plurality of processes of becoming that interact with each other in complex ways shifts the discussion of temporality from an insistence on the sharedness of *now* (as well as implicitly of *then* and *will be*) toward a consideration of what constitutes a temporal formation and how such formations might engage with and alter each other without becoming—or being plotted on—a singular timeline.

This approach undermines the idea of a given physical "now," a self-evident contemporaneous copresence among people(s). Johannes Fabian argues, "The very notion of [the] cultural construction [of time] ... implies that cultural encoding works on some precultural, i.e., 'natural' or 'real' experience of Time."[54] Insisting on the shared "intersubjective time" of the present works to avoid casting some people(s)—nonwhite, non-Western—as residual or anachronistic as well as to highlight the role of forms of contemporary force (like colonialism) in shaping extant relations.[55] However, treating participation in an inherently mutual now as a given approaches relations in the present as themselves transparent, as if all that mattered was their occurrence in a particular slice of time rather than either the process by which persons, peoples, movements, and institutions come to be active in that moment and in that place or the horizons toward which they move.[56] Asserting the intrinsic unity of time homogenizes all of the trajectories that supposedly intersect in the present. This conception of temporality implicitly suggests that the orientations borne from the past that shape movement toward the future are somewhat irrelevant when compared with a notion of temporal copresence, of simultaneity. For example, Miranda observes of her people, "Much of our culture was literally razed to the ground. I refused to believe that the absence of language meant my culture was nonexistent, but since even other Indians thought 'all you California Indians were extinct,' it's been a tough road. Along the way, I've learned a lot about stories, their power to rebuild or silence."[57] Focusing on the encounter *now* privileges the account of "California Indians" as gone because of the absence of clear signs of "culture," such as Native language use, at a given moment in time. Taking the necessary coevalness of the present as a starting point would seem to require manifesting evidence of the current content of Native "culture" as a means of registering a people's existence as such, rather than seeing the articulation of relevant timeframes and of means of connecting the past to the present as part of the expression of Indigenous sovereignty and self-determination. From the perspective of a "natural" now, stories of Indigenous survival in California cannot testify to the existence of distinct processes of motion and change that affect collective ways of inhabiting the present—how histories of violence, practices of survival, and the stories that encompass both might orient action and meaning in ways irreducible to a set of relations within an ostensibly shared slice of "real" time figured as the present.

To what extent, then, does the notion of mutual participation in the "real" time of now help in understanding interactions between persons, polities, institutions, and so on? More specifically, does referring the experience of the present to "natural" time provide ways of addressing distinctions in that experience? How can we account for the relation of experience in the present to what's come before, the pace and rhythms of how events unfold, the sensations and inclinations of moving toward what's to come, the implicit itinerary/ies in which one is immersed, the modes of temporal cross-referencing through which moments gain relative significance, and the stories that guide one's ways of being and becoming? Does the supposed physical self-evidence of *now* offer ways of encompassing the multiplicity in lived temporalities? Or does the ostensibly inevitable sharedness of "real" time function as an orienting background that normalizes particular spatiotemporal formations (such as the settler state), foreclosing or silencing countervailing stories and sensations? Indigenous narrations and sensations of time may not accord with dominant settler accounts or models in a variety of ways, including the following: modes of periodization; the felt presence of ancestors; affectively consequential memories of prior dispossessions; the ongoing material legacies of such dispossessions; knowledges arising from enduring occupancy in a particular homeland, including attunement to animal and climatic periodicities; knowledges arising from present or prior forms of mobility; the employment of generationally iterated stories as a basis for engaging with people, places, and nonhuman entities; the setting of the significance of events within a much longer timeframe (generations, centuries, or millennia); particular ceremonial periodicities; the influence and force of prophecy; and a palpable set of responsibilities to prior generations and future ones.

"Natural" time appears as if it were a singular, neutral medium into which to transpose varied experiences of becoming, such that they all can be measured and related through reference to an underlying, "real" continuity—a linear, integrated, universal unfolding. Chakrabarty notes that European historicism employs a conception of time that "is godless, continuous, and, to follow [Walter] Benjamin, empty and homogeneous": "the assumed universal applicability of its method entails the further assumption that it is always possible to assign people, places, and objects to a naturally existing, continuous flow of historical time," such that one "will always be able to produce a timeline for the globe, in which for any given span of time, the events in areas X, Y, and Z can be named"— "put[ting] them into a time we are all supposed to have shared, consciously or not."[58] The supposedly objective givenness of simultaneity, of an unmediated mutual now, depends on a historicist conception of time as an unfolding, universal line of development. Within that frame the idea of a shared present overrides the possibility for conceptualizing discrepant temporal formations, or, at best, such formations are reduced to forms of consciousness that can all be situated within the physical reality of a supervening, homogeneous flow.[59] Certain events simply are simultaneous or are objectively separated from each other by calculable units of time in ways that define possibilities for meaningful relation or causality. Further, any sense of discontinuity or disjunction between temporal figurations and formations can be explained as a result of a failure to understand their relation to this singular commonsensical medium. Positing a "natural" time that underlies any "cultural construction," then, implicitly casts non-Euro-American forms of temporal experience as a form of belief, rendering them less real than dominant accounts of a shared, linear time.[60]

To avoid this kind of culturalization, other temporalities need to be understood as having material existence and efficacy in ways that are not reducible to a single, ostensibly neutral vision of time as universal succession. The concept of frames of reference provides a way of breaking up this presumed timeline by challenging the possibility of definitively determining simultaneity while still holding onto the potential for thinking about collective experiences of time—temporal formations.[61] Within Einsteinian relativity, simultaneity depends on one's perspective based on one's frame of reference.[62] As Peter Galison observes in his cultural history of the emergence of relativity, within Newtonian conceptions of absolute space and time there is a "universal background of a single, constantly flowing river of time," but within Einstein's theory of special relativity (originally published in 1905) "there was no place for such a 'universally audible tick-tock' that we can call time.... Time flows at different rates for one clock-system in motion with respect to another: two events simultaneous for a clock observer at rest are not simultaneous for one in motion." Put another way, "the clock systems of every inertial reference frame were equivalent in the sense that the time of one frame was just as 'true' as any other."[63] An *inertial frame of reference* refers to objects moving in uniform motion with respect to each other. The classic example is the difference between people on a station platform and people on a train passing by the station: those on one are moving at a uniform speed with respect to the other, so the train and the platform are each their own frame of reference.[64] If asked whether two events were simultaneous (such as a ball hitting the floor on the train and a clock striking twelve), the people on the train and the people on the platform would have different answers. The people on the platform will see the ball hit the floor on the train at a different time (a different point on the clock) than the people on the train will. When did the event happen? Who is right? The answer is both and neither. In *Time and Space* Barry Dainton explains special relativity by noting, "Spatially separated events that are simultaneous from the perspective of one inertial frame are not simultaneous from the perspective of all inertial frames, and since the perspectives of all inertial frames are equally valid, there is no sense in the idea that the events in question are 'really' simultaneous or not."[65] Both accounts of when the ball hit the floor are equally valid, since there is no inherent reason to privilege one frame of reference over the other.

Those in different frames of reference will offer varied accounts based on the point (the time shown on a clock) at which they register something as having occurred. If there is no absolute time against which these discrepant measurements can be reconciled, then *now* has no meaning outside of the frame of reference in which it is articulated. As Steven Savitt observes, "There are (at least) as many nows as inertial frames, and there are a non-denumerable infinity of such frames."[66] For two people to inhabit shared time or to partake in a common present, they would need to occupy the same frame of reference. Following this logic, we cannot really speak of a global "coevalness"—the absolute time of Euro-American historicism—in the sense that such a concept presumes a singular timeline in which everyone moves in synchrony, rather than attending to perspectively relevant frames of reference that provide the basis for understanding lived temporalities. When addressing the relations between Natives and non-natives, then, scholars should not presume that Indigenous "identity and community," in Miranda's terms quoted earlier, can be plotted into an account of time defined by the coordinates of settler governance and sociality, which does not encompass Native stories of both fragmentation and reinvention on their own terms (rather than in terms of a settler frame of reference).

While ruling out the possibility of a frame-independent sense of simultaneity, and thus of a singular and universal time that encompasses everything, relativity still is able to situate frames of reference in relation to each other, making them mutually intelligible, through various forms of quantification. Within relativity, *time* refers to the regimented regularity of ticking clocks. As N. David Mermin explains, "While it is commonly believed that there is something called time that is measured by clocks, one of the great lessons of relativity is that the concept of time is nothing more than a convenient, though potentially treacherous, device for summarizing compactly all the relationships holding between different clocks."[67] In this way, time is, in West-Pavlov's terms quoted earlier, an external "measuring stick" separate from the contents and processes within any given inertial frame of reference. Additionally, the distinctions among frames are themselves defined in terms of their velocity with respect to each other. Thus, all of these relationships are presented in terms of numerical values that can be made commensurate through mathematical operations. Furthermore, the varied perspectives offered by different frames of reference can be triangulated through reference to the speed of light.[68] Since the speed of light does not change, it offers a constant by which to calculate the "interval" between frames of reference, their relative movement in space and time. As Brian Greene observes, "Special relativity declares a . . . law for all motion: *the combined speed of any object's motion through space and its motion through time is always precisely equal to the speed of light*," which leads to the following proposition: "Since all inertial frames are equivalent, there is no fact of the matter as to the correct decomposition of the interval into spatial and temporal components, and so the only objective (frame independent) fact about the events is the magnitude of the spacetime interval that separates them."[69] In doing away with a notion of absolute time, special relativity replaces it with a system of reference in which mathematics and the speed of light enable translations among disparate frames of reference.[70]

This process of quantifying and mathematizing time, though, runs up against the problem of whether the ticking of clocks and the calculation of the relation among points and trajectories in a four-dimensional spacetime can account for the forms of lived experience also encompassed by the concept of time. Such questions were raised perhaps most forcefully in the early twentieth century by the philosopher Henri Bergson. In all of relativity's frames of reference, time is defined as the ticking of a clock, a mechanistic process of dividing temporality into successive, homogeneous units. As against this uniformity of division, Bergson presents *duration* as the transition among qualitatively differentiable sensations such that they permeate each other in ways that defy enumeration. In *Time and Free Will* he argues, "It seems . . . that two different sensations cannot be said to be equal unless some identical residuum remains after the elimination of their qualitative difference"; "we may conclude that the idea of number implies the simple intuition of a multiplicity of parts or units, which are absolutely alike." Rather than seeking to divide time into discrete, homogeneous units, Bergson conceptualizes it as "a continuous or qualitative multiplicity with no resemblance to number."[71] Approaching time as a quantity that can be infinitely divided into equivalent units denudes temporality and the experience of time of everything beyond the "identical residuum" that supposedly can be found within (and can be commensurate with) "different sensations" as they slide into and through each other. The idea of simultaneity as a physical property of time, then, suggests that one can cut up time into equivalent units and that all the events that are spatially copresent on that temporal plane are simultaneous with each other (treating space itself as an infinitely divisible grid of equivalent units).[72] However, if duration is a "qualitative multiplicity" that is "continuous," the idea of simultaneity cuts into the continuous experience of time in order to declare that a set of events that are ostensibly spatially copresent with one's "sensations," yet do not necessarily impact them, all have an inherently shared existence, an insistence that reduces those sensations to a set of equivalent units that can be plotted along a (time)line. What does such spatial copresence at a supposed moment of time mean in terms of thinking the relation between those "simultaneous" events? What does that spatial copresence have to do with the flow—processes of unfolding, becoming—within the experience of duration? What does presence *at the same time* mean in these terms except being able to be plotted on a grid such that events occupy an "identical" temporal plane? From within Bergson's analysis, an insistence on "natural" time—that everyone occupies a singular present—looks like a mathematizing abstraction that effaces the experience of duration. In this vein, to what extent does the notion that Natives and non-natives necessarily occupy a shared temporality rely on homogenizing space and time such that ostensible copresence in space (sizing that grid at whatever scale—a particular region, the territory of the nation-state, the globe) on a slice of time (however wide—an instant, a year, a decade) is taken to mean a common inhabiting of "modern time," regardless of how Natives and non-natives enter into each other's sensations and experiences of duration?

Additionally, seeking to enumerate time—to make it determinate and calculable (and as such also convertible into a coordinate axis within a spacetime grid)—runs into the problem of its resulting frozenness.[73] If time can be plotted like a series of points on a graph, what happens to the movement between those points, "what takes place in the interval between two simultaneities?" Bergson suggests that in dividing up time into equivalent units, "as for the interval itself, as for the duration and the motion, they are necessarily left out of the equation."[74] Similarly, in *Matter and Memory* he observes, "While the line AB symbolizes the duration already lapsed of the movement from A to B already accomplished, it cannot, motionless, represent the movement in its accomplishment nor duration in its flow," later adding, "In a space which is homogeneous and infinitely divisible, we draw, in imagination, a trajectory and fix positions: afterwards, applying the movement to the trajectory, we see it divisible like the line we have drawn, and equally denuded of quality."[75] In order to understand time, according to Bergson, we must treat it as constantly in motion and, thus, not divisible into discrete units. Without such standard units, though, there is no way to determine simultaneity, except inasmuch as it is experienced as part of a lived trajectory—a qualitatively shifting process of becoming (like, in Miranda's terms, "a river where no gallon of water is the same gallon it was one second ago. Yet it is still the same river").[76] If a trajectory can be decomposed into a number of points, each one occupying a slice or plane of infinitely divisible (because homogeneous) time, there can be no movement. Reciprocally, the continuity of duration cannot be broken into units that would make experiences of time *simultaneous* with other events that lie outside a given trajectory. A given point of time cannot be separated from the orientation, momentum, and dynamism of the trajectory—and the attendant "quality" of motion—without compromising a sense of flow. Approached in this way, the representation of Native becoming as contemporaneous with that of nonnatives in "natural" time (the grid of homogenized space and time) reduces the immanent trajectories of indigeneity (processes of Native becoming) to a set of points—the supposedly shared now of the present, modernity, national history, and so on. This abstract configuration (situated within a settler grid of intelligibility) itself is treated as the neutral, natural, self-evident frame for understanding time's unfolding.[77]

If one reinterprets relativity in light of Bergson's insistence on continuity, movement, and qualitative multiplicity, the notion of frame of reference can be reconceptualized in less quantitative terms as a means of talking about collective regularities (shared backgrounds and orientations) in how time is experienced. In an unacknowledged Bergsonian turn, Lee Smolin, himself a theoretical physicist, critiques post-Einsteinian physics for its tendency to reduce time to a segment of a four-dimensional spacetime whose properties and development can be apprehended as a singular block. In *Time Reborn* he suggests that in removing the dynamism of time from its equations, physics seeks to measure the universe through conceptual tools that are treated as themselves unaffected by the processes they seek to describe. Such conceptual procedures, which he characterizes as "the background," provide "the terms that give meaning to the motion described.... A distance measurement implicitly refers to the fixed points and rulers needed to measure that distance; a specified time implies the existence of a clock outside the system measuring the time." He adds, "These background structures are the unconscious of physics, silently shaping our thinking to give meaning to the basic concepts we use to imagine the world. We think we know what 'position' means because we are making unconscious assumptions about the existence of an absolute reference."[78] What happens if, hearkening back to Ahmed, we approach "the background" not as a limitation (a set of unexamined presumptions that hamper something like proper measurement or an adequate understanding of "real" time) but as the conditions of emergence for particular temporal sensations, orienting the qualitative dynamics of duration as a collective experience of time?

If one suspends the use of the homogeneous successions of clock time as "an absolute reference," frames of reference would refer to qualitatively differentiable processes of becoming that have no inherent, neutral means of being articulated to each other, instead requiring complex processes of translation in order to be made mutually intelligible.[79] Within such an analysis, the emphasis lies on how social formations achieve substantive, if shifting, cohesion through their backgrounds, orientations, and trajectories (including the influence of nonhuman entities and forces on them).[80] Cordova suggests, "Each of us occupies a world that is made by our predecessors. We are given 'reality'; we do not *discover* it," further indicating that "there are no individual *realities*, only communal ones" and that "we reinforce our communal sense of reality. . . . We act on it. In it." She later observes that "when two people from different cultures come together," "they find it difficult to communicate with one another—their frames of reference do not meet."[81] In addition to indicating the potential for moving beyond the idea of disparate sets of *beliefs* about time that can be triangulated against the *real*, Cordova indicates the imbrication of physical reality and collective modes of perception by characterizing those "communal" orientations (trajectories of *action*) as "frames of reference." Furthermore, noting the problems of communication raised by having nonidentical frames of reference, Cordova implies that such difficulties evidence the need to *translate* among varied forms of temporal experience. Rather than engaging directly in a mutual and self-evident now, relations across temporal formations would entail, in Chakrabarty's formulation, "translations that do not take a universal middle term for granted," including the putative physical givenness of the present.[82] Each collective frame of reference might be understood as having its own forms of continuity, flow, trajectory—processes of becoming—that cannot be segmented into slices of time (in Bergson's terms "simultaneities") so as to be made commensurate with moments in other frames of reference (such as in settler-endorsed forms of historicism).

In the absence of a mutual frame of reference (a common background) between Natives and non-natives, non-natives engage in forms of translation, not primarily to understand Native temporalities but to insert them within settler timescapes. That process of interpellation is not acknowledged as such, and through it, Indigenous experiences appear as exception (an alarming rupture in time, as in the persistence of Native sovereignty or the use of violence to defend it, discussed in chapter 2), absence (the need to engender proper forms of ambition and life rhythms, as in allotment policy, addressed in chapter 3), and superstition (the denigration of complex, noncontiguous relations across time, particularly as instantiated by prophecy, taken up in chapter 4). However, recognizing such temporal inscription or conscription as part of the dynamics of settlement can open room for articulating forms of being-in-time—in terms of territoriality, politics, everyday socialities—that do not need to take settler formations as the implicit standard for what constitutes the present or historicity. Discussing the ways Indigenous philosophies and modes of living remain alien to dominant narratives of settler governance, Turner argues, "The asymmetry arises because indigenous peoples must use the normative language of the dominant culture to ultimately defend world views that are embedded in completely different normative frameworks."[83] Attending to Indigenous temporal sovereignty, then, draws attention to the ways in which settler superintendence of Native peoples imposes a particular account of how time works—a normative language or framework of temporality that serves as the basis for forms of temporal inclusion and recognition. Settler time reduces the unfolding and adaptive expressions of Indigenous peoplehood to a set of points—the supposedly shared now of the present, modernity, and national history—within a configuration that is positioned as the commonsensical frame in ways that deny the immanent motion of indigeneity. Native peoplehood gets plotted in ways that deny the movement inherent in its ongoing emergence.

In contrast to the insertion of Native peoples into settler time, the ongoing history of a people's becoming can be understood as providing an orienting background and momentum in engaging with non-native persons, practices, material culture, infrastructures, and institutions. Rather than instantiating a break into the modern, which provides the implicit framework for mutual engagement in the present, those encounters, including the experience and memory of modes of settler dispossession, will themselves become part of a people's experience of their own duration.[84] As noted earlier, Miranda suggests of Native peoples in California, including her own, that they "lost so much culture, language, land, identity—and yet still have an identity and community, albeit often fragmented and/or reinvented," and she further connects her experiences of her father's abuse to this history: "Flogging. Whipping. Belt. Whatever you call it, this beating, this punishment, is as much a part of our inheritance, our legacy, our culture, as any bowl of accorn mush, any wild salmon fillet.... More than anything else we brought with us out of the missions, we carry the violence we were given along with baptism, confession, last rites."[85] The violence of the missions here marks less a break in Indigenous temporalities than a reorientation of them, one that informs experiences of "identity and community"—an "inheritance" that comes to serve as part of the background for action in the present and toward the future. The legacies of the missions become part of Indigenous frames of reference, even as the attempted decimation of peoplehood through missionization fails. The losses do not themselves eliminate a sense of indigeneity even as they affect its texture and trajectory. In the introduction to her text, Miranda asserts, "Culture is ultimately lost when we stop telling the stories of who we are, where we have been, how we arrived here, what we once knew, what we wish we knew; when we stop our retelling of the past, our imagining of the future, and the long, long task of inventing an identity every single second of our lives."[86] The missions and their ongoing effects play a crucial role in the story of "where we have been, how we arrived here," and as such, they neither merely interrupt a sense of peoplehood nor simply supplant prior *retellings* and *imaginings*, which would generate something like a disjunction in temporality itself. In this vein, discussing the effects of forms of state violence in India, Das asks, "Are there other paths on which self-creation may take place, through occupying the same place of devastation yet again, by embracing the signs of injury and turning them into ways of becoming subjects?"[87] Mission stories, and those that follow, enter into existing frames of reference, becoming part of the totality of stories, altering them, offering new challenges and struggles, while still participating within an ongoing process of Indigenous becoming—of *invention*—as peoples shaped by lived stories and sensations.[88]

How, though, can one physicalize such sensations? How might we understand the plurality of Indigenous temporalities as having material efficacy, irreducible to "belief"? If frames of reference cannot be determined and measured against "objective" criteria, and the attempt to do so can be understood as itself an imposition of settler time, then we might turn to perception as a way of approaching Indigenous temporal formations. In *Phenomenology of Perception* Maurice Merleau-Ponty argues that perception operates holistically, taking in an environment as a "whole perceptual context" rather than a piecemeal set of sensations about particular objects or relations, and, in affectively apprehending his or her existing circumstances, a "normal person *reckons with* the possible," meaning that perception is guided by the potential for action in the world.[89] Determinations about what is possible unfold from a history of engagements with shifting environments, and such ongoing interactions work less as a set of propositions that are verified or falsified than as a continuous enmeshment in the world in which feeling, response, and judgment arise out of sensory connections to one's surroundings. In this sense perception connects a person to his or her environment, operating not as a separate consciousness or screen distinct from the actual but as an encompassing capacity for relation influenced by existing conditions and circumstances that are themselves changing—what Merleau-Ponty describes as "the momentum of existence." That process of "reckon[ing] with an environment," engaging it as a "field of possibility," further involves drawing from past encounters in order to make sense of present experience. Merleau-Ponty suggests that "since sensation is a reconstitution, it pre-supposes in me sediments left behind by some previous constitution," and he later notes, "The *person who* perceives is not spread out before himself as a consciousness must be; he has historical density, he takes up a perceptual tradition and is faced with a present." Engagement in the present is shaped and made possible by the "historical density" of a person's accreted experiences, a legacy that bears on the current moment by necessarily situating it within an extant "perceptual tradition" that is formed by the life one has lived up to that point and without which current stimuli would have no appreciable meaning.[90] As Das suggests, in a somewhat Bergsonian vein, "The simultaneity of events at the level of phenomenal time that are far apart in physical time makes the whole of the past simultaneously available." Das later adds, "There is also the process of rotation in which, independent of my will, certain regions of the past are actualized and come to define the affective qualities of the present moment" as part of the process of perception.[91] Here we also might recall Miranda's discussion of the role of the missions in affecting forms of feeling and action among contemporary Native peoples in California. In *The Memory of Place* Dylan Trigg observes, "*We carry places with us,*" further noting, "We are never truly 'in' place without already having been in another place, and that other place is never merely left behind.... Rather, coming into a place means inserting that lived history into the present," and he later refers to such accretions and projections of sensation as an "embodied hermeneutics."[92] The process of contextualizing, or orienting, new sensations within an already active set of tendencies, memories, and histories (themselves based not simply on beliefs about the world but on the accretion of material interactions in it with all sorts of entities, human and otherwise) extends beyond the present into the future. More specifically, my anticipation of the future and acting in ways that reach toward it (consciously and not) shapes the texture, contours, and dynamism of my engagement with the present, such that there is no now that can be treated in isolation from a momentum toward what will be. As Merleau-Ponty suggests, "In every focusing movement my body unites present, past and future, it secretes time."[93]

Following Cordova's caution about needing to understand experience as enmeshed within communal processes and formations, we can understand the notion of a "perceptual tradition" as itself exceeding individual sensory fields. More than indexing a given person's idiosyncratic experiences and the ways they affect the apprehension of the present and movement toward the future, perception draws on collective histories and anticipations in ways that include the following: long-standing inhabitance in a given place, or regular return to that place, and exposure to its physical and social landscape; shared material circumstances that engender common sets of concrete situations and potentials for response and agency; memories and stories of such experiences that generate feelings of belonging to a group and that inform future action; histories transmitted within and across generations that offer ways of conceptually and emotionally understanding the relation between the past, present, and future and the horizon toward which one moves as a member of the group; and the legacies of past actions by and toward members of the group that contour the "field of possibility" in the current moment (as a practical matter, not solely one of belief). These suprapersonal dynamics orient the ways one affectively encompasses, assesses, navigates, and engages the "whole perceptual context." In this way forms of perception, and experiences of duration, indicate not just generic human sensory capacities but socially mediated formations of becoming that develop and inculcate their own ways of experiencing time—what I have been characterizing as frames of reference.[94] Moreover, if perception emerges out of the materiality of one's current and previous experiences, what constitutes a "material" part of the environment, and thus a potential causal agent within it, can also be distinct among varied social formations.[95] If collective dynamics and histories shape individual forms of temporal experience, incorporating the present into the trajectory formed by such de facto belonging (chosen or not), then they provide the nonconscious background against which to register the potential for present and future action. As Merleau-Ponty suggests, "Whatever I think or decide, it is always against the background of what I have previously believed or done."[96] However, since such forms of collectivity are themselves not static, instead taking part in processes of becoming (such as Miranda's image of the flow of a river), we might characterize their role as an active and shifting process of *backgrounding*. Thus, while, in Ahmed's terms, "a background is what explains the conditions of emergence" for what appears in the foreground, the background itself is subject to change but helps shape frames of reference for temporal experience.[97] Backgrounds and modes of backgrounding provide a means by which to distinguish between temporal formations.

Within such an account, the practices, knowledges, and forms of collective identification often characterized as *tradition* can be understood as distinctive ways of being-in-time. They emerge from material processes of reckoning with an environment and are open to change while helping provide an orientation and background for everyday Native experience. At one point Turner notes, "The first difficulty is to know how we ought to characterize the distinct forms of knowledge embedded in indigenous communities. Phrases like 'traditional knowledge' and 'indigenous ways of knowing' have become commonplace in both mainstream and indigenous cultures, yet we are not at all clear about what they mean *in relation to the legal and political discourses of the dominant culture*."[98] Turner focuses on the need to engage in a process of translation whereby Indigenous self-articulations and knowledges can be made intelligible within the dominant frameworks and discourses used by the settler state, but that problem of institutional intelligibility can be reversed to suggest that the use of *traditional* to characterize Indigenous knowledges, experiences, and lifeworlds already tends to situate them within normative settler temporality. Traditional serves as the opposite of modern, indicating not simply chronological dating but qualities that belong to a different epoch—that do not fit the contours of the present. From a settler perspective, the present entails, in Mehta's formulation, "a thicker set of social credentials" that are implicitly cast as if they were simply a neutral description of now.[99] To the extent that the givenness of state sovereignty provides a significant part of the background for non-native historicism(s), the use of tradition and modern as paired and contradistinguished ways of conceptualizing Native processes of becoming often ends up translating Indigenous experiences of time into settler paradigms in ways that powerfully constrain possibilities for envisioning and realizing self-determination. Rather than arguing for Native access to modernness, which I have characterized as temporal recognition, then, I have been suggesting that the development of a notion of temporal frames of reference can provide ways that Native and non-native trajectories (as well as modes of backgrounding) might be distinguished without resorting to a notion of shared time (almost always skewed toward non-native framings), thereby opening up room conceptually for the expression of varied forms of temporal sovereignty.

In particular, "the land question," as Paul Chaat Smith puts it, can be understood as helping to generate a background that orients Native temporalities in multifaceted ways.[100] Native territorialities provide a sense of direction, regularity, and historical density for the continuing emergence of peoplehood and for figuring self-determination.[101] One might think of a people's accreting connection to a given place and neighboring peoples over generations (and the ways life in that space is affected by the interweaving climatic, vegetative, animal, social, and diplomatic dynamics at play there) as creating an experience of being and becoming whose textures, regularities, and negotiations cannot be captured through reference to a universal chronology.[102] Modes of emplacement and enduring relations to homelands—even if such zones shift (as in removal, addressed in chapter 2), are fractured (as in allotment, addressed in chapter 3), or are more extended geographically (as in movement to urban sites elsewhere, addressed in chapter 4)—shape Indigenous peoples' becoming, powerfully influencing the felt dynamics of Native being-in-time. With respect to dispossession by settlers, Miranda observes, "The loss of land is a kind of soul-wound that the Ohlone/Costanoan-Esselen Nation still feels; a wound which we negotiate every day of our lives," adding, "The loss of land clearly presaged intergenerational trauma with the accompanying loss of self-respect and self-esteem."[103] Collective temporal dynamics, then, also include histories of displacement and dispossession that inform contemporary Native sensations and self-understandings, potentially lived as forms of bodily affect ("a wound").

Furthermore, quotidian ways of living Indigenous peoplehood have their own rhythms and momentum, giving rise to ways of inhabiting time that endure even as they remain open to alteration. In the process of, in Simpson's terms, "refusing to go away, to cease to be, in asserting something beyond difference," Indigenous people(s) testify to a persistently reactivated continuity that is not the replication of the unchanging same.[104] The project of addressing Indigenous temporality, therefore, becomes a matter of attending to peoples' own frames of reference for their experiences of time: not just as beliefs set within a supervening or underlying "natural" timeline but as a basis for understanding the materiality of their ways of being and becoming. Doing so reverses the tendency to assess contemporary Natives in terms of their deviation or declension from a putative origin (their *aboriginality* in Barker's terms).[105] As Cordova observes, "the goal of persons who envision themselves in a world of motion, change, and complexity is to create and work on maintaining stability in the face of all that," and the emergence and regeneration of such stability (not to be confused with sameness or unanimity) over generations entails a continuing engagement with changing circumstances, including responding to the violence of settler impositions and displacements as well as incorporating once-alien ideas, practices, objects, and modes of institutionality.[106] In this way, becoming needs to be thought of not as a break with what's come before but as an inherent dynamism within being, in which continuity is itself an effect of activity rather than being treated simply as stasis or inertia.[107]

Miranda addresses this complex interplay of historical density, collective orientation, and engagement with current conditions and potentials in her account of Indigenous futurity in California:

> I've been thinking about the shattering and fragmentation of California Indian communities since Contact.... Sometimes something is so badly broken you cannot recreate its original shape at all; you will always compare what your creation looks like with what it used to look like. As long are you are attempting to *recreate*, you are doomed to fail! I am beginning to realize that when something is that broken, more useful and beautiful results can come from using the pieces to construct a mosaic. You use the same pieces, but you create a new design from it.[108]

What *was* does not provide a set pattern, like a mold, for what *will* or *could be*. Rather, the exertion of temporal sovereignty in the face of a history of settler violence and displacement consists in an ongoing *re-creation* oriented by an engagement with the historical density—the "pieces"—of collective identity and experience. The feeling of belonging to such "communities" provides a frame of reference and a trajectory for the effort to move forward in ways that are neither equivalent to nor simply disconnected from the past, generating "a new design" that can engender livable forms of stability. In this way, the dynamics of peoplehood take part in a continual process of creation that responds to the force of settler colonialism, as well as taking part in other forms of historical change, while not being reducible to the "shape" of peoplehood at any particular prior moment in history.

One concern about a concept like Indigenous frames of reference (or a particular people having a frame of reference) is that it will be homogenizing, effacing the variability of experience and the diversity among peoples as well as among persons who are part of the same people (Esselen, Osage, Dakota, etc.). I have been using the term *momentum* as a way of characterizing how processes of becoming carry their own immanent tendencies and directions, and, in doing so, I am suggesting less something like inertia (a simple continuance of a trajectory produced by an initial, activating external force) than the ongoing effects of patterns of regularity (which is not equivalent to sameness) that give cohesion to Native social formations. In other words, the various interanimating dynamics that sustain the collective experience of peoplehood produce forms of regularity that do not simply exist at a given moment in time but influence how a people moves toward the future.[109] Yet the boundaries between temporal formations are not inherently clear, and, like Bergson's "qualitative multiplicity," they can be understood to shade into each other, remaining dynamic and thus open to change and reorientation. An active question for this way of conceptualizing time, then, would be, what kinds of change lead to individuals, groups, or peoples diverging such that they no longer share a frame of reference? Also, a person might live in and move among varied spatiotemporal formations, and the process of doing so would be affected by the relative similarity of those formations, the conditions of that transition (especially in terms of the institutional force at play in it, such as needing to make oneself and one's people intelligible to the U.S. government), and the kinds of translation (acknowledged or not) at play in such transitions. Reciprocally, occupying a shared frame of reference is not the same as agreement, although it would entail having some shared set of experiences and modes of perception: in Miranda's imagery, having in common a history of "shattering and fragmentation," and an impetus toward re-creation even if there is disagreement about the contours and content of the "new design." Moreover, if the idea of Indigenous temporal formations or orientations raises such questions, these problems are not avoided by presuming the inherent mutuality of the present. Instead, doing so, in Mehta's terms, posits an "anthropological minimum" that serves as the necessarily shared background against which to register and reconcile, in Chakrabarty's terms, "diverse lifeworlds and conceptual horizons about being human."[110]

Adopting sovereignty and self-determination as normative principles guiding the approach to time opens the potential for thinking Indigenous temporalities—temporal multiplicity—in ways that exceed the forms of presentness posited and imposed through dominant modes of settler time. Doing so entails engaging with the profound effects of colonialism without understanding such force, struggle, and negotiation as yielding a singular kind of temporal experience that would dictate a shared present with a particular content (being *modern*, as opposed to a remnant from the past). Modes of settler invasion, intervention, regulation, dispossession, and occupation become intimate parts of Indigenous temporalities, but they do so as part of Native frames of reference, meaning that they are encountered through a perceptual tradition and a set of material inheritances that includes ongoing Indigenous legacies of landedness, mobility, governance, ritual periodicities, social networks, and intergenerational stories.[111] Together, these various aspects of being and becoming give historical density to the engagement with settler policies and everyday presence, orienting Native perception and action.

*Queer Times, Storied Landscapes, and Indigenous Duration*

This account of Indigenous temporalities requires rethinking the meaning of continuity. From within a conventional settler perspective, Indigenous continuity means the persistence of particular kinds of ritual belief and performance, modes of land use, and forms of collective decision making that have remained relatively unchanged since the period of early contact with non-natives—often characterized as the persistence of tradition or culture. These ideas about Indigenous endurance are institutionalized in various ways as part of state policy, including the mechanisms through which official recognition occurs (legal and administrative determinations of what constitutes an Indian tribe, a land claim, a sacred space, a cultural practice, etc.). Given such narratives of continuity, we should be careful that any effort to address Indigenous temporality/ies be able, in Barker's formulation, "to claim cultures and identities that are conflicted, messy, uneven, modern, technological, mixed," avoiding the quite limited visions of authenticity so often championed by the settler state and those whom Lyons has termed "culture cops."[112] How might we think about Native temporal frames of reference as allowing for continuity as well as for complex and varied change without reinstalling the notion of a singular, neutral present? Miranda begins her account of her people's history, and the enmeshment of her own life within it, by insisting, "Human beings have no other way of knowing that we exist, or what we have survived, except through the vehicle of story."[113] In this sense, temporal experience itself might be understood as intimately imbricated with story. Miranda's discussion of the work of story suggests that it functions as a crucial part of processes of becoming. Stories help provide the background for Indigenous experiences of time, shaping perceptual traditions while also influencing sensations of what's possible. Attending to story as a constitutive element of perception emphasizes the variability and changeability of Native experiences while also addressing the ongoing (re)construction of collective frames of reference, suggesting less the transmission of static narratives than active and ongoing dynamics of perceptual (re)orientation.[114] Moreover, if one thinks about Indigenous storying through the prism of recent work on temporality in queer studies, such scholarship can help highlight how stories enact relations across time that cannot be encompassed through conventional notions of tradition and that defy easy translation into the terms of Euramerican historicism.

Often referred to as the *oral tradition*, Indigenous patterns of making and circulating stories could be construed as a set of relatively authoritative texts through which peoples' histories and philosophies are transmitted across generations. Approaching them in this way, though, can freeze such stories, suggesting that they must be of a certain vintage while also making them into something like a primer rather than a dynamic and embodied part of ordinary experience. As Julie Cruikshank suggests in *The Social Life of Stories*, "Meanings do not inhere in a story but are created in the everyday situations in which they are told": "If we think of oral tradition as a social activity rather than as some reified product, we come to view it as part of the equipment of living rather than a set of meanings embedded within texts and waiting to be discovered."[115] Repertoires of shared stories of all sorts, transmitted and added to across generations, provide a means of engaging with extant circumstances in ways that generate continuity while remaining open to addition, revision, and adaptation. In *Remember This!* Waziyatawin Angela Wilson notes that "suggesting that people living today are outside an oral tradition ... assume[s] that the contemporary person is not part of a living tradition that can incorporate new information," and being part of such a legacy entails drawing on stories from long ago in addressing contemporary happenings while also contributing to the body of stories by adding recent events and dynamics.[116]

More than a kind of object inherited from the past, stories contribute to one's phenomenological frame of reference. If we recall Merleau-Ponty's point that "since sensation is a reconstitution, it pre-supposes in me sediments left behind by some previous constitution," we can conceptualize the ways that having bodies of stories in common functions as such a previous constitution, helping orient perception in the present as part of a people's ongoing processes of becoming.[117] Such an embodied sense of belonging as lived through story affects how one situates the present in relation to the past and to future possibilities. Wilson observes, "The power of these stories stems from the connection created between the shaped historical understanding and living within the present. The oral tradition, in all its forms, has the potential to cultivate thoughts, worldview, and to dictate a pattern for living."[118] As Dian Million argues in "There Is a River in Me," "The stories, unlike data, contain the affective legacy of experience. They are a felt knowledge."[119] This felt knowledge provides a background for Indigenous trajectories and temporalities, and such affective legacies connect to other forms of feeling and sensation, *cultivating* sensibilities that might be abstracted as a "worldview" but that operate in quotidian ways as modes of being-in-time. Such stories can entail transmitting memories of devastating loss (discussed in chapter 2), contextualizing social practices such as dancing and ritual (addressed in chapter 3), or conveying prophecy through social networks that exceed the inheritances of nuclear family homemaking (examined in chapter 4). In other words, stories can be understood as playing a significant role in orienting enduring forms of Native collective feeling, providing momentum for shared sensations of time—even as such feelings are themselves complex, shifting, and engaged with the specificity of varied situations. Discussing the legacies of land loss for Native peoples in California and the felt sense of continuing connection to those places, Miranda insists, "The stories still exist, and testify that our connections to the land live on beneath the surfaces of our lives, like underground rivers that never see the light of day, but run alive and singing nonetheless. The stories call us back."[120] Stories here are not just isolated narratives but themselves register relations between persons and places as well as forms of collective belonging. Such "connections to the land" are lived as forms of bodily sensation, intimately part of the flow of temporal experience ("like underground rivers"), suggesting the ways stories reciprocally inculcate modes of perception and give expression to feelings that "live on beneath the surfaces" and help shape conscious action. Das suggests of the resonance between the actions of particular rape survivors in India and a story from Hindu sacred texts, "It was as if the past had turned this face toward them—not that they had translated this past story into a present tactic of resistance," and in this way Native stories can be thought of as immanently emerging from and influencing current perceptions and practices rather than as being consciously deployed.[121] The work of storying, then, can be thought of less as the act of telling a story than as the immanent dynamism in the ways stories move through the world, the kinds of qualitative relations they generate as part of producing collective experiences of duration. Further, the process of attending to stories—acknowledging the significance and effects of the forms of temporal relation they both reflect and bear—could be characterized as a mode of temporal sovereignty.

However, drawing on story and its intergenerational transmission as a means of detailing Indigenous temporalities might seem to presume a lineage-based model dependent on a heteronormative vision of family. In describing the hardships of trying to survive amid the violences of ongoing settler occupation (specifically in terms of the life of her great-grandfather Tomás Santos Miranda), Miranda suggests, "Sometimes our bodies are the bridges over which our descendants cross, spanning unimaginable landscapes of loss." Stories and peoplehood might appear as a straight line of descent through familial inheritance, or, at least, such straightness seems to be the case if one understands kinship in linear or lineal terms (which Miranda does not).[122] Accounts of temporality within queer studies offer other ways of addressing forms of influence and identification that do not follow a linear timeline while also charting the ways that particular notions of continuity become normalized, even as such scholarly work also tends to push against the very kinds of collective continuity necessary to sustaining peoplehood.

The notion of a generic life cycle organized around conjugal union and reproduction functions as perhaps the most prominent way of envisioning the everyday meaning of continuity. Such an account positions marital couplehood as necessary for procreation itself, and thus the survival of the human species appears to depend on bourgeois family formation and homemaking. J. Jack Halberstam refers to these "conventions of family, inheritance, and child rearing" as "reproductive temporality," suggesting that "queer uses of time and space develop, at least in part, in opposition to the institutions of family, heterosexuality, and reproduction."[123] From this perspective, queer experiences of time run athwart of a projected life course organized around heterocourtship, conventional marriage, and the generationality of the nuclear family.[124] More than imposing a particular vision of proper desire and kinship dynamics, this conception of a regular life weds personal development to a universalizing account of the movement of time. That process can be described as *chrononormativity*, which Elizabeth Freeman defines as "a technique by which institutional forces come to seem like somatic facts" and in which "historically specific regimes of asymmetrical power" appear as "seemingly ordinary bodily tempos and routines." Furthermore, "these teleologies of living, in turn, structure the logic of a 'people's' inheritance: rather than just the transfer of private property along heteroreproductive lines, inheritance becomes the familial and collective *legacy* from which a group will draw a properly political future."[125] The heteronormative presumption that the nuclear family and its privatized domestic arrangements serve as the basis for human futurity per se casts the legal, political, economic, and spatial dynamics necessary to sustain that social formation as simply the immanent basis for the unfolding of time itself, as inevitably providing the framework for thinking the past, the future, and their relation to the present. Here Freeman builds on Dana Luciano's notion of *chronobiopolitics*, developed in *Arranging Grief*, which Luciano defines as "the sexual arrangement of the time of life."[126] In addition to being institutionalized in various ways, this specific developmental path comes to serve phenomenologically as part of the perceptual tradition through which people reckon with the possible, and heteronuclearity provides the background against which other modes of making a life appear as queer deviations or perverse orientations.[127] By seeking to challenge the legitimation and proliferation of straight time (which itself can be understood as a denial of, in Bergson's terms, the "qualitative multiplicity" of temporal experience), queer critique helps draw attention to how ordinary experiences are influenced by the momentum of dominant formulations of time as well as how such experiences might run in another direction, opening onto forms of temporal feeling that do not fit officially endorsed inheritances and trajectories.[128] In other words, queer analyses help open ways of registering the imposed *straightness* of time while also highlighting alternative kinds of temporal experience.

These kinds of questions about how one conceptualizes the proper shape of a life certainly resonate with the ongoing subjection of Native peoples to projects of assimilation that seek to inculcate ostensibly civilized ways of being-in-time. In fact, the imposition of heteronormative social dynamics has been a key part of the U.S. government's efforts to supplant Native modes of collectivity, casting extant Indigenous forms of association, occupancy, household formation, and governance as merely vestiges of a bygone time.[129] Many of the initiatives within Indian policy have worked to reorient everyday forms of Native feeling and practice, seeking to alter the experience of time so that U.S. legal geographies and claims to sovereignty provide the background. That chronobiopolitical project depends on an encompassing *chronogeopolitics*, implicitly positing the givenness of U.S. territoriality and jurisdiction as the self-evident basis for understanding the movement of time. In particular, the allotment program employs reproductive temporality in ways that justify the jurisdiction of the settler state (chapter 3), and a similar aim can be seen at play in the reduction of Native peoplehood to quantities of procreatively transmitted Indian "blood" (chapter 4). Queer theorizations of temporality, then, aid in understanding Native opposition to such policy framings. From this perspective, such resistance appears not as a refusal of the modern but as an expression of alternative experiences of time that persist alongside settler imperatives, and are affected by them, while not being reducible to them.

This kind of queer scholarship further challenges the implicit developmentalism of notions of a universal *now*, placing under significant pressure the historicist presumption that the past is an alien space separated by an unbridgeable gulf from the present. Carolyn Dinshaw has suggested the need to move beyond a notion of history as straight, as an unfolding "causal sequence" that, as such, rules out "an expanded range of temporal experiences—experiences not regulated by 'clock' time or by a conceptualization of the present as singular and fleeting; experiences not narrowed by the idea that time moves steadily forward, that it is scarce, that we live on only one temporal plane."[130] The possible range of ways of being-in-time is radically limited if one envisions temporality as singular and linear, as replacing what has come before in its steady forward movement (the kind of synchronous slices of time that Bergson displaces through the notion of duration). Moreover, this notion of time as an unending succession—in which the present unfolds out of the past while supplanting it—can be understood as itself relatively new. As Valerie Rohy observes, "historical alterity is, after all, a recent invention; the conviction that past ages are noncontiguous with modernity is a hallmark of modernity," and in "Queering History" Jonathan Goldberg and Madhavi Menon ask, "Why has it come to pass that we apprehend the past in the mode of difference? How has 'history' come to equal 'alterity'?"[131] The positing of the past as on the other side of a great gulf contributes to the sense of the present as something of an integrated whole against which to juxtapose historical events or dynamics (*our* understandings now versus *theirs* then), and doing so replaces the coexistence of divergent experiences of time with the difference between the contemporary moment and that which it supposedly has surpassed (such as the antinomy of the modern versus the traditional). The idea of a singular, linear unfolding in which the present supersedes the past might be thought of as a form of "compulsory heterotemporality" in which the understanding of time "mimes the heteronormative demand for proper sexual sequencing," replaying conceptions of proper individual life sequence at the level of time itself.[132] Such a vision of history can be seen at play in the imagining of certain national events, like the Civil War, as moments of transition in which the country breaks away from a degraded past (slavery in the case of the Civil War), as opposed to tracing the regularities of settler violence in which the past appears less as a space of alterity than in a relation of continuity with the present (discussed in chapter 2). In this sense, processes of settler temporal recognition and inclusion might be understood as themselves largely enacted through forms of compulsory heterotemporality that depend on treating the straightness of time (and the ongoing transcendence of the past) as given.

If historicism gains legitimacy through its implicit alignment with straightness, deviations from that experience of time can appear as queer. Temporal orientations that do not fit dominant Euramerican frames of reference can be interpellated as abnormal fixations on the past, translated as aberrant tendencies toward anachronism (as opposed to being seen as alternative ways of being-in-time). Within Euramerican discourses, the Indian becomes the paradigmatic figure for these kinds of nostalgic inclinations. Discussing the emergence of protocols of bourgeois grieving in the nineteenth-century United States, Luciano observes, "The life-world of the Indian, exterior to the new nation's modes of ordering, could only be incorporated into its historical timeline through its construction as permanently anterior," later adding, "The progressive substitution of Indian melancholia, the ultimately fatal embrace of the past, by white melancholy, the reflective look backward that enabled one to continue moving forward, thus bespoke, to whites, their own more sophisticated comprehension of the 'true' nature of time's passage."[133] The Indian serves as a symbol of backward relations to time, of insurmountable melancholic investments in the past in contrast to the putative straightness of time's passage. The supposed anteriority of Native lifeworlds provides a model of perverse fixity, and, thus, Indigenous experiences of time seem as if they are a deviant way of remaining caught in the past. From this perspective Indigenous duration can be only the carrying forward of what properly should be past, an inversion of "real" time or "natural" time which implicitly is that of Euramerican historicism.

Conversely, taking queer insights into account can enrich the meaning of historical density when approaching forms of everyday Native perception, storying, and processes of becoming. Rather than being seen as either a function of straight time (heterotemporal transmission) or simply a deviation from dominant settler linearity, storying can be treated as oriented by its own trajectories, giving rise to fields of possibility that cannot be measured within or through settler frames of reference. Conceptualizing time as not only plural but sensuous, as an expression of affective orientations, directs attention toward the need to consider how quotidian forms and feelings of continuity emerge as part of, in Cordova's terms, the work of "maintaining stability" amid ongoing processes of transformation and change.[134] Shared material conditions can engender forms of perception in common, providing a frame of reference through which individuals reckon with their joint environment. However, such an understanding of perceptual tradition can rely too much on the regularity of those shared circumstances and the group's long-term containment within a fairly circumscribed area. If they were ever applicable to Native peoples, those kinds of consistency do not necessarily characterize a good deal of Indigenous experience in the United States since the mid-nineteenth century, given dispossessions, dislocations, privatizations, programs of detribalization, urbanization, and various other mobilities (chosen and coerced). Story offers a means of understanding how collective histories can be immanent within everyday interaction and perception, generating kinds of continuity and connections across time that do not necessarily require immediate contiguity of experience (either geographic or generational). As Miranda suggests, having "an identity and community" is possible even in the absence of a legally recognized land base and amid other forms of fragmentation, and part of what enables the sustaining of peoplehood in conditions of dispersion or diaspora is the felt presence of shared (hi)stories amid the circumstances of ordinary life, stories that intimately animate and orient ongoing collective practices of becoming (of *re-creation*, *reinvention*, and *resurgence*).[135]

However, in challenging presumptions about time's singularity and developmentalism, as well as exploring the variability of kinds of temporal feeling, queer analyses tend to talk about affective connections (usually individual) that cross the apparent gulf between the past and the present, instead of addressing distinct forms of temporal orientation. For example, we might consider Freeman's notion of "temporal drag." In developing this idea of "plastering the body with outdated rather than just cross-gendered accessories," Freeman takes to task notions of performativity focused on repetition for their tendency to privilege "novelty" rather than "anachronism" in ways that suggest that "whatever seems to generate continuity seems better left behind." Freeman, though, seems less interested in investigating the potential for alternative kinds of continuity that are at odds with chrononormative modes of progress than in emphasizing the affective movement across periods, especially inasmuch as it recuperates the "cultural debris" of previous "incomplete, partial, or otherwise failed transformations of the social field."[136] Similarly, articulating a notion of "queer spectrality," Carla Freccero suggests that "doing a queer kind of history means... an openness to the possibility of being haunted" by the ways "the past is in the present," engendering "survivals and pleasures that have little to do with normative understandings of biological reproduction."[137] In insisting on the possibility of having experiences that are temporally indeterminate and/or mixed, this scholarship seeks to undo the chronobiopolitical imperative to live time in ways that line up with various dominant forms of straightness and extant modes of social reproduction. Such work aims to proliferate the possibilities for approaching historicity as dependent on forms of (cross-temporal) feeling instead of as a progressive chronology. Yet the methods developed within this kind of queer analysis seem ill equipped to account for collective frames of reference and experiences of duration. Emphasizing the idiosyncratic, the ephemeral, the spectral works as a way of creating room for other forms of being-in-time amid the insistence on heterotemporality, with its clear inheritances and uninterrupted modes of generational succession. However, such figurations work less well as a means of addressing the temporal robustness of Indigenous modes of self-understanding: the duration and renewal of connections to place and peoplehood, processes of intergenerational storying, their role in orienting everyday phenomenologies of time, and the ways such modes of continuity might serve as the basis for experiences and expressions of temporal sovereignty.

Queerness, then, cannot itself name all that lies outside of normative conceptions of time. Or, rather, using *queer* in such a way ignores how the particular kinds of temporal relations marked by many queer studies scholars may still occur within a settler frame of reference. How might nonheteronormative temporalities, for example, still participate within the life of the settler state and depend on its jurisdictional structure? How might non-natives who deviate from straight time still situate themselves (explicitly or implicitly) as participants within national history, taking the territorial and jurisdictional coordinates that orient that history as their frame? For example, many non-native members of sexual minorities envision themselves as inheriting Native peoples' supposedly traditional tolerance of sexual and gender nonnormative people, whom non-natives treat as their own queer Native progenitors. This kind of cross-time identification does not unsettle non-native political geographies or narratives of Native disappearance.[138] As Beth Brant suggests, "We have learned that a hegemonic gay and lesbian movement cannot encompass our complicated history.... Nor can a hegemonic gay and lesbian movement give us tools to heal our broken Nations."[139] Conversely, to be committed to queer critiques of normalization does not necessarily entail challenging settler frames, nor does it inherently involve a commitment to engaging Native sovereignties and struggles for self-determination.[140]

Deviations from straight time need not inherently mean that they are disjunct from the chronogeopolitics of the settler state, in which the nation-state and its coherence serves as the background against which to register forms of temporal feeling and fields of possibility. However, I also want to avoid a homogenizing dichotomy between Natives and non-natives (such as in the notion that time for one is necessarily circular whereas for the other it is linear). Moreover, I've been using the term *non-natives*, but in doing so, I've almost entirely been addressing white perspectives, narratives, and perceptions. Simply lumping in the experiences of non-natives of color with those of whites erases the ongoing dynamics of racialization and white supremacy, including the ways that people of color have experiences of time that differ from the chrononormativities of whiteness. For example, Marlon Ross has addressed how certain prominent ways of narrating histories of sexuality do not work well across the color line, suggesting that the object-choice-based definitions of "modern sexuality" do not fit the ways nonwhites were understood as perverse regardless of their object choice, and that such modernness can be understood as itself "constructed over and against the premodern present of traditional... sexual practices being engaged in by those not privy to Europe's progress." These differences, then, can be seen as giving rise to "alternative sexual modernities."[141] In a related vein, Afro-pessimist work has suggested that the legacies of the history of enslavement and related modes of antiblackness continue to be determinative for Afro-descended people in ways that mark a clear distinction between black and nonblack social lives, presumably including with respect to experiences of time.[142] Thus, non-natives of color, whom Jodi Byrd has characterized as "arrivants," can be understood as having their own complex relations to dominant modes of temporality.[143] However, aligning Natives with other racialized groups as fellow people of color also can efface the specificities of indigeneity as well as displace the question of to what extent arrivants' experiences of time draw on settler colonial frames of reference, even as they remain outside of the phenomenology and privileges of whiteness.[144]

Furthermore, scholarship on pre-twentieth-century conceptions and experiences of time in the United States suggests the unevenness of the emergence of linear time as the dominant model, also raising questions about the extent to which it should be taken as paradigmatic. For example, Thomas M. Allen suggests, "Temporal heterogeneity... becomes central to the experience of modern collective belonging," adding, "These heterogeneous temporalities are not marginal or resistant to the nation, nor do they represent forms of collective affiliation that will emerge after the demise of the nation. Rather, they are themselves the threads out of which the fabric of national belonging has long been woven."[145] In a related vein, Lloyd Pratt argues, "Conflicts between different modalities of time... forbid the homogeneously linear time whose emergence has sometimes been associated with early American nationalism," later observing with respect to U.S. regionalism, "Modernity's distanciation of time and space produces these figures, but it does not reembed them in a single new order of time synchronized with broader translocal norms—quite the opposite. The distanciation of time and space leads them to inhabit several different orders of time."[146] Yet while these scholars offer differing ways of interpreting the interactions among discrepant temporal modalities, they posit an inherently shared frame of reference—"national belonging" or "modernity"—in and through which these varied times can be brought into meaningful conceptual and causal relation. While displacing linearity as such, then, these accounts posit a particular formation that serves as the background for thinking processes of becoming.[147] To what extent, though, can these frames engage with Native experiences? To what degree do they translate such experiences, and Native social (and temporal) formations, into non-native terms? How might non-native "temporal heterogeneity" or "modalities of time" remain distinct from each other while also taking part in forms of settler expansion and occupation? With respect to settler colonialism and Native sovereignty, what difference do these non-native differences make? Moreover, within such framings, do Native experiences of time appear as parts of a larger formation whose contours and operation remain disconnected from the dynamics of Native peoples' exertion of sovereignty and self-determination (such as in Mignolo's account of "coloniality," discussed earlier)?

Rather than trying to resolve these questions or tensions, my aim is to explore the possibilities that might be opened by conceptualizing Native peoples as occupying temporal formations that are not reducible to non-native ones. I explore how such an approach might facilitate moving past certain intellectual and political impasses that arise when positing a necessarily shared now between Native and non-natives, while leaving open the question of what such an approach might mean for other groups. In particular, my way of approaching Indigenous duration, orientations, and storying aims to undo the tradition/modernity bind by offering an alternative account of continuity. As discussed earlier, the notion of tradition gains meaning by being juxtaposed with the modern, the current, the new. Representing Native stories, knowledges, and experiences as traditional casts them as residual of some other, older time instead of characterizing them as participating in a present whose frame of reference differs from that of the chrononormativities of settler governance. Conversely, understanding the work of story in the present as something like temporal drag emphasizes the leap from the past to the present, the uncanny reappearance of the former in the latter, instead of highlighting the diffusion of stories through networks of relationships that provide the basis for living peoplehood as an ongoing process (the collective "retelling of the past" and "imagining of the future" that Miranda addresses). This kind of continuity produced through the everyday materiality of storying is neither that of the reproductive temporality of familial relation nor the historicist logic of successive unfolding.

Story engenders ongoing forms of connection that are not necessarily about an unbroken chain of possession or inhabitance, an uninterrupted line that can be traced from the present into the past. In this vein Miranda repeats an observation from a lover's letter to her: "You do have stories.... Those stories your dad tells are connected with older stories, stories that might not have been passed down to you, but which existed and maybe even still exist in a world that isn't this one.... It is a fragment in one way, but like the shard of a pot that can be restored."[148] The stories that have not "been passed down" can be understood as both temporally continuous and discontinuous in different ways. Their effects on the orientations of older generations may become part of a younger generation's ways of being in the world, even without the transmission of the stories per se or with the communication of only some or part of those stories. In addition, while the stories here are passed through Miranda's father, her text suggests the possibility of receiving stories from people other than one's heteronuclear forebears. Further, the stories may themselves be recovered, reconstructed, or remade from the "fragment[s]" that remain, generating forms of temporal relation that are not those of continuous succession (such as in Miranda's return to the stories of her great-grandfather and the stories contained in the notes of the anthropologist John P. Harrington—including those of her relative Isabel Meadows). A sense of peoplehood is conveyed as a felt knowledge that itself gives momentum to the conscious search for such stories, as in Miranda's figure of "underground rivers" that run "beneath the surfaces" of individuals' conscious perceptions.[149] Instead of necessarily following the lineality of familial inheritance, storying accretes cross-references, resonances, and recollections, giving historical density to everyday Native perception by endowing it with collective forms of temporal breadth. As Kimberley Blaeser suggests, "When we invoke teachings and tell ourselves into communities, we build a genealogy of story."[150]

To return to "the land question," story provides ways of connecting peoples and places that encompass territoriality as a key part of the sense of time's unfolding.[151] Contrasting the legalities of reservation territory with more expansive and shifting Indigenous relations to place (including urban centers), Mishuana Goeman notes, "Stories teach us how to care for and respect one another and the land. Responsibility, respect, and places created through tribal stories have endured longer than the Western fences that outlined settler territories and individual properties that continue to change hands."[152] Being in place entails having collective stories that provide orientation with respect to that place's relation to other places, its ongoing participation in a shared history and futurity, and the ethics that guide how one connects to the land and to other people. Such emplaced and emplacing stories (what Coulthard refers to as "grounded normativities") generate a frame of reference for relation across time, but less like an inheritance passed generationally—something akin to an heirloom—than a potentially open-ended way of (re)connecting to social and physical landscapes.[153] As Keith Basso says of Apache practices of place naming, "the place-maker's main objective is to speak the past into being, to summon it with words and give it dramatic form, to *produce* experience by forging ancestral worlds in which others can participate and readily lose themselves." Later he observes, "By virtue of their role as spatial anchors in traditional Apache narratives, place-names can be made to represent the narratives themselves, summarizing them, as it were."[154] Stories, then, give meaning to current and former occupancy in particular places while also conjuring the specificities of those places, producing kinds of experience and forms of relation that cross apparent temporal gulfs but do not arrive as an uncanny or spectral remainder. These connections to place exceed the terms of individual affect and transect the chronogeopolitics of settler policy and popular narratives. Everyday participation within such storying produces emotional and sensory investments in placemaking that give shape to and help animate collective processes of becoming and ways of being-in-time that can be understood as expressions of temporal sovereignty.

Conceptualized in this way, Indigenous duration operates less as a chronological sequence than as overlapping networks of affective connection (to persons, nonhuman entities, and place) that orient one's way of moving through space and time, with story as a crucial part of that process. In this way, storying helps engender a frame of reference, such as by providing a background against which to perceive motion, change, continuity, and possible action in the world in ways that cannot be encompassed within dominant modes of settler time. Stories become part of, in Cruikshank's phrase, the "equipment of living," furnishing density to everyday forms of perception, informing the direction of individual and collective trajectories, and giving them momentum.[155] More than representing events, stories, in the words of Heidi Kiiwetinepinesiik Stark, "*do things*, like provoke action, embody sovereignty, or structure social and political institutions" and, in doing so, they open up alternative temporalities to those institutionalized within Indian policy.[156] Approached in this way, storying can be understood as remaking the potentially rupturing effects of settler colonial violence (like removal, allotment, and termination) into part of the affective repertoire through which indigeneity persists as such despite the force of non-native occupation. Such occurrences clearly have profound effects on everyday experience, yet they need not be understood as a kind of epochal rupture. Rather, they become, via story, part of the perceptual tradition through which the present is experienced—through which to reckon with the contemporary field of possibility. Such stories connect the current moment to other sites and sensations in ways that may be messy, multiple, and conflicted but that remain extensions (rather than disruptions) of the complex temporality/ies of Indigenous peoplehood(s).

The project's turn to textual analysis as a way of engaging with these dynamics reflects a conceptual and political investment in storying as a mode of world making (as well as my own training and inclinations). If story has the ability to realize modes of perceiving and living time, then that potentiality can be enacted through Native forms of writing and cultural production. As Goeman observes, "the literary maps of Native people presented in oral stories, or later in writing" offer "subversive or alternative geographies," later adding that "Native literature provides a mechanism to see the limits of territory, as it is legally interpreted from original treaties, and give sustenance to Native people's relationship to the land" and that such narratives offer "examples of a writer's ability to disrupt the 'truths' of settler colonialism."[157] The Native literary texts I engage can be understood as themselves engaged in a reverse process of translation, seeking less to make Native modes of becoming intelligible to non-natives than to mark that distance and disjuncture in ways that highlight the violences entailed in normalizing settler time. Attending to Native texts opens up possibilities for envisioning and engaging with alternative temporalities, ones that do not fit within official and ordinary non-native accounts.[158] Reciprocally, settler "truths" about time can be understood as conveyed through non-native texts, governmental and popular, making them valuable as objects of analysis in order to investigate the contours and limits of dominant forms of historicity.

While these various kinds of texts can be approached as instantiations of temporal formations, an analytic procedure I perform at various points throughout the book, I also should note the queerness of my own intellectual aims. As mentioned previously, I'm less interested in demonstrating the accuracy of my claims about time by marshaling proof that it *actually* functions in such ways than in offering this intellectual account of time as a set of interpretive possibilities—as a hermeneutic. How might these texts be read in ways that highlight the potential for alternative experiences of time to those normalized within non-native articulations of *the* past, *the* present, and *the* future? How might we think against the sense of time's unity and coincidence with settler interests and imperatives, and what might doing so yield? How might emphasizing such alternatives aid in conceptualizing and living forms of Native self-determination? My orientation to the materials gathered in this study, then, might be characterized in Eve Kosofsky Sedgwick's terms as "reparative," trying to contest the inevitability of time's singularity in ways that sketch possibilities for imagining and feeling otherwise.[159]

## TWO. THE SILENCE OF ELY S. PARKER

I'm watching *Lincoln* (2012) in the movie theater, and in the middle it cuts from a scene about the machinations of political operatives paid by the Lincoln administration to secure House votes for the Thirteenth Amendment to a scene of General Ulysses S. Grant in Virginia meeting with Confederate commissioners. One of the men in the frame appears to me to be Native. Later, at the end of the film, in a scene set at the Appomattox courthouse, the site of General Robert E. Lee's surrender, that man returns, and I suddenly realize that it must be Ely S. Parker.[1] A chief of the Tonawanda Senecas who played a central role in their fight to reclaim their reservation in the wake of the supplemental Treaty of Buffalo Creek of 1842, Parker was Grant's secretary and aide from his enlistment in the army in 1863 through Grant's election as president (Parker had originally met Grant in 1860 while serving as a civil engineer in Galena, Illinois), later serving as commissioner of Indian Affairs during the Grant administration.[2] The published screenplay confirms the historical identity of this figure on the screen, including Parker in the official cast of characters.[3] In the scenes in which Grant appears, though, it never designates Parker by name, instead including him as one of the general's other unremarked-upon "aides," "officers," or "staff."[4] Furthermore, at no point in the movie does Parker speak.

The film's screenwriter, Tony Kushner, clearly felt it necessary to represent Parker's existence for the sake of accuracy, but even as the movie captures his presence, it vacates that fact of any real significance, with Parker functioning more or less as a historical prop. The film's simultaneous awareness of his presence and failure to address it can be read as a kind of allegory for the ways Native sovereignties, histories, and struggles with the U.S. settler state cannot meaningfully enter into a historiographic imagination in which the Civil War is taken as indicating the immanent trajectory of national time, especially one in which emancipation serves as the key moment for casting the nation as always in the process of becoming democratic. Onto what other conceptualizations of time might Parker's presence open? If we were to approach *Lincoln* as something like a condensation of how the Civil War continues to be envisioned in the United States, how might Parker's unnamed and silent appearance suggest the ways that such accounts of the U.S. past, present, and future foreclose an engagement with Indigenous polities and geographies as well as the ongoing history of state assaults on them? How does the focus on the Civil War help orient national temporality, indicating a particular set of background assumptions about what constitutes continuity and change in the United States as well as the continuing efficacy of non-native modes of perception in which Indigenous struggles and self-determination appear as marginal (when registered at all)? Parker's passing and uncommented-on entry into the film's imagination can be read as a trace, pointing toward temporalities of indigeneity and settler colonialism that remain unintelligible within the narrative of the Civil War as an epochal and/or redemptive break.[5] Attending to *Lincoln*'s way of envisioning the war and its meaning for U.S. history allows for a broader discussion of how periodization participates in the storying and unfolding of national time, giving shape to the present and the sense of its potentials by reactivating, in Merleau-Ponty's terms, an inherited perceptual tradition.[6] More specifically, within the tradition embodied by the film, in which the national union provides the de facto frame of reference for marking the movement of time, Native people(s) appear anomalous, as if they emerge from nowhere in ways disjointed from other processes of becoming.

Clearly a product of contemporary perspectives and sensibilities, *Lincoln* suggests the role the Civil War plays in current forms of national self-representation, particularly as they seek to cast the United States as a multicultural nation committed to modes of antiracist inclusion.[7] Within this vision of history, emancipation serves as a crucial marker of a national commitment to equality, but one that requires a break in the usual operation of the nation in order for the promise of liberty to be actualized. In *Lincoln* the Civil War appears as a transformative caesura in the time and space of the nation, an exception in which the usual legal and political order of the country is suspended in the process of realizing national ideals. Giorgio Agamben has theorized "the state of exception" as "the legal form of what cannot have legal form." He describes the exertion of state sovereignty as a process that places normal law in abeyance while both claiming to be acting in the spirit of the law and using that deferral as a means of enacting what otherwise would be considered illegal forms of violence (such as detention, torture, and murder)—usually against subjects of the state who would nominally have recourse to forms of protection under the regular legal regime.⁸ With respect to the Civil War, though, the exceptional recourse to state violence is narrated as a means of endowing previously abject subjects with personhood. Those who were held as slaves become full participants in national progress through the suspension of normal jurisdiction, such that the war less interrupts the unfolding of U.S. history than serves as a paradigmatic moment in its realization. A national norm of universal equality that has no precise legal status (and that directly contravenes the institutionalized structure of chattel slavery) ostensibly serves as the basis for a suspension of law in the war and the implementation of a new law that puts into legal form that which was always supposed to be, but never actually was, the legal order of the nation. This story, reiterated in *Lincoln*, presents the ideal of increasing freedom, personified in the dechattelment of African Americans, as immanently animating the nation's history and moral life (regardless of actually existing institutional structures at any particular moment). The film, then, invokes the Civil War as the ever-present possibility of a break with the past but one that marks less a rupture in national evolution than the means through which the inherent democratic tendencies of the nation can be materialized—a dynamic I describe as the *emancipation sublime*.⁹ *Lincoln* predicates the resolution of the legacy of slavery on the ratification of the Thirteenth Amendment, which putatively remediates antiblack racism and institutionalized white supremacy, and narrating the Civil War in this way circulates it as something of a symbol of national promise, as well as the potentially ambiguous relation of the actual legal order to such putative ideals at any given time.¹⁰

However, when national history is cast as pivoting around the Civil War (itself presented as the vehicle through which emancipation can be realized), what happens to the ongoing processes through which Native peoples and lands are made domestic? What role do they play in the nation's ongoing becoming? Or, more to the point, how does the emphasis on the war as the means of progressing toward greater freedom illustrate the ways the experience of national history, and its division into meaningful periods, depends on treating the union as the necessary background? How can the film be thought of as engaging in a process of backgrounding whereby the national union comes to appear as necessary to the movement of time itself? The union provides, in Sara Ahmed's terms, "the conditions of emergence" for antiracist futurity.¹¹ What room is there for thinking about how the union's development—and the need to resort to exceptional violence in its defense—depends on the persistent and continuing subjection of Indigenous nations to settler governance? As Jodi Byrd argues, "The generally accepted theorizations of racialization in the United States have, in the pursuit of equal rights and enfranchisements, tended to be sited along the axis of inclusion/exclusion," figuring oppression as being "redressed by further inclusion into the nation-state" in ways that present "the foundation for U.S. participatory democracy" as something other than "the colonization of indigenous peoples and lands."[12] The notion of U.S. history as itself oriented by a never-ending process of democratic development provides historical density to contemporary forms of perception. Within this frame of reference, Native peoples, histories, and sovereignties appear not simply as irrelevant but as aberrant, as if they could not exist in time because they do not form part of the trajectory through which the present emerges from the past. The problem, then, does not lie in the failure to recognize Native histories such that they could become part of the process of national storying but rather in the ways the temporality of the union—the crucial role it plays in orienting settler experiences of time—remains disjunct from Indigenous temporalities and antagonistic to the potential for Native temporal sovereignty.

Parker's mute unintelligibility within the film—his highly visible yet meaningless presence—suggests how *Lincoln* relies on certain commonsense ways of apprehending the national past, present, and future.[13] His indigeneity ill fits the vision and experience of national evolution that *Lincoln* enacts, the itinerary of national progress it implicitly traces by invoking the Civil War in the ways it does. Parker's evident yet silent Indianness marks the movie's inability to engage with the particular history of the Seneca people, and the dynamics of Indian policy writ large, in which Parker was enmeshed before and after the war. The film speaks to certain persistent ways of inhabiting and storying national time that orient non-native sensations of history, shaping the possibilities for engaging with Indigenous experience while helping legitimize settler incursions by casting them and their effects as the immanent unfolding of the democratic and civilizational promise of the union. The kind of temporal framings at play in the film can be seen as also at work in nineteenth-century Indian policy. Specifically, focusing on Parker's role in Indian affairs both before and after the Civil War as well as during the Dakota War of 1862, which occurred just weeks before the issuing of the Emancipation Proclamation, highlights the roles played by non-native narratives and experiences of time in the process of dispossessing Indigenous peoples.

If one understands that the violence of settler colonialism need not be primarily understood through exceptional events of spectacular violence (like the Civil War), one can begin to narrate U.S. national history as an accumulation of mundane, state-sanctioned processes. The insistent pressure, imposition, and dispossession enacted by settlers backed (sometimes implicitly, sometimes explicitly) by the U.S. government over the course of generations can be characterized as a "quasi-event," in Elizabeth Povinelli's terms. Povinelli defines a quasi-event as circumstances in which "nothing rises to the level of an event let alone a crisis"—"legal forms of bad faith" or "not being killed by the state in any way that would be recognizable as state killing."[14] One might describe such relations as a temporality of attrition, a relentless assault on Indigenous occupancy and placemaking conducted largely through nonmilitary, bureaucratic means, justified on the basis of a story of continuing advancement—a chronogeopolitics of progress. Treaties serve as the mechanism for such lawful dispossession.

Expressions of Native self-determination are figured as aberrance (as anomalous deviations from the liberatory baseline of national union), rather than as politically meaningful expressions of Indigenous duration and peoplehood. Such struggles over the meaning of temporality were very much at play in the crises faced by Dakotas and Senecas throughout the nineteenth century.[15] Moreover, as Dian Million argues in *Therapeutic Nations*, colonialism operates "as a *felt*, affective relationship," and these examples bring such affects to the fore by emphasizing how Native "oral traditions[,] and ... literary and historical voices, are suppressed by western knowledge that denies its own affective attachments to certain histories."[16] This chapter, then, explores the ways Indigenous peoples' felt sense of time is supplanted by official narratives that are rooted in settler phenomenologies and policy imperatives but that are cast as simply neutral recountings of a shared history.

If nineteenth-century Indian policy seeks to impose a particular trajectory of national becoming onto Native peoples, in which their insistence on their own futurities appears as perverse recalcitrance, the question of how Native modes of peoplehood survive and develop amid this colonial onslaught seems key. How does peoplehood persist as a mode of being and becoming in time in the wake of such profound settler violence? Furthermore, how does such violence affect Native temporal frames of reference without simply displacing them (or creating some kind of existential rupture in Native identity and continuity)? The writings of Charles Alexander Eastman illustrate how the dynamics and dislocations of this period become part of the self-understanding of the next generation. Himself driven into diaspora by the Dakota War of 1862, as well as serving as the doctor on the Pine Ridge reservation during the Wounded Knee massacre (1890), Eastman was steeped in the effects of Indian policy in the late nineteenth century—the brutalities enacted through treaty making, combined with the employment of overwhelming military force against Native movements for survival deemed deviant by the government. Yet, while registering the proliferation of settler invasions and assaults, his writings explore the continuities of Indigenous history (Dakota, in particular) across these acts of occupation, and the ways they become part of the background for Native experiences of time, without producing something like a temporal break into a shared modernity. He does suggest that Native people(s) can become national subjects, but, in doing so, his account of what that belonging might mean, and of the significance of Indigenous presence and continuance, differs greatly from the perception of national time at play in dominant accounts of U.S. history. Addressing what happens in the wake of the mounting quasi-events of settlement, Eastman's texts illustrate how stories and legacies from the nineteenth century help orient early twentieth-century modes of Indigenous becoming.

*The Civil War as the Horizon for Emancipation*

In the foreword to the published screenplay for *Lincoln*, Doris Kearns Goodwin suggests, "In an age when we are cynical about politicians and frustrated by our political system, Tony Kushner's screenplay is a vivid testimony to the ultimate strength of our democratic form of government—the revolutionary idea, designed by our founding fathers and secured by the Civil War, that ordinary people can govern themselves without kings or queens, dictators or tsars."[17] Viewed in this way, thinking about the Civil War allows Americans in the present to reinvest emotionally in the nation because that conflict helped realize the "revolutionary idea" ostensibly embodied in the existence of the U.S. government—namely, rule by the people. This gloss suggests the kind of perceptual tradition through which the war is remembered: the war makes possible the persistence of the "democratic form" of the nation and the union, serving a crucial role in perpetuating the ideal represented by the country's existence. The vast violence of the Civil War does not mark the failure of the juridical-political order but its reinvigoration. From this perspective, emancipation as a specific set of political actions and their particular effects—in terms of the executive order by the Lincoln administration during the war or the ratification of the Thirteenth Amendment—takes part in the narration of the Civil War as a triumph of democracy. As A. O. Scott says in his review of the film in the *New York Times*, the film offers "a civics lesson that is energetically staged and alive with moral energy."[18] Or, flipped over, such storying of the Civil War depends on its fusion with emancipation, which itself serves as a sign of the nation's tendency to become ever more democratic. From a slightly different angle, the linkage of the Civil War with emancipation as its internal truth allows the work of preserving the union, and thus the union itself, to appear as a necessary precondition for the achievement of democracy. The structure of the U.S. state appears as the political "form" through which that "revolutionary idea" will be immanently materialized, even in the absence of a clear indication of how. In this contemporary vision of the nation and its past, the Civil War *secures* the union as the precondition for emancipation, as the a priori vehicle that makes democracy possible. The war, then, represents not a rupture but an exception in which the norm of democracy can be cast as (increasingly) animating the juridical structure of the U.S. state. Furthermore, with its inherent democratic trajectory, the union provides the background against which to register the becoming of national history in its increasing materialization of liberty. This dynamic, the ways *Lincoln* positions emancipation as a key moment in the unfolding of national time, can be characterized as the *emancipation sublime*.

The film casts the war as most certainly about the struggle over slavery, and in doing so, it presents the conflict as itself a return to first principles. At the beginning of the movie, after encountering two black soldiers at the battlefield of Jenkins' Ferry in Arkansas who question his commitment to integration and African American civil rights, Lincoln comes upon two white soldiers who quote to him from the Gettysburg Address. One declaims, "Our fathers brought forth on this continent a new nation, conceived in liberty," followed by the other saying, "Now we are engaged in a great civil war, testing whether that nation or any nation so conceived and so dedicated can long endure."[19] These parallel encounters, the one coming right after the other, raise the question of the extent to which the "liberty" in which the nation was "conceived," and for which a "great civil war" is being fought, applies across the color line. The movie answers this question somewhat decisively when Lincoln, talking to workers in the White House telegraph room, observes in commenting on Euclid's theorems that "it is a self-evident truth that things which are equal to the same thing are equal to each other. We begin with equality." The promise of "equality," which appears as the substance of the "liberty" that the "new nation" of the United States was created to achieve, not only extends to African Americans, but the film depicts the effort to secure racial equality through the abolition of slavery as the normative aim of the war itself. At one point Lincoln asks Elizabeth Keckley, Mary Todd Lincoln's African American dressmaker, what might follow "once slavery's day is done," and she responds, "I never heard anyone ask what freedom will bring. Freedom's first.... My son died, fighting for the Union, wearing the Union blue. For freedom he died. I'm his mother. That's what I am to the nation, Mr. Lincoln. What else must I be?"[20] In this scene the film offers a syllogistic series as a moral certainty: to die in the war on the Union side is to die fighting for the national union; to fight for the national union entails a battle for an end to the institution of chattel slavery; and the elimination of that institution through a process of emancipation means the achievement of "freedom." While freedom has no clear referent other than the cessation of lawful enslavement, it functions as a self-evident aim that can provide significance for the deaths incurred in the war.

The term freedom, treated as more or less synonymous with equality, appears as if it were semiotically self-sufficient, as naming a "truth" that might be reflected *in* law but that transcends the definitional parameters *of* law. The question of what specific legal capacities, entitlements, and responsibilities might attach to freedom seems irrelevant to understanding its contours or promise, especially inasmuch as freedom animates national time (providing its beginning and its driving impulse or organizing principle).[21] In a meeting with his cabinet, Lincoln asserts, "I can't accomplish a goddamned thing of any human *meaning or worth* until we cure ourselves of slavery and end this pestilential war. . . . *This amendment is that cure!* . . . Blood's been spilt to afford us this moment!"[22] The Thirteenth Amendment both instantiates something new (realizing the "cure") and merely reflects an existing ("first") principle of liberty—what "our fathers brought forth" in the Revolution. Ending slavery, then, gives the war, in all its carnage, "meaning" and "worth," such that emancipation provides the (somewhat catachrestic) guiding principle that stretches across the rupture that is the conflict itself. In a meeting at Fort Monroe in Virginia, the vice president of the Confederacy, Alexander Stephens, confronts Lincoln, saying, "How've you held your Union together? Through *democracy*? How many hundreds of thousands have died during your administration? Your Union, sir, is bonded in cannonfire and death," and Lincoln responds, "Say all we done is show the world that democracy isn't chaos, that there is a great invisible strength in a people's union? Say we've shown that a people can endure awful sacrifice and yet cohere? Mightn't that save at least the *idea* of democracy, to aspire to? Eventually, to become worthy of?"[23] This exchange highlights the gap between the union as ordered around a particular set of legal structures and the union as an attempt to realize the "*idea* of democracy," as well as underlining the ways the potential disjunction between the two generates extraordinary violence in the effort to make them "cohere."

In this way the film partakes in the contemporary ideological formation that Chandan Reddy has described as "freedom with violence." Reddy defines this notion as the ways that "socially and institutionally produced forms of emancipation remain regulatively and constitutively tied to the nation-state form," further noting, "The irony, of course, is that the social history of the excluded community is now dependent for its conditions of representational existence on the popular affirmation of the norm from which it was excluded."[24] One might add that such inclusion also relies on a temporal frame of reference in which emancipation either was always already on the way or has been definitely proclaimed, such that to be part of that history is to have no experience of time in which racist violence constitutes the union and national policy as such.[25] As part of a speech addressing the legality of the Emancipation Proclamation in a conversation with his cabinet, Lincoln makes an argument for the necessity of the Thirteenth Amendment, observing, "I decided that the Constitution gives me war powers, but no one knows just exactly what those powers are. Some say they don't exist. I don't know. I decided I needed them to exist to uphold my oath to protect the Constitution."[26] Here the film represents the Constitution as less a juridical framework than a temporally expansive idea of the union and its role as a vehicle for the unfolding of democracy, liberty, freedom, and/or equality, despite the limits to the actual "powers" granted and distributed in the Constitution as a functioning legal document.

To the extent that ending slavery can be imagined as the aim of the war and as the realization of the liberty for which the nation was founded, this goal helps legitimize the defense of the union, as well as the powers invoked and invented in doing so. *Lincoln* casts the union as inherently bearing the ideal of racial equality. Moreover, the Thirteenth Amendment functions as the "cure," formally enshrining in the Constitution the idea of democracy and/or equality for which the war ostensibly was fought. The necessity of holding together the union gains normative meaning and justification from its association with the movement for black freedom. In his review A. O. Scott suggests that *Lincoln* functions as something of a rejoinder to *The Birth of a Nation*, correcting such earlier "distortions" of the historical and national record. Roderick Ferguson suggests in *The Reorder of Things*, with respect to the discourses of racial equality since the civil rights era, "The state manages diversity through political emancipation," engaging in a "promotion of diversity without ensuring equal opportunity" in ways that allow the state (and inclusion within it) to be cast as a vehicle for securing freedom and liberty without the prospect of a more far-reaching "antiracist redistribution" of material resources.[27] Reciprocally, the bloodshed of the war and the potential violation of the law in its conduct look like temporary oddities when seen in light of the immanent relation between law and equality that supposedly characterizes the nation's governance. In this way the film enacts a genealogy in which the Civil War operates as a signal moment in the becoming of American freedom (characterized by reviewers as "a pivotal moment in America's history" and as engaging "the nation's central political and moral debate"), in which cohering the state's jurisdiction functions as a moral, antiracist necessity. Through this prism, the expansion of governmental authority and the eruption of state-licensed violence work in the service of an antiracist extension of the potential for democracy that dwells within the idea of the nation—if not, perhaps, its actual laws and juridical structure.[28]

Not only does the film's approach to national time displace antebellum and postbellum Indigenous struggles over sovereignty and territory, for which the war is not an inherently meaningful marker of change or development, it also casts Parker as Indian flesh. As Alexander G. Weheliye suggests, drawing on the work of Hortense Spillers, "flesh" alludes to "forms of political death" enacted through a "biocultural stigmatic apparatus in which ideas are literally and figuratively deformed into racialized assemblages ... that invest human phenomenology with an aura of extrahuman physiology."[29] Flesh appears as the racialized excess beyond the "body," itself defined with reference to normative modes of (presumptively white) personhood. The film's figuring of Parker as a generic, unnamed Indian can be understood as an expression of a broader chronobiopolitical dynamic whereby the continuities of Indigenous landedness and governance, as well as persistent dispossession by the United States, become perceptually condensed as the racially marked presence of Native persons. Indians' incongruity against the available background signals their nonimportance to the process of national becoming. Through this phenomenology of Indianizing enfleshment, the temporalities of Indigenous self-determination and settler occupation disappear. Weheliye's project, though, entails drawing on blackness, and theorizations of it, as a way of thinking other "genres of the human" than the post-Enlightenment personhood that he, following Sylvia Wynter, terms "Man," and he suggests the need to explore "the social life found in those bottomless circles and circles of sorrow around political violence."[30] In addition to bringing into relief the limits of the emancipation sublime as a way of figuring the history or horizon of black freedom struggles, Weheliye's analysis further points to how the flesh, which from a dominant perspective signals merely absence (of personhood, capacity, civility), opens onto various kinds of nondominant sociality. In a complementary fashion, although in terms quite different from those Weheliye uses, one can interpret Parker's presence as inadvertently indexing Indigenous spatiotemporal formations that do not fit within the film's frame of reference.

The film, then, can be understood as enmeshed within settler colonial temporalities. It identifies the union as the political and normative fact that must be defended against all threats to it, including Native peoples whose claims to territoriality, sovereignty, and self-determination potentially challenge the legitimacy of the United States and its jurisdiction. Moreover, the film licenses state-organized violence, in excess of any constitutional or legal framework, as the means of securing that union, while positioning national law as the ultimate vehicle for remedying racial inequity. From such a perspective, the incorporation of Native peoples into the state/union constitutes a promise of liberty rather than an act of imperial subjection.[31] In "Contract and Usurpation" Robert Nichols observes, "Even though antiracist discourses may serve a critical and destabilizing function in one context, in another (specifically settler-colonial) context, that *same* discourse may serve a totalizing and hegemonic function," and he further suggests, "The normatively favored solution to the problem of racism" is often "a more expansive, universalist redescription of personhood or humanity, realized through a deeper integration of racialized subjects under the legal protection of (unproblematized) colonial sovereignty," adding that "one of the ways the settler contract continues to function today is through the hegemonic reproduction of its universality through antiracist discourse."[32] The universality claimed by the settler state as a harbinger of democracy—and the vision of antiracist inclusion marshaled to legitimize this conception of the state as *the* vehicle for the achievement of freedom—depends on a form of temporal totalization. Not only is the nation-state (or, in the case of *Lincoln*, the union) the a priori political form in which to remediate racism by incorporating racialized subjects as nationalized persons, it also serves as the only basis for marking the passage of time. As the immanent bearer of antiracist progress, the national union serves as the singular frame of reference through which to perceive the unfolding of justice and the possibilities for its realization. Kinds of political collectivity that do not take the form of the union, and that actively contest the legitimacy of its jurisdictional and administrative claims, can have no being-in-time or must be residues from a past that precedes the formation of the union—one that does not contribute to the becoming or extension of liberty. In this way the exertion of settler sovereignty emerges through a process of temporal backgrounding whereby Native people(s) are not perceived as acting in time, except to the extent that they participate as subjects in the "revolutionary" legacy of national founding, identity, and transcendence. The silent figure of Ely S. Parker—the mute facticity and fleshliness of his visible Indianness, which has no other meaning within this sense of history—testifies to the colonial force that orients national time.

*Inexplicable Violence*

The Civil War functions as a key figure in a dominant narrative of national history (as in *Lincoln*), but its exceptional status relies on an unstated exceptionalization of Native peoples. That process is less particular to the contemporary moment than a sustained feature of U.S. policy from the Revolution onward. As Aileen Moreton-Robinson suggests, "Slaves were brought to America as the property of white people to work the land that was appropriated from Native American tribes.... Thus, the question of how anyone came to be white or black in the United States of America is inextricably tied to the dispossession" of Indigenous peoples.[33] If one foregrounds Indigenous self-determination, U.S. history appears as an endemic and irresolvable legitimacy crisis, in which the prior claims of Native peoples to the domestic space of the nation makes the assertion of U.S. jurisdiction utterly incoherent except as a colonial imposition. However, the U.S. government at all levels has persistently cast Native peoples as anomalies within the normal operation of law and politics. Doing so, up through the present moment, allows for the structural challenge they pose to national identity, authority, and morality to be bracketed. They are perceived as aberrant figures who seem to appear out of nowhere, with no clear roots in the past or future. As noted in chapter 1, Ahmed suggests, "The figure 'figures' insofar as the background both is and is not in view," and in this way the figure of the Indian figures as such against the background of the chronogeopolitical normalization of U.S. jurisdiction and settler occupation, which orient the sensation of time's passage.[34] Within this frame of reference, armed Native action in response to these expropriations and the conditions of destitution created by Indian policy seems deviant, appearing as discontinuous with circumstances before and afterward such that the outbreak of violence is experienced as (Indian) oddity rather than as symptomatic of broader, long-standing, and ongoing patterns of settler invasion. Such counterinsurgent storying, in which Indigenous resistance appears as the exceptional atrocities of barbarous Indians, functions as the dominant account of the Dakota War of 1862, effacing the persistent violence that engendered the uprising as well as the experiences of time and sense of historicity that endowed it with meaning for Dakota people.[35]

After ceding millions of acres to the United States through treaties in 1837, 1851, and 1858, Dakota peoples in 1862 were confined to a seventy-mile strip along the southern bank of the Missouri River. As Waziyatawin suggests, "By the mid-nineteenth century, most of the treaties had simply become a form of legalized land theft. Nowhere is this more apparent than the treaties negotiated with the Dakota people of Minnesota."[36] In light of the vast constriction in their access to hunting resources and other possibilities for securing subsistence, Dakota peoples had become increasingly dependent on treaty-guaranteed annuities in order to purchase needed goods from traders, who preyed on Dakota vulnerability while charging heavily inflated debts against federal funds. In 1862 the annuity had not arrived by late spring, when it had been expected, and Dakotas at both of the agencies assigned to them were starving.[37] The urgency of the situation was illustrated by the armed Native demand for access to the food stores of the Upper Agency (at Yellow Medicine) beginning on August 4, which lasted until four days later, when the recently appointed agent, Thomas J. Galbraith, finally capitulated. The event that immediately precipitated the war directly points to this broader context of destitution. Beginning as an attempt on August 17 by a small group of Mdewakanton warriors to steal eggs from Robinson Jones (a local settler with whom they had traded previously), the raid quickly escalated into the murder of Jones and his family. After retreating to consult the leaders of their village, Rice Creek, they and the rest of their band quickly decided to seek shelter with Little Crow, an important leader who had been a signatory to the treaties of 1851 and 1858.[38] The warriors from Rice Creek pressured him and his band to join them in hostilities, and on the morning of August 18, he took part in an assault on the Lower Agency (at Redwood) in which twenty Americans were killed and ten captured. Over the next week or so, Dakota forces attacked Fort Ridgely and various settlements in the area, particularly New Ulm. Upon hearing of the conflict on August 19, the governor of Minnesota, Alex Ramsey, appointed the former governor and longtime Indian trader Henry H. Sibley as a colonel to lead state militia units.[39] When Sibley arrived on September 26 at the camp in the vicinity of Yellow Medicine occupied by the "peace party," which had established its own site separate from Little Crow and others who desired to continue the war, he took into U.S. custody over twelve hundred Dakotas, augmented by almost eight hundred more who came in over the next several weeks—including many who had left to travel westward with Little Crow but later decided to return or were captured in U.S. military raids to the west.[40] By October 17 Sibley had almost four hundred prisoners, and the dependents of those men numbered around sixteen hundred.

On September 28 Sibley ordered the creation of a five-member military commission to try anyone who had participated in the attacks and had committed violence against settlers. After the conflict had already begun, General John Pope was reassigned from his previous station in northern Virginia to take command of the situation in Minnesota, largely as a punishment for his failures during the Second Battle of Bull Run in late August, and by early October Pope was in communication with his superior officers and the War Department about the trials occurring under Sibley. Although Pope had instructed Sibley to execute those found guilty and to send the remainder of the prisoners and their dependents to Fort Snelling, that order was countermanded by a message Pope received in mid-October indicating that President Lincoln wished to evaluate the findings of the military commission. The trials continued until November 2, and by the end of the process, 392 Dakota men had been tried, 323 convicted, and 303 sentenced to death. Those convicted were sent to Mankato, while all the other Dakotas held in custody were sent to Fort Snelling. Based on an assessment of the trial transcripts and the evidence presented in each case, Lincoln commuted the sentences of the vast majority of those convicted, and on December 26 thirty-eight Dakotas were hanged at one time as the penalty for their participation in the war. In the spring of 1863, those who had been convicted were sent to Camp McClellan in Davenport, Iowa; their dependents were removed from Fort Snelling to the Crow Creek reservation; and Congress passed a statute to remove all Dakota peoples from Minnesota. The remaining 177 Dakota prisoners who had survived their internment of over three and a half years were pardoned by President Andrew Johnson in March 1866.

Perhaps the clearest and most extensive official description of the war is offered by Thomas Galbraith in his report as Indian agent, filed on January 27, 1863. He cannot conceptualize the events of the previous fall as a war, illustrating how his Indianization of the Dakotas enacts a chronobiopolitics in which their actions appear as expressive of innate bodily tendencies rather than part of a historical trajectory. He describes the Dakota turn to violence as "the recent and, although smothered, yet existing rebellion, or murderous raid," "atrocities and savage outrages of the Indians," and an "outbreak."[41] Characterizing the conflict in these ways denies it the kind of political significance or intelligibility that would accompany its designation as a military clash between contending sovereigns. Instead, the language here suggests something like an act of treason, an attempt at theft accompanied by bloodshed, wholesale unmotivated slaughter, and/or a natural disaster on the order of an epidemic. Galbraith observes, "In the beginning it was the intention of [Little] Crow to make regular 'war' after the manner of white men, but his 'braves' having tasted of blood and plunder became wild and unmanageable and again yielded to the popular current, and 'Crow's war' degenerated into a savage, barbarous, and inhuman massacre."[42] The intent to engage in "war," performing violence in a proper political mode like "white men," quickly *degenerates* into a "wild and unmanageable" orgy of carnage driven by an insatiable "taste" for "blood." This depiction presents efforts to interpret such struggle in the ways one would if it were conducted by civilized nations as ridiculous, thereby forestalling an engagement with the questions of sovereignty at play in Dakota grievances and choices. Galbraith declares, "Whenever Indians on a large or small scale commit crimes, they should be promptly punished."[43] Indian violence cannot constitute an act of *war*, and therefore an expression of Indigenous sovereignty, because, instead, it needs to be explained and adjudicated as *crime*—the inappropriate and unmotivated actions of Indian persons (presumed to be subjects of the state) rather than the performance of a collective politics arising from generations of inhabitance and decades of escalating non-native incursion.[44]

The interpretation of Native uses of force as criminality depends on a framing in which Indigenous aims remain temporally unintelligible. While the Civil War can be felt and remembered as a signal moment in the process of national becoming, particularly inasmuch as the war is fused with emancipation, the Dakota War appears as a rupture in time owing largely to the participants' *Indianness*. In seeking to answer the question, "What was the cause of the outbreak?" Galbraith insists that first "it will be necessary to strip the Indian of the filigree coloring of romance, which has been thrown around him by sentimental poets and lovesick novelists, and present him as he is, a matter of fact being."[45] Freeing this "fact" from "romance" entails recognizing that "the Sioux" "regard most of the vices as virtues. Theft, arson, rape, and murder are among them regarded as the means to distinction." Furthermore, "ignorance, indolence, filth, lust, vice, bigotry, superstition, and crime, make up the *'ancient customs'* of the Sioux Indians, and they adhere to the code with a tenacity and stoicism indefinable."[46] From Galbraith's perspective, a particular population needs to be subjected to the legal order, but its very nature proves resistant to such inclusion.[47] As Byrd argues, "The United States has used executive, legislative, and juridical means to make 'Indian' those peoples and nations who stand in the way of U.S. military and economic desires," later suggesting that "all who can be made 'Indian' in the transit of empire, can be killed without being murdered." In this vein, official accounts surrounding the Dakota War can be thought of as turning Indigenous *peoples* into *populations* by suspending the question of sovereignty, a process, in Byrd's terms, "of making racial what is international."[48] Describing Native "customs" as "crime" (including the legal charges of "theft, arson, rape, and murder") presents Native actions as an accretion of forms of "vice," having no relation to collective claims that have a history and for which Dakotas are struggling to secure a future. Indian agency figures only insofar as it signifies in relation to the background of U.S. law, against which it is experienced as an eruptive violation—foreclosing the possibility of recognizing the existence and persistence of an autonomous Dakota legal and political order (a "code") that would be at odds with the assertion of U.S. authority over them and their lands.[49] The report, and other accounts of the conflict by non-native officials, converts the potential understanding of Dakotas as outside the normal legal order due to their indigeneity (their endurance as a people in this place) into an attribution of anomaly to them as Indians that makes them subject to prosecution, such as in the trials that led to the mass execution of thirty-eight Dakota men (to which I will return shortly).[50]

Even as officials tend to cast Dakota actions as criminal rather than military, and thus as lacking political content and as disruptions of the ordering structure of national time, their discussions register the ways that the frame of reference for Native people(s) might differ considerably from their own, specifically in the accrued experiences arising out of being subjected to the promises, deferrals, and denials of the treaty system. When faced with accounts of the self-evidence of settler futurity, one is left with, in Vera B. Palmer's terms, "the radical vacuum where an account of grief and mourning should be."[51] While rejecting the idea that the "outbreak" was caused by anything but the Indians' savage inclinations, Galbraith does address prior treaties as an "inciting" element:

> From the best information which I have been able to obtain, it seems that at the time of the treaties of Mendota and Traverse des Sioux, in the year 1851, in order to induce the Indians to sign the treaties, very liberal, if not extravagant, promises were made to them—promises for the occasion, without regard to consequences.... This, I must say, however, that the alleged non-compliance with "promises" made "at the treaty" was the text and conclusion of nearly every Indian orator's speech which I have had the fortune to hear, (and I have heard not a few).[52]

Dakota people can engage in treaty making as a diplomatic activity, in which formal promises are made by the U.S. government in exchange for territory over which the Dakota exercise authority. The existence of the treaty system and the uneven history of recognizing Dakota sovereignty push against the storying of Native participation in the conflict as simply lawbreaking, suggesting that this indictment and the rhetorical and emotional force that animates it arise from an existing perceptual tradition in which U.S. jurisdiction provides the self-evident context for registering continuity and change. The need to insist that Dakota actions do not constitute warfare as such implicitly points toward the place of diplomacy in the prior cession of lands and the ways Dakota uses of force might be interpreted as formal military engagement emerging out of their cumulative fury at the slow-motion invasion of their homeland. During the conflict Pope observes in his commands to Sibley, "It is idle and wicked, in view of the atrocious murders these Indians have committed, in the face of treaties and without provocation, to make treaties or talk about keeping faith with them": "They are to be treated as maniacs or wild beasts, and by no means as a people with whom treaties or compromises can be made."[53] The turn to violence by "these Indians" seems to license the suspension of treaties, both prior ones and the prospect of treaty making as a vehicle of securing peace, and in committing "atrocious murders" the Dakota morph from "a people" into an aggregation without any cognizable political status—"maniacs or wild beasts."

However, the supposed violations for which Dakota combatants are tried by the military commission remain unclear. When commuting the death sentences of the hundreds of Dakotas marked for execution by Sibley's commission, Lincoln cites a distinction between "participation in *massacres*" and "participation in *battles*," implying that participants in the latter could not be tried since battles were a regular part of combat.[54] In listing the people to be executed, though, that same message indicates that taking part in battles serves as one of the charges against seven of them and as the principal charge against two of them.[55] The president tries to frame actions as punishable because they violated the laws of war, rather than casting all uses of force by the Dakotas as inherently punishable, but that distinction collapses in the effort to provide a semilegal means of disciplining Dakota insurgency, instituting a regime of domestic governance predicated on disavowing Native peoples' claims since they substantively challenge the coherence of settler sovereignty.[56] Sibley's message to General Pope announcing the creation of a military commission notes that those appointed to it will see "if there are guilty parties among them" that "can be arrested and properly dealt with," two days later indicating, "If found guilty they will be immediately executed." Of what precisely are they "guilty"? In the middle of his campaign, Sibley asserts with respect to Dakota actions, "This system of plunder must be suppressed and the criminals punished," and he observes in a letter three days later, "I will do all in my power to chastise the miserable savages who have devastated the frontier." Furthermore, Sibley characterizes those who took part in the conflict as "bad Indians," and both Sibley and Pope consistently refer to the entirety of the conflict as "the late outrages."[57] These comments do not distinguish between battle and massacre, casting all use of arms by the Dakota as a violation of some unstated norm. In this vein the criminal character of their "plunder" seems to lie in its disruption of settler temporalities: to "devastate the frontier" makes one a "bad Indian" since doing so claims land to which whites have developed attachments. The "outrage," then, lies in challenging white possessive expectations that form the basis for their engagement with the present and movement toward the future.

Official narratives treat the immutability of non-native occupancy and settler development as given, and, in this context, Native peoples appear as a dislocation of time itself. These accounts offer a genealogy of settlement that serves as the backdrop against which to assess the actions for which the Dakota prisoners are tried. At one point during the war, Sibley insists, "There is no use to disguise the fact that unless we can now, and very effectually, crush this rising, the state is ruined, and some of its fairest portions will revert for years into the possession of these miserable wretches," and Pope notes, "I have proposed to the government to disarm and remove entirely from the state all the annuity Indians, and all other Indians now within its boundaries, and to place them where they can no longer impede the progress of settlements nor endanger the settlers."[58] Armed conflict becomes the occasion for removing Dakota peoples from Minnesota, and while that dislocation might be cast as a response to recent "outrages," it points toward "the progress of settlements" and the access of non-natives to the "fairest portions" of the state as the horizon for policy making.[59] More than indicating a yet-to-be-fulfilled intent, the fear that the state will be "ruined" by "this rising" suggests a process already well in motion whose momentum will be catastrophically redirected should overwhelming force not be used to "crush" everything that blocks the path of white territorial becoming. Galbraith observes in his report that, in response to Dakota actions against the "peaceful and industrious frontier people," "if the sufferers are promptly compensated, the Indians removed, and the frontier secured against the reasonable probability of future raid [sic] of the kind, then the *effects* of the outbreak will soon comparatively disappear, and the frontier will, in a short time, resume its wonted prosperity."[60] The trial of combatants, then, operates as part of a broader process of dispossession that makes land available to "peaceful and industrious" people who will bring "prosperity" to the "frontier." That connection appears perhaps most blatantly in a letter to Governor Ramsey from Morton S. Wilkinson, one of Minnesota's U.S. senators, in which he asserts, "I have done all in my power to induce our President to have the law executed in regard to your condemned Indians," noting at the end of the correspondence, "If the people will be patient we will be able, I think, to dispose of those condemned, and will also succeed in removing the Sioux and Winnebago Indians from the state."[61] To fully *execute the law* entails not only punishing those Dakota pronounced guilty for their actions during the war but removing all Native peoples from Minnesota, even those who played no role in the conflict. The invocation of non-native *patience* expresses the sense that the conflict functions less as a justifiable response to ongoing and expanding dynamics of settler expropriation than as a pause in the self-evident unfolding of settler time.

When considered in the context of policy before the war, this push for removal functions as a projection forward in time of the established pattern of increasingly exerting control over Dakota lands, largely accomplished through the treaty system. While scholars have addressed the legality of the trials of the Dakota combatants, tracing the categorical and procedural errors that undermine their legitimacy, the trials' failure to distinguish between regular criminal law and martial law indicates less a categorical error (interpreting Dakota actions as "crime" instead of "war") than a continuity between peace and war in the extension of settler authority and inhabitance.[62] Separate from reference to particular statutes, "the law" as a way of designating a normative order gains meaning here by employing Indianness as a biopolitical category through which Dakota challenges to U.S. jurisdiction are translated as an innate propensity toward backwardness. In his postwar report Galbraith observes of the "immediate causes of the outbreak" that the previous agent had instituted "a new and radical system" that was put into place by "the treaties of 1858."[63] The theory, in substance, was to break up the community system which obtained among the Sioux; weaken and destroy their tribal relations; individualize them by giving each a separate home, and having them subsist by industry—the sweat of their brows; till the soil; to make labor honorable and idleness dishonorable; or, as it was expressed in short, *'make white men of them,'* and have them adopt the habits and customs of white men."[64] For Natives, to participate in the futurity of the region, and thus of the nation, would mean subjecting themselves to a legal order that enables, protects, and further incites non-native property ownership. For non-natives, Indianness functions as a category through which Dakota dissent from that sensation of time is experienced as an ingrained and racialized form of temporal aberrance. In *Firsting and Lasting* Jean O'Brien addresses the notion that Indians eventually will and must disappear as a function of "the temporalities of race," but she further observes that within non-native historical accounts "Indians reside in an ahistorical temporality," having a certain quality of "timelessness."[65] More than presenting Native peoples as an anachronism, Indianness suggests they have no proper role within the unfolding of time. It takes part in a chronobiopolitics in which the spatiotemporal formations of Indigenous sovereignty appear from within settler frames of reference as a mode of racialized embodiment with particular antisocial tendencies—criminality, destructiveness, irrationality, viciousness—that threaten to divert the proper unfolding of development.

This rendering of Indigenous presence as temporally exceptional in both everyday and official ways creates the potential for non-native place and history. The issue lies less in a failure of recognition than in the way settlers are oriented in time, or, more to the point, how the way their sense of time emerges through particular forms of orientation. As Ahmed suggests in *Queer Phenomenology*, whiteness may be conceptualized "as a straightening device" that "gets reproduced through acts of alignment," such that "whiteness allows bodies to move with comfort through space, and to inhabit the world as if it were home."[66] The phenomenology of whiteness entails aligning a particular possessive relation to place with an experience of time as the necessary and immanent unfolding of settlement over the terrain of the nation. From within this felt experience, Native landedness and duration appear as something of a queer temporal deviation.[67] The sense of settler momentum appears regularly in the annual reports from agents to the Office of Indian Affairs. The report for 1851 indicates, "A review of the history of this nation presents no fact so striking as the noiseless, multitudinous movements of its people westward. This is not a local or transitory accident. We see wave following wave in endless succession," and the next year's report observes, "The Indian must retire before the wave of the Anglo-Saxon race, in his onward march north and west."[68] Beyond offering standard visions of Native disappearance, these statements suggest an immanent process of unfolding, in which white occupation defines the terms of temporal "succession" itself.

What, then, would it mean for non-natives in the period, either before or after the conflict of 1862, to recognize Dakota claims, histories, experiences?[69] It would require a fundamental reorganization of the terms and trajectories of perception. If engagement with the world follows from a sense of what's possible based on prior engagements, whites' substantive engagement with Dakota landedness and governance would entail a significant change in their phenomenal processing of historicity and futurity, what is experienced as real against the background of what has been and what could be. Such a profound shift in ordinary frames of reference would extend beyond interacting with Dakotas and their demands in the present by altering the affective sense of what the present is in terms of the continuities from which it emerges and the potentialities onto which it opens. Merleau-Ponty suggests, "Our perception in its entirety is animated by a logic which assigns to each object its determinate features in virtue of those of the rest, and which 'cancel out' as unreal all stray data; it is entirely sustained by the certainty of the world," providing a sense of "the primordial constancy of the world as the horizon of all our experiences."[70] The fact of non-native possession and the extension of settler territorial authority function as certainties, as means of *aligning* place and time. The self-evidence of white claims provides a sense of constancy that guides perception and against which Dakota presence and opposition appear as "stray data," as a perverse and impossible divergence from the course of history itself.[71] As Waziyatawin Angela Wilson observes, "By the 1840s the Dakota were facing incremental land dispossession through treaties, which were nonetheless repeatedly violated," and "the ethnic cleansing [in the wake of the war] was so complete that no public memorials dedicated to the forced removals exist."[72] Officials' accounts, then, go beyond justifying actions taken during and after combat. They are indicative of the operative temporal sense guiding settler activity, the normalization of that trajectory of becoming as the movement of time itself.

Attending to the decades of treaties that preceded the conflict, however, points toward a very different temporal experience for Dakota people(s), one shaped by escalating land loss, dislocation, and extended periods of uncertainty.[73] The Mdewakantons and Wahpekutes in 1837 ceded "all their land, east of the Mississippi river, and all their islands in the said river." In 1851 they, along with the Sissetons and Wahpetons, relinquished "all their lands and all their right, title and claim to any lands whatever" in what would become the state of Minnesota.[74] While the treaties of 1851 originally had provided for reserved lands on both sides of the Minnesota River, that provision was struck out by the Senate, replaced with a supplemental article giving the president authority to designate lands elsewhere, but the president allowed all four peoples to remain on the lands that had been reserved in the provision that was removed from the treaty. In 1854 Congress passed an act authorizing the president to confirm Dakota rights in perpetuity to that territory, but he never formally did so. The treaties of 1858 affirmed Dakota claims to the land south of the Minnesota River that had been promised in the provision omitted by the Senate in 1851, and they offered compensation for the territory north of the river should the Senate determine that the Dakotas held a right to that land.[75] These repeated renegotiations of the boundaries of Dakota territory, coupled with the persistent ambiguities as to the exact legal character and extent of Dakota landholding, produced a profound sense of insecurity. This mounting series of, in Povinelli's term, quasi-events created a temporality of anxiety that hampered Dakotas' ability to project a clear future for themselves based on past events and patterns.[76] Yet, even as extant forms of awareness and engagement were challenged by these shifting conditions, ongoing struggles with the United States became part of the historical density of Dakota perception. These conflicts engendered their own complex affects that influenced choices going forward and that certainly contributed to the antagonisms unleashed in the conflict. In this vein the supposed "outrages" of the war appear as expressions of Native outrage. Glen Coulthard suggests that "what implicitly gets interpreted by the state as Indigenous peoples' *ressentiment*—understood as an incapacitating inability or unwillingness to get over the past—is actually an entirely appropriate manifesting of our *resentment*: a politicized expression of Indigenous anger and outrage directed at a structural and symbolic violence that still structures our lives, our relations with others, and our relationship with the land."[77] As noted previously, Agent Galbraith observed that "alleged non-compliance with 'promises' made 'at the treaty' was the text and conclusion of nearly every Indian orator's speech which I have had the fortune to hear," indicating the centrality of treaty making and the violations of those promises within Dakota memory, and such memory serves as an incitement to future action. Thus, General Pope's conclusion that "in the face of treaties" Dakotas acted "without provocation" appears as an utter evacuation of Dakota temporalities, in which the treaties and their abuse function as key parts of the perceptual tradition that animates Dakota "anger and outrage."[78]

To what extent, then, can the Dakota War and the patterns of negotiation and abandonment that preceded it be understood as "a shared past," in Philip Deloria's terms addressed in chapter 1?[79] From one angle, Dakotas and whites in the 1830s through the 1860s may be said to have been acting simultaneously, responding in their own ways to the same set of circumstances, which were overdetermined by the actions of U.S. officials and institutions. Yet, from another angle, if one considers the frames of reference at play in their varied ways of situating events within temporal narratives and orientations, and the conceptual and perceptual backgrounds against which they made sense of occurrences, their perspectives are so disparate as to constitute incommensurate experiences of time. Looking at the terms of the treaties and officials' commentary on them as contrasted with evidence of Dakota responses reveals prominent disjunctions in how the different parties contextualized the treaties within their extant temporal trajectories. The texts of the treaties of 1851 position federal discretion as the basis for Indian occupancy and governance. In them Dakota peoples divest themselves of all territorial claims, such that they officially cease to possess any specific land rights based on their indigeneity in what would become the state of Minnesota. Furthermore, the Senate inserts into the treaty a clause giving the president the power to choose "such tracts of country" as he shall deem proper, suggesting a diminished, contingent inhabitance in which Dakota peoples do not exert a determinate political authority over that space.[80] That impression seems confirmed by the provisions in Article 7: "Rules and regulations to protect the rights of persons and property among the Indian parties to this Treaty, and adapted to their condition and wants, may be prescribed and enforced in such manner as the President or the Congress of the United States, from time to time, shall direct."[81] More than gesturing toward the management of Indian-white relations, as in the regulation of non-native presence on Native lands in the various federal Trade and Intercourse Acts, this clause seems designed to allow for U.S. superintendence of intrareservation dynamics among Native persons, and it envisions that the peculiar "condition and wants" of Indians license such intervention. Governor Ramsey, who helped negotiate the treaties, said of Article 7 that it "introduces an entirely new relation between these Indians and the federal government. It disposes at once of the fanciful pretensions and artificial rules of construction to which the assumed *sovereignty* of Indian tribes has so often given rise."[82] This "new relation" extends the premise he articulated the previous year, when he referred to "the suppositious independence of the Indian" as "but another of the anomalies... of which the general subject of the relative rights and duties of a civilized and barbarous people is so fruitful."[83] The supposed innovations of the treaties of 1851, then, emerge out of the extant projection of settler futurity in which Native sovereignty functions as an *anomaly*—a temporary fiction to be overcome.

Even as the legal contours and status of Dakota lands become increasingly ambiguous and contingent over the 1850s, Dakota peoples' geographies continue to fail to match those formally instituted through policy. Despite federal declarations, Dakota rhythms of occupancy and movement do not become decisively oriented toward government-approved territory, suggesting a phenomenological horizon that projects an alternative field of possibility to that which serves as the basis for non-native perception (at least in bureaucratic terms). The annual report in 1855 indicates, "I have still to regret that part of the Medawakantoan and the whole of the Wahpekuti bands have failed to perform their promise to come on to the reserve," adding, "The same complaint must be made of those Sisetons and Wahpetons who have been accustomed to plant below the reserve." The report in the following year observes, "Three of the upper bands still remain off the reserve, and in the midst of the white settlement, giving occasion to constant complaints. Their chiefs at the last council held with them, say positively they will not come to the reserve, but are willing to do so whenever the United States government performs it [*sic*] promises to them."[84] These moments of administrative pique suggest the ways that Dakota expectations and regularities continue to defy U.S.-instituted patterns, illustrating what might be termed their exercise of temporal sovereignty. In asserting themselves as the arbiters of how to interpret the meaning and terms of the treaties (assessing when the government has "perform[ed] it[s] promises"), the Dakotas also position themselves as the ones who know how the treaties should be contextualized within ongoing practices of occupancy, economy, and subsistence (such as choice of hunting areas)—how the treaties should be situated within the continuities of Dakota history and life.

When viewed as an isolated event, the Dakota War of 1862 can be treated as having no cause other than ingrained Indian propensities toward lawlessness.

Native warriors' employment of force can be attributed to an inherent tendency toward criminality that explains their violation of a social order organized around the terms of settler governance. Within this frame, the Dakota use of violence to oppose increasing dispossession appears as an incongruous break in the unfolding of a history shaped by the enhancement and expansion of settler occupancy—a history that is also that of extending and perfecting the national union. The chronogeopolitics of state and federal policy in Minnesota before the war posited a progressive understanding of time whereby Native modes of inhabitance cease to disrupt the exercise of settler sovereignty. In this vein, the "outrage" of the war lies less in the specific actions of combatants than their refusal to accede to the staging of Dakota peoples as a spatiotemporal problem in need of resolution. The abrogation of former treaties with Dakota peoples and their statutory expulsion from Minnesota by congressional fiat in February and March 1863, then, can be understood as less an immediate response to the recent conflict than an expansion of existing aims justified as the spatial amplification of an inevitable settler futurity.[85]

Speaking of current Dakota experience, John Peacock notes, "Dakota descendants of the war ... tended to think in terms of how the war has, in fact, *never* ended," and, starting in 2002, a biannual tradition has developed of retracing the 150-mile walk from the site of Fort Snelling to the Lower Sioux reservation in commemoration of the forced removal of Dakota people along that route in November 1862.[86] Contemporary references to the Dakota War suggest not just collective memory or trauma but temporal orientation, the ways the war affected material relations and possibilities for becoming going forward. In this way it provides part of the frame of reference for Dakota experiences of time. However, Peacock also observes that "few Dakota people were raised knowing about the war."[87] The war may not be actively remembered by all, but it still can help provide the background for Dakota perception through the ways the effects (and affects) it generated are conveyed intergenerationally, whether or not they are explicitly named or conceptualized as arising from the experience of the war.[88] Also, the invocation of the war in contemporary processes of storying becomes a way of reconnecting current feelings, sensations, and experiences to the war. As Wilson observes of Dakota oral tradition and history more generally, "The power of these stories stems from the connection created between the shaped historical understanding and living within the present."[89] The war functions as a crucial context for understanding Dakota responses to ongoing forms of colonial occupation, and drawing on the memory of it serves as a tool for reinvigorating the phenomenology of Dakota duration—the felt sense of having a collective history as a people (or related peoples).

Unlike the Civil War, the Dakota War does not serve for non-natives as a building block for national history, in the sense of being experienced as such at the time or being seen afterward as having been such an event.[90] Rather, the Dakota War appears as an interruption, an anomalous diversion from the straight line of settler progress; it does not have periodizing effects and does not come to function as an important episode in the storying or experience of national time. Dakota presence is cast as Indian aberrance within non-native impressions of historicity and futurity, and attending to such perceptions highlights how the use of the national union as the frame of reference for understanding movement in time actually enacts forms of settler colonial force. We can see in the archive of governance before and after the Dakota War the kinds of Indianization at play in *Lincoln*, in which Parker's flesh marks him as an anomaly in the unfolding of national history. For most viewers, his presence functions as "stray data" that has no place in their sense of history and continuity as U.S. subjects. Within the film, as in accounts by officials before and following the Dakota War, the chronobiopolitics of Indianness reinforces the chronogeopolitics of the state, in which the shape of national jurisdiction serves as a key element in the perception of time itself.

*Parker in Peacetime*

Ely S. Parker's silent cameo in *Lincoln* does not point directly toward the Dakota War or other spectacular moments of state violence against Native peoples during the period (such as the Sand Creek massacre or the Navajo Long Walk), nor does it substantively gesture toward Parker's own role in Indian affairs before and after the Civil War.[91] His presence, though, operates as a trace of possibilities that arise when the unfolding of national development represented by the war—the experience of U.S. history as an expansion of freedom via the emancipation sublime—does not serve as the background against which events figure (or don't figure) as meaningful. As Ahmed suggests in *Queer Phenomenology*, "Histories shape 'what' surfaces: they are behind the arrival of 'the what' that surfaces."[92] *Lincoln* helps reveal the limits of "the what" that can surface when the jurisdiction of the state is held constant in one's way of perceiving time/history. However, focusing on the figure of Parker opens onto a different kind of history, one that is set against the background of Indigenous sovereignties and the accretions of escalating, everyday settler encroachments. His role in the defense of Tonawanda lands and his service as the commissioner of Indian Affairs during the Grant administration highlight temporalities of Native experience that do not conform to those of national crisis and resolution with respect to the Civil War. Drawing attention to Parker's career emphasizes how the political struggles generated by the effort to manage Native presence provide a continuity that straddles the war. Indigenous polities and landedness persistently trouble U.S. jurisdictional and administrative mappings in ways that challenge the chronogeopolitics of settler time. Attending to the antebellum and postbellum regularities of dispossession enacted through the treaty system foregrounds the significance of settler impositions and interpellations in the absence of armed conflict—when there is no explosive event like the Dakota War. Addressing the feelings that attend the failure of revolutionary movements, or the collapse of regimes set in place through such movements, David Scott suggests that there is a "sense of temporal *rupture* and collective *disorientation*" that arises as part of the "phenomenology of an intractable temporal experience." However, what occurs when there is no such "intractable" moment of calamity? How might we think about Native phenomenologies that emerge from banal relations of attrition, the pileup of quotidian and nonspectacular modes of invasion and expropriation that characterizes the operation of U.S. Indian policy over the course of the nineteenth century? What experience of time emerges amid the "exhaustion of life" generated by these routine and persistent dynamics of dispossession that do not constitute historical events within ordinary settler frames of reference?[93]

Parker came to his role as the representative for the Tonawanda Senecas in the early 1840s in the wake of the supplemental Treaty of Buffalo Creek in 1842 (first as the principal interpreter in 1844 and eventually as a Seneca sachem of the Iroquois League in 1851).[94] The initial Treaty of Buffalo Creek in 1838 had ceded four Seneca reservations (Buffalo Creek, Cattaraugus, Allegany, and Tonawanda) in exchange for lands in the Kansas Territory, but after extensive Seneca protest about the validity of the process through which that treaty was negotiated, a new agreement was reached four years later, brokered by the Society of Friends with virtually no Seneca input.[95] That treaty agreed to cede Buffalo Creek and Tonawanda in order to retain Cattaraugus and Allegany. However, the Tonawandas refused to assent to this trade and continued to insist on the illegitimacy of both treaties, especially given the absence of Tonawanda consent. After over fifteen years of struggle, in 1857 the Tonawandas finally were able to secure a reduced version of their reservation by agreeing to buy the land from the official property holders, selling the territory in Kansas they putatively possessed through the treaty of 1842 in order to do so.[96] Their central antagonist in this conflict was the Ogden Land Company, which as of 1811 held preemption rights to Seneca lands.[97] This chain of prospective title reached back to an agreement in 1786 between New York and Massachusetts that resolved a boundary dispute by declaring a particular stretch of territory to be under the jurisdiction of the former while giving the latter the right to retain the funds generated from selling it to private purchasers. Preemption rights to Seneca lands passed through a series of hands before 1811, providing much of the impetus for the Treaty of Big Tree in 1797, in which the Senecas ceded all of their lands in New York with the exception of eleven reservations (including the four that would be at stake in the Buffalo Creek treaties). Once it acquired the preemption right, the Ogden Land Company began its decades-long campaign to gain access to the rest of Seneca territory, as well as that of other Haudenosaunee peoples in what had become central and western New York.[98] The claims of the Ogden company to territory they did not actually own, the federal government's acceptance of this supremely fantastic form of legal fiction, the collusion of numerous U.S. officials in facilitating the realization of this phantom title-in-potential (including many on the payroll of the Ogden company), and the suffering caused by the resulting displacement of Seneca people and flouting of Seneca sovereignty together constitute a slow-motion invasion over more than half a century.

That dynamic does not take the form of, to recall the terms used to characterize the Dakota War, an "outbreak": there is no moment of "war," "insurgency," or "emergency." Rather, there is a series of quasi-events enacted through the treaty system, which sanctifies relations of force as consensual.[99] While not seeking to portray the treaty process as completely unilateral or as a fall from a sense of prelapsarian Indigenous wholeness, I want to emphasize how the coercive influence of state imperatives does not cease with the signing of a single treaty and how later possibilities for Native agency remain affected by the momentum of earlier and ongoing impositions of settler law and administrative mappings.[100] As Dale Turner suggests in his discussion of policy in Canada, "there are intellectual landscapes that have been forced on Aboriginal peoples.... These intellectual traditions, stained by colonialism, have created discourses on property, ethics, political sovereignty, and justice that have subjugated, distorted, and marginalized Aboriginal ways of thinking." As part of this list, one might add discourses on *futurity*. These institutionalized temporal narratives, made routine in everyday forms of sense making, treat settler legal and political norms as given, such that Native peoplehood appears as an aberration whose endurance indicates less sovereignty than oddity, a perverse deformation in the otherwise even alignments of settler becoming. Povinelli's discussion in *Economies of Abandonment* of how liberal temporalities engender and sustain racial precarity seems apt to describe the situation faced by Native peoples within the treaty system's frame of reference: "They are used to hearing that the harms in their present lives should be bracketed. They are used to being aggressively abandoned within a temporal horizon of a future perspective: a future from whose perspective their present suffering has already been mourned and buried."[101] Moreover, having to negotiate such settler orientations generates accreting effects over time, potentially producing forms of rage, despair, and/or exhaustion, which can be seen in both the Dakota War and Parker's later career in Indian affairs. Such effects on Indigenous processes of becoming do not register as meaningful within settler experiences and accounts of time, including the periodization of national history.

Official accounts of the treaty of 1838 present supposed Seneca acquiescence as a temporal Rubicon, and they further situate the effects of that assent within another story of the necessary progress of Seneca people toward civilization. After having worked out a deal with the Ogden company and the federal government for the Senecas to retain the Cattaraugus and Allegany reservations, the Society of Friends met with Seneca representatives in April 1842 to explain the situation as they saw it. Describing the compromise that had been reached without Seneca input, the Quaker representatives indicate, "The well-known policy of the government to remove all the aboriginal race beyond the Mississippi forbids the idea that the treaty would be suffered to remain a lifeless form," adding that "the expulsion of the Senecas at the point of the bayonet is a circumstance which could not be contemplated without horror."[102] While taking some form of removal (and the larger aim of displacing Native peoples as such) as a necessary frame for considering possibilities for the future, the Quakers also disavow violence as a vehicle for enacting that policy imaginary. What remains unclear, though, is how this self-evident unfolding of non-native occupation will occur without force given clear Seneca opposition. In response to this implicit question, the Quakers make the potential for continued Seneca inhabitance on any part of their lands contingent on their ceasing to be aberrant with respect to settler geographies and temporalities. In that same council in 1842, the Quakers observe, "We have seen that from the day when the *white men* first set their feet on your land *they* have been *increasing*, and the *red men* have been *decreasing*," and they add, "We believe it is not too late to reform. If you will take our advice now," which entails adopting Euramerican heterodomesticity, private property, and commercial agriculture, "then will your nation grow and increase, and become strong," whereas the alternative is "extinction."[103]

In this vein, treaties signal the exceptional character of Indigenous peoples within the scheme of U.S. law and policy, indicating less the survival of an alternative sovereignty than a noncoercive process by which Native peoples will merge into a future defined with respect to settler jurisdiction and patterns of land use. Even as they suggest a diplomatic relation, the treaties are situated by non-natives within a frame of reference oriented around settler processes of becoming. In the council with federal treaty commissioners at Buffalo Creek in May 1842, the state representative from New York observes, "You are under the protection of the laws of this State, and to a degree you are liable to their exactions and restrictions, like our own citizens. Ours is a government of laws, and not of force. It is impossible to protect our own citizens against improvident contracts of their own making."[104] The treaty of 1838 functions as a "contract" to which the Senecas, ostensibly including the Tonawandas, have agreed, and by that fact they have submitted to the "laws" promulgated by the "government." After 1842, when the Tonawandas insist that the treaty is invalid with respect to their reservation given the absence of their consent to it, various governors tell them that state officials have no authority to act. In 1844 the governor of New York, William C. Bouck, insists, "I had no power as Governor of the state of New York, legally to interfere in your difficulties with the Ogden Company. The treaty under which they claim your land was with the President & Senate of the United States. It is not within my province to inquire into the legality of this treaty," and two years later Governor Silas Wright observes, "I have examined the subject carefully and do not find that I have any power to interpose on your behalf and prevent the execution of the Treaties. They were made with the authorities of the United States, and are as binding upon us as they are upon you," adding, "If I were to exert the authority of this state by force to prevent the execution of these Treaties, I should be guilty of insurrection under the laws of the United States."[105] The status of the treaty system as a federal matter supposedly binds the hands of state officials, even though the state previously had engaged in numerous treaties with Haudenosaunee peoples absent any federal oversight, and the specialness of federal-Native relations means that treaties must de facto be considered lawful and binding, as to do otherwise would constitute "insurrection."[106] Similarly, in the same letter, Wright warns the Tonawandas "to do nothing in violation of the law, or to the interruption of the public peace," noting, "If the Treaties have conveyed away your rights in the reservation, you cannot get them back by acts of violence." Given their posited prior assent, Tonawanda resistance to their dispossession now constitutes a "violation of the law" and can be constellated with "acts of violence" as a mode of aggression against the sociopolitical order instituted through treaties. The quasi-events of transgenerational invasion disappear behind the narrative of Native acquiescence, via treaties, to the legalities and mappings of the settler state. That process of geopolitical transformation apparently has nothing to do with the exertion of "force" (such as would be the case in the deployment of the military), instead appearing as simply the neutral, natural unfolding of national time.

However, Tonawanda accounts in the period between 1842 and 1857—largely translated, delivered, and/or authored by Parker—refuse that story of amicable and nonantagonistic development and the attendant citation of treaties as signs of the absence of violence. In a memorial from numerous Tonawanda chiefs and warriors to the president and Senate, they observe that they have been "informed that the legal tribunals cannot look behind or below the outward face of treaties which have been ratified, and inquire into the *manner or means* whereby they were obtained. The courts of law, therefore, cannot reach the evil or do us justice," and they respond by noting, "The United States government, which authorized a commissioner to make these treaties, can authorize another commissioner to unmake them, and we on our part, as a nation, will most gladly assent."[107] "Law" does not equal "justice," and to the extent that treaties stand beyond the jurisdiction of "legal tribunals" in which questions might be raised about their validity, they function as an "outward" sign of legitimacy for a fundamentally corrupt process of obtaining Native territory. Moreover, the Tonawandas inquire why federal authority with respect to treaties somehow recedes in the wake of their ratification, especially when one of the putative signatories challenges the authenticity of the agreement. In a letter to the commissioner of Indian affairs in 1848, Parker asserts, "The Tonawanda Senecas will not surrender to the Ogden Land Company the Tonawanda reservation, because they are no parties to the Treaties under which the Ogden Company claim their lands, and by which they propose to remove them."[108] What prevents such treaties from being *unmade*? The apparent answer is the presumption that Indigenous lands eventually will and must be given over to non-native forms of inhabitance and jurisdiction, such that moves in that direction cannot be undone.

As against that temporality of "civilized" progress, Tonawandas assert a different frame of reference, one that takes into account the duration of their relation to their lands as well as their decades-long struggle to retain them. In a memorial to Governor Wright in 1845, Tonawanda leaders insist, "The justness of our cause we think to be approved by the Great Spirit, who has given the lands we occupy to our forefathers, who gave it to us in trust for our children, and we do not wish to violate that trust which is so sacred to us."[109] Enduring Seneca occupation of this territory suggests a kind of claim, one "approved by the Great Spirit," that cannot be reduced to the terms of U.S. law, indicating a "sacred" connection passed down from "forefathers." That intergenerational belonging offers a very different sense and scale of time from that at play in Quaker articulations of Seneca prospects or in the treaty system.

In addition, the memorial draws attention away from the scene of consent that appears in the events immediately surrounding any given treaty and toward the broader pattern of settler intrusion over time that shapes the circumstances of that negotiation. Awareness of the persistence of non-native aggression provides historical density for the Tonawandas' perception of the current environment of political engagement and struggle. After reminding Governor Wright of the Treaty of Canandaigua (1794), in which Haudenosaunee peoples were promised the possession of their existing lands in perpetuity, the chiefs observe:

> Citizens of the United States hav[ing] the name of the Ogden Company have for many years past, in direct violation of this provision of a Treaty, harassed us in the quiet possession of our lands and homes. They have sold our lands at public auction against our consent and the consent of the people we represent; and we did publicly protest against the sale of our lands time after time, but seemingly with no effect, for the purchasers have now come and settled upon our lands under the title of the Ogden Company, and we do not wish to remove them by force, because we should then violate the treaties of peace we have made[.]

The prior and ongoing violations of the Treaty of Canandaigua by "citizens of the United States" do not signify for U.S. officials as a prism through which to interpret the Treaty of Buffalo Creek and its aftermath. However, that continuing legacy of *harassment* orients Tonawanda interpretations of the events of the late 1830s and 1840s. Not only did they not consent to those later "treaties," but the possibility of meaningful consent is vitiated by the persistent patterns of settler action—seemingly unregulated and at times actively supported by the government—in disturbing Tonawandas' "quiet possession" and thereby breaching the prior treaty. Further, Senecas have not responded to such acts of aggression (ultimately countenanced by the United States under the cover of legal "title") through recourse to violence, due to their commitment to upholding the terms of extant "treaties of peace."[110] The Tonawandas situate the treaty system as it actually functions within a historicity of dispossession in which the law operates as a retrospective projection to validate forms of appropriation enacted by U.S. subjects. In this way Seneca opposition to removal in the 1840s and 1850s refuses the legalized notion of Indian anachronism, instead presenting the ongoing history of settler lawlessness as the principal challenge to the potential for just U.S.-Native relations predicated on treaties as vehicles of diplomatic good faith.

When considering the accumulating affects produced by the struggle against removal, though, what would it mean to characterize these dynamics as part of a shared history? In the effort to highlight Seneca being-in-time, can such a formulation do justice to the profound asymmetries in Seneca and non-native experiences of time, especially with respect to the ways policy developments gain meaning when figured in relation to (disparate) temporal orientations and the persistent dynamics of attrition faced by the Tonawanda Senecas? As a central figure in the Tonawanda struggle to retain their lands, Parker took part in shaping their response and played a crucial role in the process of negotiating the repurchase of their reservation. How does Parker's later career register this earlier moment? We can approach the complexities of his postbellum perspective on Indian affairs by viewing them in light of the exhaustion engendered by the decades-long battle against the Ogden company and the sanctioning of its claims by the state through the treaty system. More than reflecting an acceptance of the need for assimilation, or of the need to abandon prior Native lifeways in favor of something more "modern," Parker's later disappointment with the use of treaties can be interpreted as expressive of the cumulative fatigue of working against official efforts to materialize a futurity in which Native sovereignty has no place. In Parker's career after the Civil War we can see how the national union that is championed as the index and herald of freedom in *Lincoln* functions for Native peoples as a mode of imperial force through which the settler state works to realize a story of inevitable national progress, perceived as an immanent tendency within time itself.

Parker's official communications on Indian affairs after the Civil War note the ways the treaty system helps validate a state-sanctioned process of non-native annexation, but he also seems to accept the temporal narrative of settler development and expansion as the frame through which to approach U.S.-Native relations. In his recommendations in 1867 for rethinking Indian policy, requested by the Secretary of War, Parker emphasizes how the treaty process has served as a license to white aggression. He observes of the system's ostensible aims, "The plan of removal was adopted as the policy of the government, and, by treaty stipulations, affirmed by Congress; lands were set apart for tribes removing into the western wilds, and the faith of a great nation pledged that the homes selected by the Indians should be and remain their homes forever, unmolested by the hand of the grasping and avaricious white man."[111] From this perspective, Native peoples accepted new lands in exchange for the ones they currently occupied, with the proviso that those to which they would be removed would be held by them "forever," and such a promise by the federal government was meant to permanently hold at bay "grasping and avaricious" U.S. citizens. However, the repeated use of treaties as the vehicle for accomplishing transfers of this kind ended up sanctioning non-native incursions: "as the hardy pioneer and adventurous miner advanced into the inhospitable regions occupied by the Indians, in search of the precious metals, they found no rights possessed by the Indians that they were bound to respect. The faith of treaties solemnly entered into were totally disregarded, and Indian territory wantonly violated."[112] When nonstate actors would assert claims to treaty-guaranteed territory, the government would either look away or negotiate another treaty to cover this most recent wave of dispossession, such that "the Indians" effectively have "no rights" that settlers are "bound to respect." Drawn from the infamous *Dred Scott* case in 1857, which denied African Americans national citizenship, this phrasing intimates the ways that the recursive dynamics of non-native invasion faced by Native peoples exceed the terms of emancipation and reconstruction through which *Dred Scott* legally was superseded.[113] Parker adds that "if any tribe remonstrated against the violation of their natural and treaty rights, members of the tribe were inhumanly shot down and the whole treated as mere dogs. Retaliation generally followed, and bloody Indian wars have been the consequence."[114] In this alternative account of national time, the eruption of armed conflict, such as the Dakota War, needs to be understood as part of a cycle of settler intrusion backed by violence that leads to Native response but in which only the latter becomes visible as a disruption of the extant legal order, as an exception that needs to be handled as an emergency, crisis, or "outrage." While such white actions often violate treaty terms, Parker implies that they are driven by a broader temporality of treaty making that projects future cessions as the means of retroactively legitimizing these putatively illegal forms of trespass by citizens.

By the time he becomes the commissioner of Indian Affairs in 1869, Parker's perspective on the treaty system has become even bleaker. In his annual report for that year, he states, "Arrangements now, as heretofore, will doubtless be required with tribes desiring to be settled upon reservations for the relinquishment of their rights to the lands claimed by them and for assistance in sustaining themselves in a new position, but I am of the opinion that *they should not be of a treaty nature*." If treaties have functioned as a means of validating the transfer of territory from Native control and making it part of the regular jurisdictional hierarchies of the United States, often as a retrospective measure in the wake of existing settler encroachments, Parker argues that this legal fiction should be suspended due to the falseness of its premises. He asserts, "A treaty involves the idea of a compact between two or more sovereign powers, each possessing sufficient authority and force to compel a compliance with the obligations incurred. The Indian tribes of the United States are not sovereign nations," and "Great injury has been done by the government in deluding this people into the belief of their being independent sovereignties, while they were at the same time recognized only as dependents and wards." In light of the fact that Indian tribes are deemed "dependents and wards" in federal law, narrating them as "sovereign nations" simply for the purposes of concluding treaties that ease the exchange of land can only be a vicious pretense, one that is especially pernicious given that it trades on the inability of Native peoples to enforce the terms of these agreements against the United States and, thus, to prevent continuing breaches of them before the next round of treaties and cessions.[115] Parker adds, though, "In regard to treaties now in force, justice and humanity require that they be promptly and faithfully executed, so that the Indians may not have cause of complaint, or reason to violate their obligations by acts of violence and robbery."[116] Fulfillment of treaty obligations appears as something of a cynical means of deferring Indigenous "complaint," so as to prevent the outbreak of the kinds of activity deemed criminal by U.S. officials (such as in the Dakota War).

The sort of peace secured by treaties, or whatever "arrangements" might replace them, gains meaning within a vision of futurity defined by the inexorable movement of non-natives onto Indigenous lands.[117] In his recommendations in 1867, Parker suggests that if the reasons they should concentrate themselves on much smaller land bases and take up "agricultural and pastoral pursuits" (as well as "the habits and modes of civilized communities") were explained to Native peoples, "they could probably be made to comprehend that the waves of population and civilization are upon every side of them; that it is too strong for them to resist; and that, unless they fall in with the current of destiny as it rolls and surges around them, they must succumb and be annihilated by its overwhelming force."[118] The chronogeopolitics of settlement—the "waves" of development—here provides the background against which to assess the (im)possibilities for the continuance of Native polities. In his recommendations in 1867, Parker observes, "Originally their greatest desire was to be left undisturbed by the overflowing white population that was quietly but surely pressing to overwhelm them, and they have been powerless to divert or stem the current of events," further noting that "naturally many of them at times have sought by violence the redress of what they conceived to be great and heinous wrongs against their natural rights."[119] In contrast to the characterization of Natives' violent response as savagery or crime, like in officials' accounts of the Dakota War discussed earlier, Parker insists that such actions are attempts to "redress" white "wrongs." However, even if understood as a "heinous" violation of "natural" law, the expansion of white occupancy, and the attendant extension of U.S. modes of governance, possesses an *overwhelming* momentum that fundamentally (re)orients the relation between the present and the future—"the current of events."

While presenting Native peoplehood as something of a vestigial anomaly, and at times arguing for the necessity for Indians to move away from communalism, Parker also highlights the importance of taking existing forms of Indigenous territoriality as the frame of reference for developing future policy, even if such recognition can function in somewhat carceral ways.[120] Discussing the general terms of federal Indian policy, he observes in his report of 1869 that Indians "should be secured their legal rights; located, when practicable, upon reservations; assisted in agricultural pursuits and the arts of civilized life; and that Indians who should fail or refuse to come in and locate in permanent abodes provided for them, would be subject wholly to the control and supervision of the military authorities, to be treated as friendly or hostile as circumstances might justify."[121] Although reserving the military as the insurance that Native movement, grievance, and/or warfare will not disrupt non-native geographies of property, transit, and commerce, Parker concedes that Indigenous peoples have specific "legal rights," largely derived from extant treaties, and that they should not be deprived of tribally specific land bases, even while they are being trained out of such collective attachments via education in "agricultural pursuits" and the individualized, privatized cartographies of "civilized life."

If the experience of the Senecas—and the Tonawandas in particular—illustrates for Parker the potential violence of the treaty system, including the ability of the U.S. government to utilize such apparent consent as a means of legally validating an ongoing process of expropriation and displacement, Seneca struggles also indicate the persistent commitment of Native peoples to retaining separate governance over homelands that remain apart from the regular jurisdiction of the settler state.[122] Parker often stages his own engagement with this dynamic as a confrontation in the present with a residual formation that, for better or worse, cannot be sustained in the face of non-native advancement (whether cast as invasion or progress) and the increasing integration of the national union as such. His statements as the Commissioner of Indian Affairs, though, should not be taken as necessarily typifying his perspective on Native landedness in his later years. In a letter to a friend in 1887, he observes, "The tenacity with which the remnants of this people have adhered to their tribal organizations and religious traditions is all that has saved them thus far from inevitable extinguishment; when they abandon their birthright for a mess of Christian pottage they will then cease to be a distinctive people," and his speech on the twenty-fifth anniversary of the Gettysburg Address features discussion of the Delaware chief Tammany and the violence of the history of white expropriations.[123] The seemingly residual here appears as a strategy for projecting forward a "birthright" through which Natives can produce an alternative future oriented around "distinctive" peoplehood, one at odds with the de facto "extinguishment" envisioned by administrators and missionaries. In this vein Parker's insistence, albeit uneven, on the need to honor extant treaties and to provide a legal status for collective Indigenous occupancy through reservations suggests that, whatever may occur in the future, Native peoples as landholding, self-governing entities exist now and must be reckoned with as such. The legacy of Seneca struggles, then, generates a complex frame of reference in which peoplehood remains under threat but itself serves as the background against which to figure Native movement forward in time.

*Legacies for the Modern*

For Parker the treaty system provides both the basis for legally recognizing the ongoing continuity of Native territoriality as peoples and the most effective legal mechanism for the, ostensibly peaceful, annexation of Indigenous lands. Such complex treaty temporalities do not fit the impression of U.S. history as the perennial expansion of democratic possibility and inclusion, a way of perceiving national time that I have characterized as the emancipation sublime. Reckoning with Indigenous sovereignties and the effects of Indian policy entails refusing to normalize U.S. national jurisdiction as the de facto container in which time happens. Instead, that very presumption serves as the means of imposing settler time by narrating the dispossession of Native peoples as simply the inevitability of progress while casting Indigenous peoples' continuing inhabitance in their homelands as an anomaly—an anachronistic residue. From this perspective, arguing for belonging to the nation would suggest that Native peoples had capitulated to a settler-ordered modernity in which U.S. authority serves as the uncontested framework in which to envision the future. Yet in the generation following Parker, many of the most prominent Native intellectuals, including Charles Alexander Eastman, actively argued for the right to U.S. citizenship and sought to cast themselves as properly national subjects. Considering those born in the late nineteenth century, and Eastman in particular, Gerald Vizenor wonders in *Manifest Manners*, "What did it mean to be the first generation to hear the stories of the past, bear the horrors of the moment, and write to the future? What were tribal identities at the turn of the last century?"[124] Eastman was among very few Native people who gained great fame as writers and intellectuals among non-native publics in the early twentieth century, and as David Martínez reminds us in *Dakota Philosopher*, "When wondering where Eastman's heart may lie, one must bear in mind that he lived through the Dakota exile and the Wounded Knee Massacre."[125] For these reasons, he serves as a compelling example through which to explore the question of how nineteenth-century stories of dislocation, which play no role in dominant periodizations of national history, become part of the process of assessing current possibilities and projecting the potential for movement forward. How do these events provide the background for Eastman's sense of time? How is his outlook on the policies and the potentials of the allotment era, including his endorsement of U.S. citizenship, oriented by having been witness to such spectacular violence, as well as the quasi-events of invasion and occupation that surround such explosive episodes? In contrast to Parker, Eastman tends not to stress the legacy of treaty making as a means of emphasizing the coherence of Native peoplehood and its role in Indigenous futurity, but this relative absence signals less a renunciation of Indigenous self-determination than an articulation of it that has been given shape and direction by the circumstances faced by Dakotas over the prior century. Scholarship on Eastman tends to take the polarity of "tradition" and "assimilation" as the framework for tracing the itineraries and implications of his writings, positing a vexed temporal threshold Eastman must cross in becoming "modern" that gives rise to his work's complicated mediations. For example, Drew Lopenzina observes, "Eastman sought to reconcile the divisions of ethnicity by balancing the need of an authentic Indian identity against the need for adaptation to modern life," adding that "the question then arises, how much of Eastman's memories are a result of his full immersion into white standards of narrative and representation, and how much is emulsion, a resistance defined by the cultural integrity of his memories and experiences that retains its separateness despite having been swept up in the currents of an alien culture?"[126] To be "authentic," Indian identity remains separate from "modern life," such that "integrity" appears to reside in priorness: retaining what was before they were "swept up" in the tidal wave of "white standards." In addition, critics tend to contextualize the putatively assimilationist elements of Eastman's work as expressive of "their historical moment," "of their times," and part of "negotiat[ing] . . . through a complex modern world."[127] To be in the present, then, means being potentially cut off from indigeneity, such that the available responses are "resistance" or "adaptation" to a contemporary "world" that is self-identical and whose modernness is, by definition, disjunct from Indian identity as such. As Vizenor suggests, though, "the tribal real is not an enterprise of resistance."[128] In this vein, Eastman's texts register the effectivities of non-native force (military and otherwise) while exploring what it means to put together a life as a Native person amid them.[129]

The forms of temporal experience Eastman addresses, therefore, can be understood not as strung between a separate Indigenous pastness and the "historical moment" of the modern present but, rather, as expressive of a continuing and shifting Indigenous duration in the context of ongoing (and in many ways intensifying) forms of settler pressure, constraint, aggression, and fraud. In *Bad Indians* Deborah Miranda asserts, as discussed in chapter 1, "Culture is ultimately lost when we stop telling the stories of who we are, where we have been, how we arrived here, what we once knew, what we wish we knew; when we stop our retelling of the past, our imagining of the future, and the long, long task of inventing an identity every single second of our lives." Seen in this way, Eastman's writings engage in the process of *retelling the past* in ways that provide an alternative account of the present than that available in dominant Euramerican visions (including those implemented through the treaty system). As Malea Powell suggests in "Imagining a New Indian," Eastman sought "to refigure the possibilities of Indianness for future generations." The issue, then, is less whether one currently would want to take Eastman's ideas or example as a model than the ways his work illustrates how trajectories of Native becoming are affected by non-native modes of temporal recognition, incorporation, and imposition without being reducible to them. What are the potentials for exercising temporal sovereignty in the context of extraordinarily diminished possibilities for sovereignty of all kinds?[130]

Eastman's texts frequently reference the Dakota War of 1862, doing so in ways that highlight its overwhelming violence while also presenting it as the condition of possibility for everything that came later.[131] That future neither redeems the horrors of dispossession and displacement nor marks a definitive break in which the character of time itself is altered irrevocably, becoming something modern that fundamentally is disjunct from something traditional. Rather, Eastman thematizes how the war functions as part of the frame of reference through which he perceives the conditions and potentials of Native life in the moment of writing. In *Indian Boyhood* (1902) he chronicles his life in Canada, where his family had fled after the war (in which his father had participated). While he and his family believed his father to have been executed by the U.S. government in the mass hanging of Dakota combatants, his father had been among those pardoned, coming to find Eastman over a decade later and bringing Eastman back to his home near Flandreau. Eastman observes just before his return to the United States, "I was scarcely old enough to know anything definite about the 'Big Knives,' as we called the white men, when the terrible Minnesota massacre broke up our home and I was carried into exile."[132] *From the Deep Woods to Civilization* (1916) largely picks up where *Indian Boyhood* leaves off, chronicling Eastman's brief time on his father's homestead in Dakota Territory, his education (including graduating from Dartmouth College and receiving a medical degree from Boston University), his service on-reservation at Pine Ridge, and his later work (particularly as an agent enrolling Sioux people for allotment). In the text Eastman describes himself and his family as having been "driven by the troops into exile" after the Dakota War, and he says of his attempt to create a private practice after leaving the Indian service, "After thirty years of exile from the land of my nativity and the home of my ancestors, I came back to Minnesota in 1893."[133] These moments suggest not only the viciousness of the displacement from their homeland but the ways the attendant feeling of decenteredness affects subsequent sensations. Dispossession, then, functions less as a discrete event than as an animating principle that shapes his later experience.

Similarly, although the war marks a decisive moment of change, Eastman situates it within a long-standing process of expropriation. In *Indian Boyhood* his uncle observes, "The greatest object of their [whites'] lives seems to be to acquire possessions—to be rich. They desire to possess the whole world. For thirty years they were trying to entice us to sell them our land. Finally the outbreak gave them all, and we have been driven away from our beautiful country," and in *Deep Woods* Eastman recounts this history in his own voice, indicating, "My people had been turned out of some of the finest country in the world. . . . The Americans pretended to buy the land at ten cents an acre, but never paid the price; the debt stands unpaid to this day. Because they did not pay, the Sioux protested; finally came the outbreak of 1862 in Minnesota."[134] The experience of exile references less a singular occurrence than the ways Dakotas affectively register the momentum of decades of settler campaigns conducted before the war, and it cannot be remediated through a simple return, as it marks more than mere distance. As Lisa Tatonetti suggests in "Disrupting a Story of Loss," Eastman's work "displays a perspective that is rooted in Native history."[135] Exile defines the horizon of possibility for Dakota futurity, indicating the accumulating and ongoing force of U.S. intervention to which Native peoples remain subject while also positioning the continuing legacy of Dakota principles and peoplehood as the background against which to figure a way forward.[136]

Eastman often speaks in ways that seem to accept evolutionary notions of advancement, appearing to adopt dominant allotment-era notions that full incorporation into the U.S. nation will emancipate Indians from confinement within the savage backwardness of the reservation, but he repeatedly underlines the fact that Native peoples' adoption of non-native practices over the preceding several decades has occurred in the context of pervasive, structural coercion.[137] Addressing the politics of Native arguments for U.S. citizenship in the late nineteenth and early twentieth centuries, Beth Piatote situates that project in the context of the stifling system of federal superintendence to which Native peoples were subjected on- and off-reservation. In "The Indian/Agent Aporia," she observes, "'Indian' had come to mean a subject administratively and legally devoid of agency and anomalously positioned in relationship to the American state and its values," and citizenship offered a possible way to be "free of wardship" while not "necessarily . . . sacrific[ing Native peoples'] status as tribes with treaty rights."[138] Rather than normalizing the story of U.S. history as a process of expanding democratic inclusion, Eastman might be seen as seeking to envision continued potentials for Indigenous becoming against the background of intensive conditions of domination. In *Deep Woods* his father notes, "It is true that they have subdued and taught many peoples, and our own must eventually bow to this law; the sooner we accept their mode of life and follow their teaching, the better it will be for us all." Eastman explains that these conclusions are based on his father's "meditations during those four years in a military prison," and he further indicates that his father saw "no alternative for the Indian," quoting him as saying, "One would be like a hobbled pony without learning to live like those among whom we must live." This articulation of necessity seems to capture the text's perspective, and while Eastman does not retreat from insisting on the importance of accommodating non-native "mode[s] of life" broadly stated, such a call clearly arises under duress—the absence of another option owing to the force of the "law" and increasing forms of *hobbling*.[139] As Martínez suggests, "In Eastman's complex view of Dakota history, his concern is for the well-being of the people, who have been forced by the growing pressures of 'progress' and 'civilization' to make not only difficult decisions but also very unfortunate ones."[140]

In discussing the options available to Native people(s), Eastman emphasizes the constraints on their temporal sovereignty, underscoring how U.S. policy works to foreclose Indigenous ways of being-in-time by heavily circumscribing (when not outright seeking to eliminate) extant rhythms of life and relations to place while proscribing Native-initiated responses to such dislocations. Addressing the circumstances surrounding the Wounded Knee massacre, he notes in *Deep Woods* that "the Sioux had many grievances and causes for profound discontent, which lay back of and were more or less closely related to the ghost dance craze," later insisting, "I have tried to make it clear that there was no

'Indian outbreak' in 1890–1891, and that such trouble as we had may justly be charged to the dishonest politicians, who... first robbed the Indians, then bullied them, and finally in a panic called for troops to suppress them."[141] If the Ghost Dance arises because of the changed conditions of Sioux life, the reaction to it by "dishonest politicians"—Eastman observes that just after his arrival at Pine Ridge, the agent asserts, "If I had my way, I would have had troops here before this"—illustrates a refusal to accept the legitimacy of Indigenous adaptations and initiatives in the face of settler-imposed circumstances (grievance-inducing *robbery* and *bullying*).[142] Instead, Native people(s) are reduced to Indian(ized) flesh, subjected to chronobiopolitical projects whose aim is to enlighten and improve them into submission. Although somewhat dismissing the Ghost Dance as a "craze," Eastman's discussion foregrounds U.S. efforts to thwart Native attempts to generate their own futures from within situations overdetermined by persistent forms of non-native assault, regulation, and deception.[143]

The reservation serves for Eastman as the most prominent sign of broader white efforts to arrest Native-led development.[144] He begins *Indian Boyhood* by noting that "*the Indian no longer exists as a natural and free man. Those remnants which now dwell upon the reservations present only a sort of tableau—a fictitious copy of the past,*" and at the end he remarks that his father, after his release from prison, "soon became convinced that life on a government reservation meant physical and moral degradation." More than functioning as a site of surveillance and containment, the reservation creates a simulacrum of Native life. As a "fictitious copy," it does not so much preserve older practices as generate a sense of the Indian as inherently, immanently "past," as outside the present moment and as having been superseded. The "degradation" inheres in the ways the extreme management of Indigenous people on reservations cuts them off from the possibility of "natural," self-organized patterns of modification and transformation in response to changed conditions. In *The Indian To-day* (1915) Eastman describes the situation faced by "the Indian of the Northwest": "One morning he awoke to the fact that he must give up his freedom and resign his vast possessions to live in a squalid cabin in the backyard of civilization," additionally observing, "He was practically a prisoner, to be fed and treated as such; and what resources were left him must be controlled by the Indian Bureau through its resident agent." He further notes in *Deep Woods*, "An Indian agent has almost autocratic power." As opposed to presenting reservations as treaty-guaranteed spaces in which to sustain peoplehood, the vision offered by Parker (albeit in rather qualified ways), Eastman perceives them as carceral sites. What most characterizes life on-reservation is not the retention of older ways of being but the imposition of a subordinate and heavily policed relation to "civilization." Native people(s) remain pinioned to a settler-defined order in which they are cast as in need of temporal aid—as perennially lagging in time and thus requiring white help to move forward.[145]

Eschewing the notion of a preset formula for development, such as the transition from tradition to a singular (settler) modernity, Eastman presents the realization of Native-authored change as drawing on existing knowledges and impressions in order to envision possibilities for current and future action. In *The Soul of the Indian* (1911), he asserts, "It is my personal belief, after thirty-five years' experience of it, that there is no such thing as 'Christian civilization.' I believe that Christianity and modern civilization are opposed and irreconcilable, and that the spirit of Christianity and of our ancient religion is essentially the same," and he later observes, "As a child, I understood how to give; I have forgotten that grace since I became civilized. I lived the natural life, whereas I now live the artificial." One might read such moments as indicative of the ways Indians in the early twentieth century come to be seen as bearers of primitive wisdom.[146] Eastman's writings do sometimes employ this idiom.[147] However, the distinction between the "natural" and the "artificial" instead might be understood in terms of his comments about the *unnaturalness* of the reservation as a site of settler force and anachronization. The imposition of "civilization" as a set of mandatory orientations and dispositions works to disjoint Native experiences of time by seeking to replace existing Indigenous frames of reference. He sketches what might be characterized as a natural affinity between Christianity and pre-reservation modes of being, suggesting that contemporary Native engagements with what previously had been exclusively non-native systems of belief and practice might be seen and felt as continuous with prior ways of reckoning with the possible.[148] That connection across time signifies a dynamic relation, a becoming in which the historical density of Native perceptual traditions shapes the encounter with new ideas and social forms such that they are incorporated into, while also themselves shifting, existing trajectories. Furthermore, Eastman indicates that non-native social formations do not operate as a totality, but rather, elements of them, such as Christianity, might be recontextualized as part of extant Native social formations. In addition to separating those elements from the more destructive and "artificial" tendencies clustered together as "civilization," that distinction suggests the absence of something like an encompassing shared modernity that functions as a unified formation into which Native people(s) enter in ways that mark a definitive conceptual and perceptual break from what came before.

While neither smooth nor uncontested, such a process of Indigenous-centered development shapes Eastman's understanding of what it would mean for Natives to participate in American life as national subjects.[149] As Penelope Myrtle Kelsey observes, Eastman's refusal of Euramerican dominance can be "seen in his strategic inclusion of a Dakota worldview in his books," and Powell suggests that "for Eastman's new Indian, being Indian and American is not a contradiction."[150] In *The Indian To-day* Eastman suggests of Euramerican sociality, "Here is a system which has gradually taken its present complicated form during two thousand years. A primitive race has put it on ready made, to a large extent, within two generations. In order to accomplish such a feat, they had to fight physical demoralization, psychological confusion, and spiritual apathy. In other words, the old building had to be pulled down, foundations and all, and replaced by the new. But you have had to use the same timber!" The insertion of Native peoples into settler time required that they *put on* a "ready made" set of orientations that had emerged for Euramericans over the course of "two thousand years," and Eastman indicates that the consequences of that interpellation were immense and quite destructive.[151] However, he insists that the "new" edifice to be built depends on the "same" materials, that it will involve not so much replacing extant modes of perception wholesale as recomposing and redirecting them in novel ways. He adds, "It has long been apparent to us that absolute distinctions cannot be maintained under the American flag. Yet we think each race should be allowed to retain its own religion and racial codes as far as is compatible with the public good, and should enter the body politic of its own free will, and not under compulsion. This has not been the case with the native American."[152] The "compulsion" to which Native peoples have been subjected has meant that the policy formulations about their entry into "the body politic" have not been guided by their own determinations, including decisions (explicit and implicit) about retaining and adapting their own "codes" of belief and conduct.

In contrast to prior official tendencies, Eastman observes, "It has come to be more and more the case that the Indian, so long and so oppressively paternalized, is allowed to take a hand in his own development."[153] Although being "allowed" some autonomy is not the same as exercising self-determination, Eastman's emphasis on a form of development driven by Native priorities and experiences marks a shift from the sense of Indigenous peoples as utterly transformed by their encounter with the modern and/or the sense that they, their histories, and their futures are largely irrelevant to American national time. He closes *Deep Woods* by asserting, "I am for development and progress along social and spiritual lines, rather than those of commerce, nationalism or material efficiency. Nevertheless, so long as I live, I am an American."[154] Such "progress" (linked to what he elsewhere characterizes as Indians' "natural life") appears as something other than dominant modes of nationalism, and perhaps as directly counter to them. Yet the qualifier *nevertheless* suggests that being "American" need not mean accepting the trajectory of development posited in the system that has been imposed on Native peoples.

Even as Eastman's texts can be read as seeking recognition of Native people(s) from non-natives, the terms of that acknowledgment require a shift in settler perceptions of time, particularly with regard to the dynamics and effects of Indian policy over the preceding half century. While Eastman's accounts take U.S. national belonging as their frame, the background against which it comes into focus is not the standard periodization of U.S. history, especially not the account of it as engendering ever-expanding spheres of freedom as in the emancipation sublime. Rather, he highlights both the centrality of settler force to the national past (including its ongoing effects and legacies in the present) and the importance of Indigenous modes of being and becoming to Native futurity, as well as that of the settler nation-state. Drawing on Dipesh Chakrabarty's critique of Euro-American historicism, discussed in chapter 1, one might characterize Eastman as displacing a narrative of Native *transition* to civilization—becoming historical in ways that take settler structures as the given frame—in favor of a process of *translation*, Native people(s) responding to non-native force from within their own sociotemporal formations. Such an account of the past and its momentum in shaping the present and future defies chrononormative accounts of national life and time. History functions in Eastman's work not as an unfolding of inclusion in which various kinds of bodies, multiple types of racialized flesh, become full participants in the becoming of the United States. Rather, for Eastman, the nineteenth century is experienced as an accretion of institutionalized modes of violence, and the possibilities for living as an "American," for Natives and non-natives, emerge out of that continuing legacy of colonial dispossession. The potential for such a perspective arises out of the phenomenology of dislocation, out of collective Indigenous sensations perceived against the background of temporalities of peoplehood, landedness, and sovereignty. This frame of reference serves as the one that Eastman seeks to share with non-native readers.

This way of articulating Native histories and experiences of time stands in stark contrast to the dialectic of national continuity and transcendence at play in *Lincoln*. In its positioning of the Civil War as the prism through which to understand the U.S. past, the film, and the ways of orienting national time it emblematizes, obscures both the dynamics of settler colonialism and the temporal narratives of exception employed to normalize such occupation and dispossession. Casting the violence of the Civil War and its challenge to national jurisdiction as a process of democratic becoming (via the war's linkage to emancipation as its immanent horizon) enacts a mode of periodization in which the preservation of the national union serves as the vehicle for realizing freedom and racial justice. Within that story there is no place for engaging with Indian policy and Indigenous sovereignties as anything but a curiosity, an oddity that does not fit. Like Ely S. Parker in the film, they may appear as a marker of accuracy—these things happened—while remaining silent and marginal with respect to the central drama of national (re)formation. However, Parker's appearance suggests the potential for a different accounting of time that runs against the grain of that national plot. Attending to official accounts of the Dakota War and to Parker's antebellum and postbellum participation in Indian affairs reveals how U.S. legal and administrative discourses have sought to manage indigeneity by coding it as a (temporal) anomaly, as an eruption within time whose aberrance is experienced and explained as a resurgence out of the past (outdated claims to land, atavistic tendencies toward violence, a failure to progress toward a modern future).

Conversely, one could trace the history of how the state generates geopolitical cohesion for itself in any given moment by projecting a futurity predicated on expansive and invasive settler inhabitance. Native peoples and sovereignties appear as a temporal aberration within a geography defined by the normalization of settler law. The projection of the inevitability of the union—the geopolitical cohesion of the United States—as the framework for temporal experience depends on a cross-referenced and mutually defining set of perceptions, sensations, and processes of backgrounding that can be described as settler (colonial) time. From this perspective, the history of settlement in the nineteenth century appears less as a series of eruptive episodes of armed conflict than as a more slow-motion temporality of expropriation—a series of *quasi-events*, largely enacted via the treaty system—through which Native nations are subjected to the union. Addressing performances of mourning in the nineteenth century, Dana Luciano suggests that they can "rearrange the dominant chronobiopolitical dispositions of the historical moments in which they were produced," engendering "a self-conscious distance from the 'official' materials of history."[155] Focusing on Parker's silence brings to the fore the histories of Indigenous struggle, survival, and self-determination toward which his presence gestures. Doing so allows for an attention to the chronogeopolitical dynamics of settler colonialism, in terms of the operation of policy discourses in the nineteenth century as well as the elision of Indigenous peoples in the process of periodizing the U.S. past—the operation of a metanarrative of Indian irrelevance. In contrast to the frame of the emancipation sublime, Parker's fleeting, mute presence provides an opening to an alternative accounting of (Indigenous) temporality, one in which Native histories, polities, sovereignties, and futures cease to be exceptional.

THREE. **THE DURATION OF THE LAND**

In 1906 Congress passed the Osage Allotment Act, extending to the Osage Nation the principles at play in the allotment program generally.[1] These include efforts to break up Native land tenure into privatized property holding, organized primarily around nuclear family units; dismantle Indigenous structures of governance, asserting greater U.S. jurisdictional authority over Native peoples and places; insert Native peoples into the cash economy and Euramerican agricultural production; and transform everyday patterns of life so that they would conform to Euramerican conventions of dress, language, religion, literacy, gender roles, and so on.[2] This policy imaginary draws on temporal figurations in order to remap and reorder spatial relations. Presented by officials and supporters as a means by which Indians could progress from a stunted and backward savagery toward civilization, allotment offered a vision of necessary development over time that enabled the struggle between Indigenous and settler geopolitical formations to be conceptually bracketed.[3] Emplotting Native governance and sovereignty as merely a moment within an evolutionary process of becoming casts Indians as moving toward the achievement of liberal modernity rather than as struggling to retain control over their extant territories and to maintain their self-determination as peoples. This turn in the discourses of Indian policy can be understood as complementing the ways of perceiving national time discussed in the previous chapter, in that allotment offers a means for Indians to cease to be temporally anomalous and to be included within the increasingly democratic potentials of American national life.[4] Allotment policy projects a futurity oriented around settler modes of being, and doing so incites and legitimizes various processes (legal, administrative, and quotidian) that work to transform Native sociospatial dynamics at all levels so as to make Indigenous lands more available for non-native expropriation, occupation, and investment.

We can understand allotment as a field of force working to reshape Native experiences of space and time, but one that does not operate in a vacuum. Rather than treating it as instituting a fundamentally new and different kind of temporality (dividing Native time between tradition and modernity), we might conceptualize allotment, in the idiom of general relativity, as exerting something like gravitational influence on extant Indigenous trajectories.[5] The question of simultaneity was a central concern in the late nineteenth and early twentieth centuries, due to the need to produce a global sense of it because of European and U.S. military, political, and commercial aims, and that process of universalization generated considerable scientific and philosophical debate over the nature or possibility of an absolute, shared "now," a process of intellectual ferment that provided the context for the emergence of theories of relativity and the notion of spacetime.[6] The concept of spacetime merges space (and its three dimensions) and time (its own dimension) into a single manifold whose shape depends on the operation of gravity. Space, then, does not function as a neutral container, and ways of ordering space—like the presumption and realization of U.S. jurisdiction over its "domestic" territory (discussed in the previous chapter)—affect the contours and texture of temporal experience. For example, allotment policy's division of Native lands into privatized units works to position nuclear family homemaking as the implicit frame for personal timescales of living, in terms of patterns and periodicities of maturation, rhythms and relations of labor, shifting and atomized attachments to place, and generationally compact connections across time. In this way the force of U.S. policy (exerted through statutes but realized through the discretion given to Indian agents, government licensing and leasing of lands, forms of taxation, mandatory schooling, etc.) can exert what might be thought of as something like gravitational pressure on existing Indigenous spatiotemporal formations, potentially shifting them in ways that give rise to collective orientations and trajectories of becoming at odds with those that previously had shaped Native perception.

However, in the absence of a clean, clear break (a "leap" into "modernity"), one formation or frame of reference does not simply replace the other.[7] How can we think about the complexities and tensions involved in living within two disparate spatiotemporal formations, in being subject to the varied and uneven forms of force that they exert on everyday self-understanding and ways of being? Put another way, becoming subject to the allotment program entailed less the immediate supplanting of one's existing sense of time (rhythms, periodicities, forms of periodization, ways of understanding and experiencing causality) than the imposition of an alien set of orientations that have effects on everyday experiences and regularities.⁸ Existing forms of Native perception in their historical density continue to shape engagements with the surrounding environment, and the patterns already at play in those environments do not simply dissipate when confronted with new non-native political, economic, and residential dynamics. Borrowing from Henri Bergson, we might characterize the persistence and momentum of those forms of collective memory, experience, and engagement as *duration*. As discussed in chapter 1, Bergson characterizes duration in terms of the sense of time's movement as immanent flow, rather than it being able to be broken up into a series of disconnected simultaneities. Moreover, he indicates that movement itself provides the basis for perception as well as the connection between recollection and action.⁹ In *Matter and Memory*, he suggests, "There is no perception which is not full of memories. With the immediate and present data of our sense we mingle a thousand details out of our past experience," enacting a dialectic in which "past experience" gives coherence and meaning to what we sense while being guided by encounters and possibilities in the present: "Our representation of matter is the measure of our possible action upon bodies: it results from the discarding of what has no interest for our needs, or more generally for our functions."¹⁰ As opposed to thinking about Native temporality as being ruptured into a new, modern simultaneity with non-natives as a function of U.S. administrative interventions, then, we can conceptualize Native perception—in this case, specifically Osage perception—as guided by shared forms of memory that affect engagement with allotment-era conditions and changes. Such collective orientation gives momentum to Osage experiences and ways of negotiating the shifting social landscape as well as influencing what is sensed as a possible action within the present. Thought of in this way, the temporalities that allotment (and U.S. policy more broadly) seeks to realize run up against the counterforce of extant Osage modes of becoming.

John Joseph Mathews's *Sundown* highlights this disjunction. The novel illustrates the imposition of U.S. legal geography, which is animated by a developmental momentum and which exerts warping effects on everyday Osage experience, and it shows the relation between such settler mappings and alternative Osage forms of sociospatiality that have their own complex temporal dynamics and that provide the basis for an experience of continuity within Osage peoplehood. Critics often have characterized the novel's portrayal of the ongoing legacy of allotment in terms of a struggle between tradition and assimilation, mediated by the figure of the mixed-blood.¹¹ However, not only does this approach overlook the ways Mathews links Osage placemaking to a pervasive sensation of time that occurs alongside that of allotment, but this interpretive frame in many ways reiterates the internal logic of allotment, in which the movement of history itself immanently reaffirms the coherence and inevitability of the transition to settler social norms and the realization of the state's jurisdictional imaginary. Instead, the novel explores the complex interactions of two spatiotemporal formations, addressing how they interpenetrate and affect each other without becoming identical. Rather than merely indicating that settler institutions employed particular discourses of temporality as part of the effort to manage Native affairs and to legitimize the seizure of Indigenous lands (as discussed in chapter 2), the novel highlights how the influence of federal Indian law and administration actually materially alters the phenomenological experience of the present (and its felt relation to the past and future) as well as the framework within which change occurs. Reciprocally, in engaging with the ongoing force of allotment, *Sundown* suggests that amid these pressures to conform to a futurity defined by the state's extension of jurisdiction over Native peoples, another way of sensing time and space is also operative; the novel continually gestures toward an Osage sense of time irreducible to U.S. history and policy.[12]

The novel traces what it feels like to be made to experience yourself and your people as a temporal anomaly as part of the imposition of an alien geopolitical order, and in doing so, it refuses to take U.S. time as the frame through which to approach the enduring complexity of Osage peoplehood. In "From Difference to Density," Chris Andersen argues that a scholarly insistence on Indigenous "difference" from non-natives creates a situation in which "Indigenous complexity [is] reductively fixed in time and space through apparently objective, logical markers used to bear the discursive weight of our authenticity and legitimacy," and he proposes, instead, "beginning with the assumption that Indigenous communities are epistemologically *dense* (rather than just *different*)."[13] In this vein *Sundown* refuses a static (anachronizing) assessment of relative Osage *difference* (from a settler standard taken as the norm) by instead highlighting the *density* of Osage experience. The novel does so by juxtaposing three modes of time: the implementation of allotment-era Indian law and policy, the felt sensations of an Osage man (Chal Windzer) coming of age during this period, and the duration of Osage occupancy in their homeland, attending to the emergence and persistence of modes of perception, experience, and memory that link Osage people to that place. In moving among these discrepant temporal frames, the novel illustrates how settler legal and administrative interventions generate everyday feelings of backwardness on the part of Native people. At the same time, Mathews offers a means of envisioning Osage modes of continuity and change without making them subject to settler legal geography and national history as their condition of intelligibility.

The text often marks the switch point between Euramerican-dominated and Osage temporalities and the conflicts generated by their lived incommensurability by indicating the presence of a feeling of *queerness*. The term signals a sense of being out of sync with Euramerican narratives of development while also referencing the ideological and institutional nexus of reproductive lineality, presumptively diminishing Indian bloodedness, and land loss constructed through the legalities of Osage allotment. In addition to marking the impressment of Osage people into federal Indian policy's heteronormative conceptions of nuclear family property holding and racial inheritance, Mathews's repeated invocation of queerness alludes to the extant linkage within sexological and popular discourses of people of color with perversity. *Sundown* plays on this set of associations to suggest how Chal's inability to fit in, including his supposed failure to be properly heterofamilially directed, might open onto a larger set of questions about how the imposition of U.S. jurisdiction becomes (chronobiopolitically) naturalized through the presentation of its reordering of ordinary life as merely expressive of the normal temporality of procreation. Conversely, the novel suggests how Indigenous modes of history and placemaking are dismissed by coding them as an enduring, racially transmitted incapacity for civilization.[14] In narrating Chal's sensation of disorientation with respect to the events unfolding around him, the novel suggests that his feeling of queerness within the social formations created by allotment indicates less an Indian inability to adapt (one attributable to degrees of Indian bloodedness) than continuing Osage connections to the land they inhabit. The novel explores how the duration of that history of occupancy provides not only a perspective from which to challenge the self-evidence of the developmental trajectory envisioned by U.S. policy but also a resource on which Osage people draw, both explicitly and implicitly, in quotidian negotiations with the accreting material effects of allotment.

*In the Time of Allotment*

When narrated in terms of U.S. history, including that of Indian policy, Osage experience will appear as a blockage, as a drag on or diversion from a trajectory shaped by the orientations and momentum of settler colonial imperatives. Yet Mathews draws on aspects of modernist style to explore how allotment's discourse of civilizational advancement forcefully comes to inhabit everyday perception as a phenomenological frame.[15] The text manipulates the reader's sense of pace by juxtaposing different representations of time in order to register the disjunction between the alterations in everyday Osage experience resulting from policy developments in the allotment era and the feeling of suspension or stasis (not yet *doing something*) expressed by (some) Osage people. The text gestures toward the significant material changes made possible by extant federal policy, including the multiplication of derricks, the vast expansion in the non-native population, the increase in direct federal regulation, the exertion of control over Osage governance, and the imposition of Oklahoma statehood. It also foregrounds Chal's and others' related sense of stasis, of an apparent inability to gather momentum toward any productive endeavor or goal. The text juxtaposes that feeling of incipiency and/or immobility with what it terms "the Great Frenzy" (266), referring to the profound effects on early twentieth-century Osage people of their wealth.[16]

Osage ways of relating to place and attendant expressions of peoplehood continue to provide the frame of reference for ordinary sensation, even as the impositions of the allotment era seek to reorder quotidian Native affects and practices by replacing the durable networks that shape them. In fact, the law passed in 1906 allotting the Osage reservation uniquely registers that friction. Unlike the Dawes Act (1887) or Curtis Act (1898), which instituted allotment for Native peoples generally outside Indian Territory and then for those in Indian Territory, this 1906 act institutionalizes continuing Osage communal claims to land, even if they are placed on the same timetable for elimination as other forms of legal recognition for a distinct legal status for Indians separate from regular U.S. property holding, policy, and jurisdiction.[17] Beyond placing allotments in trust for twenty-five years, a common procedure that was supposed to protect Native people from the effects of a market economy they supposedly did not understand, the law specifies that "oil, gas, coal, or other minerals" are "reserved to the use of the tribe for a period of twenty-five years," with "the royalty" from those resources "to be paid to said tribe" on a per capita, quarterly basis, creating what has since come to be known as "the mineral estate."[18] In addition, Osages had collective claims to the interest from the funds generated by the sale of their prior reservation in Kansas as well as "all moneys received from grazing lands." These provisions require an entity that will handle the affairs of the collectivity maintained by them, and the act creates a "tribal council," consisting of a principal chief, an assistant principal chief, and eight other members.[19] Moreover, the law allocates 160 acres each to three reservation communities (at the towns of Pawhuska, Hominy, and Gray Horse) that are "set aside for the use and benefit of the Osage Indians, exclusively for dwelling purposes," also for "a period of twenty-five years."[20] Together, these provisions—a communal claim to subsurface rights, the creation of a governing body, and the acknowledgment of communities whose land remains unallottable—indicate the impress of extant forms of Osage peoplehood on the law, even as they are translated into forms more amenable to settler governance.[21] The act's recognition of the persistence of collective Osage modes of occupancy and decision making points toward the presence of forms of emplacement that do not fit the trajectories of allotment.

From the outset, though, Mathews roots Chal's life in the history of U.S. intervention into Osage affairs. The text begins with Chal's birth in his parents' home near the agency buildings in the emerging town of Kihekah.[22] Standing over the bed Chal's father, John, asserts of his newborn son, "He shall be a challenge to the disinheritors of his people. We'll call him Challenge" (4), situating Chal in an agonistic relation with those who would seek to disinherit Osage people from their lands. However, neither John nor the narrator makes clear the terms or history of that process of expropriation, and not until later in the book does the question of allotment emerge, with John actually endorsing it. The narrator observes, "He talked about allotment more and more and said that in a few years there would be thousands of people in the new town of Kihekah which had grown out of the old Agency," and John often contrasts his enthusiasm with the resistance of other Osages: "If it hadn't been for the progressives on the council, they never would have been any allotment, if it was left up to the fullblood party" (44–45). Presenting himself as oriented toward the future, John casts the full-bloods as holding onto the past from sheer "stubbornness" (46).[23] As against John's sense "that something momentous was about to happen" that "would change the whole existence of people who lived at the Agency," the narrator notes that "something" "never quite happened," further indicating that despite the impressions of "the Progressives" that allotment resulted from their agitations, "In reality the allotment was forced upon the tribe by people outside the reservation who had no particular interest in the welfare of the tribe" (49). Here the text locates non-native desire for access to the reservation as the principal drive behind allotment, providing a more concrete sense of the character of the disinheritance to which John initially alludes while suggesting John's own misapprehension of the stakes and motivations behind the policy for which he advocates. This confusion echoes John's earlier uncertainty about the meaning of "challenge." Before naming his son, he declares, "I live as a challenge," but the narrator remarks, "He didn't know what he challenged"; "it had never been definite" (3). Like his father, Chal occupies a collective relation of disinheritance but lacks the ability adequately to name that process, to fully articulate the nature of the challenge that faces "his people."[24]

Mathews periodically reminds readers that the political economy in which the characters are enmeshed, which helps give rise to their particular structures of feeling, depends on U.S. governmental action to catalyze and sustain it.[25]

After John and other members of the Osage council created by the 1906 act have been removed by the Secretary of the Interior because of their role in seeking to negotiate oil leases, the text notes, "Chal had never thought much about the Government except that it seemed always present, like an atmosphere. But its presence had been beneficent and protective he felt. However, if it could dismiss the Council..., it must be very sinister and more powerful than they thought," adding, "He had visualized it as a great force which had overcome everything; but a force that was just and kindly.... A great bearded patriarch somewhere among the clouds, with outspread arms" (60).[26] More than simply affecting the lives of Osage people, U.S. legal and administrative interventions create "an atmosphere," a presence that may appear "beneficent" but whose very ubiquity seems to crowd out alternatives. In addition, while Chal narrates the role of U.S. policy in Osage affairs as "just and kindly," the government also exerts what he experiences as a potentially omnipresent "force," godlike in both its scope and its dictation on high from elsewhere. Later, while on break from attending a fictionalized version of the University of Oklahoma, Chal observes of his continuing habit of using the term "guv'mint," "in the old way of the reservation," "He guessed it was because that word had always been associated with authority outside of the reservation; that potent thing which controlled the destinies of Indian agents, of school children, and controlled the payments. He knew he ought to say 'our government' or 'the United States'" (165). These moments highlight the pervasiveness of Indian policy in shaping ordinary routine on Osage lands (including before allotment, although not as extensively), even as Chal also registers the alienness of that regime and the ways it seeks to solicit consent through positing a future that encompasses Native people as U.S. national subjects.[27]

The suffusion of Osage space with settler administrative mappings and directives—including the allotments themselves, leases for grazing, leases for oil production, regulation of town sites, and regulation of inheritance—creates an environment in which merging into a U.S. "our" as part of a national process of becoming comes to seem the self-evident basis for understanding movement forward in time. In addition to mentioning in passing the absorption of the territory of the Osage Nation into the state of Oklahoma in 1907 ("the reservation had become a county in the new state—'the biggest county,' John had said proudly" [63]), the novel notes that while still in the air force after the war Chal "received letters from his father; short letters telling him about what the guv'mint had done or was gonna do" and that the town "had been recognized as a city of the first class, and the oil sales were larger and larger and would soon run into millions of dollars" (233).[28] Ongoing forms of federal involvement—"what the guv'mint had done or was gonna do"—provide the impetus for impressions of progress in Kihekah. In this vein, during the process of choosing either town sites or land plots just before the formal passage of allotment, John says to Chal "something about his 'seein' history being made'" (47–48), suggesting that for John, and implicitly for Chal, history itself becomes associated with the privatization of land as part of the broader implementation of U.S.-sanctioned progress—a maneuver that resonates with the operation of the treaty system as discussed in the previous chapter. Rather than indicating a particular Osage or mixed-blood conflict with respect to modern life, something like a conflict of "values" or "cultures," Chal's emotional orientations register the impact of settler temporal narratives given the proliferation and materialization of such narratives in the government-initiated remapping of Osage space.[29]

The novel suggests how dominant discourses of time translate Indianness as a form of nondevelopment, gesturing toward the legal dynamics of "competency" and the key role it plays in the administrative architecture of allotment. Throughout the novel Chal compares himself to other Osages his age, particularly his childhood playmates Sun-on-His-Wings and Running Elk, who initially attend college with him but who leave soon thereafter (refusing to take part in a fraternity ritual that involved paddling the pledges). Chal thinks that they "lacked the spirit of the times—lacked 'get-up,' as John expressed it. They seemed contented just to sit in the village and talk, like many of the other young men" (68). To choose not to engage in pursuits associated with business and to continue to dwell in places explicitly marked as belonging to the Osage collectively ("the village") means having a somewhat recumbent relation to time—lacking "spirit" or "get-up"—such that one's activities do not in fact count as activity but as immobility, inertia, sloth.[30] Further, when Chal returns to the reservation from college, he "was disappointed in his friends [including Sun-on-His-Wings] because it seemed that they didn't have any ambition": "He was the only person not doing something except the mixedbloods and the fullbloods, but he believed that there wasn't much interest in them—he certainly didn't want to be like them. He knew that he ought to be doing something. Of course he had never in all his life done anything" (162). Turning away from Indianness provides the precondition for advancing in a version of time defined by settler interests and imperatives, explaining the otherwise logically incoherent description of twenty-plus years of life as devoid of activity ("never...done anything"). To have any degree of Indian blood seems, by definition, to indicate a pervasive and intractable inactivity due to a lack of "ambition," but this supposed turning away from opportunity and possibility appears

THE DURATION OF THE LAND · 103

as such only within the context of Chal's experience of "ought," the impulsion toward progress—toward citizenship—animated by the force of settler policy.

One of the central ways that "ought" is materialized in Osage life is through the legalities of competency. The allotment act of 1906 specifies that "at the request and upon the petition of any adult member of the tribe," the Secretary of the Interior "may issue to such member a certificate of competency, authorizing him to sell and convey any of the lands deeded him by reason of this act, except his homestead" (the first 160-acre plot allotted).[31] To be *competent* one must offer evidence of the capacity to engage in extant commercial relations of sale, money management, credit, and debt, and successfully doing so illustrates that one has crossed a civilizational threshold, has reached a state of advancement such that one can participate fully in the present of the nation.[32] Being competent, though, further means losing federal trust status for most of the land one holds (the nonhomestead part of one's allotment). Withdrawal of such recognition, cast as the wished-for achievement of fee simple ownership, makes that land fully fungible as well as taxable as private property by the state and federal governments. By the mid-1930s between a quarter and a third of the Osage reservation had passed into non-Osage hands as a result of the voluntary sale of land (often either to avoid taxes on it or to pay existing debts), inheritance by non-native spouses, and the auctioning of allotted plots to satisfy taxes.[33] Conversely, those Osages not deemed competent come under the authority of non-native "guardians" as determined and regulated under Oklahoma state law, and under an act of Congress in 1912, all of the moneys due to such "restricted" Osages would be paid to the guardians on their behalf, a policy reaffirmed in a law in 1921 that specified that those without certificates of competency or guardians could receive only $1,000 quarterly (regardless of what they actually were due based on extant Osage Nation funds).[34] The often-stifling oversight by guardians, the extraordinary banality and extremity of guardians' fleecing of their clients, and the statutory curtailment of access to the wealth increasingly generated by oil production incentivized the pursuit of competency, further driving the loss of land in the ways already noted.[35]

*Sundown* registers the multiple kinds of violence made possible by implementing this policy framework, ordered as it is around institutionalized narratives of Osage maturation into a capacity for private property holding. John's murder by bandits in an attempt to steal his new car occasions the novel's commentary on the emergence of the guardian system and the implications of dissolving Osage territory into a series of individual claims, as opposed to it being under the jurisdiction of Osage national governance. While the novel does not indicate directly whether John and Chal have certificates of competency, it suggests that they would, given their ability to access their own wealth for various purposes and Chal's inheritance from his father absent any mention of its mediation by non-natives. However, when Chal's mother narrates how his father died, since all Chal knew was the fact of John's death via a telegram he received while still away in the air force, she intersperses the events of the homicide with discussion of the appointment of guardians for "restricted" Osages. After noting, "They found his pistol in his hand," she recounts, "The agent said that white people in town could be guardians for young Osages and they will not have their money long, I believe," and after a brief pause she continues, "Your father said that the guv'mint would not let these white people cheat Indians, but they have done it all the time" (235), although she later suggests, "I believe your father did not believe this. I believe his tongue said this so that his heart could hear it" (236). The superintendence of whites over those Osages not deemed competent is part of broader patterns of *cheating*, with the government enabling this individualized but cumulatively quite sizable project of resource extraction. In addition, the distinction Chal's mother makes between John's "tongue" and "heart" suggests a disconnection between the kinds of sentiments promoted by allotment policy, with its promise to inculcate "civilized" tendencies, and the experience of being subjected to alien rule by the U.S. "guv'mint." Chal's mother does not differentiate the program of resource theft enabled by the temporalities of competency from the murderous assault on John and the taking of his car. John's protest against his wife's warning about the prominence of white bandits is that "it is a civilized country now" (237).[36] The text ironically indicates that the putative *civilizedness* of Osage territory in the wake of allotment—with its subdivision of the land into privately held (and often salable) units and the dismantling of a collective Osage capacity to protect themselves through their own exertion of jurisdiction—is what unleashes the potential for attacks on Osage people.

Beyond noting the pervasiveness of allotment's effects and the implications of its temporal narratives for Osage well-being, the novel further explores how the significant alterations in the landscape and sociality of the reservation help generate forms of temporal affect. Without specifying quite who is the subject of the feeling addressed, the text reveals, "As the years went by, the fevered expectancy seemed to increase. Nothing was certain and calm any more, but the atmosphere was a-tingle with uncertainty; a thrilling uncertainty which would some day evolve into a glorious certainty. Each day brought more news of something about to happen" (61). This permeating sensation of expectancy constitutes an "atmosphere" that echoes the atmosphere of government presence. Or, rather, the latter makes possible and secures a range of non-native investments in Osage wealth and territory that intensify exponentially owing to the vast expansion of oil leases and related income from the mid-1910s onward. As "the black derricks crept farther west" from the initial sites on the eastern part of the reservation, people had a "feeling in their hearts that the indefinite glory was not far off now": "they talked about the future . . . which was sure to be glorious, though its particular glory was vague" (74). This sense of the future emerges out of the potential for various kinds of commercial development enabled by Osage oil, whose bounty can be directed into private accumulation of various sorts because of the United States' dismantling of Osage sovereignty and constitutional governance both before and through allotment.[37]

The novel suggests, then, that this shared structure of "feeling in their hearts" can be traced to the shifted frame of reference incited by allotment policy. The sense of "fevered expectancy" among Osage people, though, signals less an accomplished transformation than the inculcation of particular ways of turning toward the future, ones that are themselves partial and vexed. While in college, Chal "decided he would be a business man and amount to something in Kihekah," further observing when he goes back to the reservation, "Everybody seemed to be doing something," and "Everyone talked about oil" (161). The felt need to do or be "something" gains meaning and momentum from the commercial networks propelled by oil production. That material reordering of life in and around the reservation generates an emotional trajectory toward a particular kind of sociality endorsed by U.S. policy. The narrator indicates that Chal "wanted to be identified with that vague something which everybody else seemed to have, and which he believed to be civilization" (281), earlier noting that "he was proud of the new paved streets and the tall buildings that had been built in his absence" while in the air force during World War I (237). Correlating this alteration of the built environment with civilization presents the changes in Osage space as progress in time, as advancement from a benighted past toward the potentials of an enlightened tomorrow. From this allotment-induced vantage point, the alterations in Osage life brought by oil production and its economies seem less a historical shift within Osage sociality to which individual Osages are more or less attracted, as with any form of change, than a break with Osage ways that appear as *uncivilized* and of the past—as a stasis against which to register the dynamism of civilized "doing." However, rather than providing a clear path by which to transition from the one to the other, existing circumstances produce a kind of affective *vagueness*, an impression of being outside of time(on the cusp of "something" "about to happen"). The novel suggests how the allotment-era reconfiguration of the geopolitics of Osage sovereignty and the topography of everyday life provides momentum for particular feelings of modernness, which are experienced as necessarily a renunciation of a sense of Osage identity even as the contours of the future toward which Chal ostensibly is moving remain elliptical at best. These feelings suggest the influence exerted on him by the "history" he was "'seein'... being made," yet whose contours remain amorphous and ill fitting within his everyday perceptions—the "vague something" about which he hears so much "talk" and which he feels as an indistinct, if palpable, presence.

However, the novel does not engage with the differential gender effects of such temporal narratives or the gendered terms of competency, presenting the sense of the need to be doing "something" in "business" as if it applied equally to all Osages rather than operated through a postallotment gendered division of labor in which women would not normatively be imagined as wage workers or entrepreneurs. Moreover, Chal's mother—who does not speak during much of the narrative, to whom Mathews never gives a name, and who stands as the only significant female Osage character in the novel—serves as a contrast to John, providing a kind of placeholder for the persistence of an alternative Osage sensibility in spaces other than the three reserved villages (which will be discussed in greater detail later). Through this character, the text extends the conventional tendency to cast women as the bearers of tradition even as Mathews raises questions about the temporality of that particular concept. Thus, the account the text provides of the sociopolitical effects of allotment and the affective dynamics that attend its implementation remains deeply masculinist by both making men's experience paradigmatic of Osage response writ large and positioning Chal's mother as a counterpoint to that story.[38]

## Emplaced Silence

Addressing contemporary Osage processes of constitution making, Jean Dennison argues for a conception of "entanglement" that "calls attention to the inherent power dynamics within the ongoing colonial context" of Osage life "without erasing [Osage exertions of] agency," "understanding settler colonial forces as having a varied, dynamic, and uneven impact across space and time" while also "negat[ing] the easy divide of colonized and colonizer."[39] However, if "entanglement" points toward the ongoing effects of settlement and their unevenness (and thus the ability of the Osage people to take up tools of settler governance, like constitutions, and mobilize them in the service of Osage self-determination), might settler colonial fields of force be understood as not simply moving "across space and time" but altering them, seeking to reorder Indigenous spatiotemporal formations? As Russell West-Pavlov observes, "alternative temporalities remain latent and active under the threshold of linear time and its all but ubiquitous stranglehold," or in Dipesh Chakrabarty's terms, challenging the Enlightenment vision of time as singular and linear involves recognizing "a plurality of times existing together" that indicate "a disjuncture of the present with itself," such that we may acknowledge that there are varied "ways of being through which we make the present manifest" that cannot be resolved into a universal, singular time.[40] In this vein Mathews draws attention to ongoing Osage relations to place that provide the frame of reference for collective (if not explicitly articulated or necessarily homogeneous) processes of becoming, and the novel uses figures of silence to illustrate that shared background, suggesting the density of everyday Osage perception and duration—including the ways connections to the land exert their own spatiotemporal force.

Osage oil wealth testifies to a collective Osage territoriality preserved in the provision of the act passed in 1906 that "reserved to the use of the tribe" all "oil, gas, coal, or other minerals" found on the reservation.[41] Chal's father, John, dismisses this clause, indicating in passing, "We had to let Running Horse and his fullblood party have that provision about the minerals, so's tu git the allotment bill through" (50). However, if Chal "has never in all his life done anything," that fact can be traced to the wealth generated by Osage lands. The text notes, "The payments in royalties to the members of the tribe on the roll became larger and larger as the oil production increased" (62), adding, "There was nothing to do except talk. Their incomes were so large now that they didn't think of working at anything; in fact, they had never worked except by spurts when some enthusiasm came over them" (75). As a result of their quarterly income, especially that generated by royalties from oil, holders of Osage headrights do not need to engage in wage labor in order to have revenue to meet their expenses and fulfill their desires. U.S. Indian policy provided the frame in which leases for production were negotiated, transposing Osage sovereignty and land tenure into a form amenable to large-scale resource extraction. Yet the Osage Nation successfully fought to retain collective rights to the mineral estate, and in the novel's representation, the funds generated by the oil economy facilitate forms of everyday action (the "talk[ing]" and forms of "enthusiasm" that do not count as "anything") that are not consistent with the kinds of interactions and modes of inhabitance envisioned and incited by allotment's privatizing imaginary.[42]

The existence of the mineral estate expresses a shared Osage relation to place that continues to provide a basis for common experiences of time, indicating the duration of peoplehood as well as the momentum of persistent affective connections that are irreducible to allotment-era initiatives. Chal "remembered that his father had said that an Indian is not a wanderer—that people said they were nomads, but that no one loved his native soil more than Indians" (234), and during Chal's time in the sweat lodge, Chief Watching Eagle, Sun-on-His-Wings's father, amplifies this sensation of an enduring connection to the "soil": "We must have time to keep our place on earth. . . . Our children must keep place on earth. If we think all time of these troublous things, we will not have time to think of other things. We will not have time to keep our place on earth" (275). Like the indefinite "something" for which Chal often longs, the project of "keep[ing] our place" also involves activity within time, not simply the absence of productive endeavor. The insistence on the time of "business" as the only activity that counts as "anything" displaces the work and ongoing engagement necessary to maintain *place*, (re)making and sustaining the spatial matrix of Osage peoplehood. While from the perspective of U.S. policy the mineral estate indexes a residual geography of peoplehood, one anachronistically retained so as to facilitate certain kinds of resource extraction, it instead appears here to condense a persistent process through which an Osage "we" is (re)constituted as a landed entity—as an entity whose regeneration in time depends on an active connection to the "earth" they inhabit.

*Sundown* presents that interdependence less as an idea or principle in which Osages believe than as the animating material context from which Osage peoplehood continues to emerge. Foregrounding the problem of "hav[ing] time" for the labor of preserving Osage continuity in their homelands suggests the difficulty of prioritizing it among contemporaneous demands, but this phrasing also gestures toward the notion that such activity entails a kind or sense of time incommensurate with the "troublous" imperatives of U.S. policy.[43] As discussed earlier, the novel plays on the disjunction between the significant, forced reorganization of Osage political economy in a fairly short period and the felt sense of (some) Osages of being unable to move toward the "something" of allotment-projected progress. The text further contrasts these temporal framings with a third temporality: the duration of Osage peoplehood experienced through inhabitance on the land. More than serving as a blankness onto which various persons and populations can project concepts and cartographies, the territory exerts pressures on those who live there, producing effects that influence the contours, character, textures, trajectories, and rhythms of human sociospatiality. The text notes of the spread of the oil derricks, "At the tip of the westward movement, half a dozen little towns grew up; not out of the earth like mushrooms, as they were not of this part of the earth; they had no harmony with the Osage. Later they were like driftwood carried in from strange lands on a high tide and left stranded when the tide went out" (303). The non-native infrastructure that emerges out of petro-capitalism ill fits "this part of the earth," appearing as something borne from somewhere else and left abandoned in an alien land. The narrator suggests that these towns do not belong because they arise suddenly and have no enduring connection to this territory, their *strangeness* bespeaking the absence of a history that would accommodate them to and integrate them within this place. Such discordance contrasts with the ongoing set of relationships generated by Osage inhabitance over (at least) hundreds of years.[44] In using the term "harmony," though, Mathews is suggesting less a state of sustained equilibrium, or something like an ability to commune with the land, than an extended process by which the multiplicity of elements of the territory (themselves changing) collectively exert influence on Osage lifeways.[45] The passage in its syntax emphasizes that process of becoming in tune with the dynamics of "this part of the earth," positioning "the Osage" as the object of *with* instead of the land. That substitution suggests less that Osage people are equivalent to or merely an extension of a generic nature than that the time frame of their habitation here has forged a connection in which they have been affected by the particular nonhuman dynamics of this region, such that Osages have a relationship to this place that settlers do not.

This storying of the connections between Osage territoriality and experiences of time, though, does not take the form of defending Native political authority per se. In *Tribal Secrets* Robert Warrior observes that the "real problem" the text addresses is "a community having severely limited ability to make choices regarding its own future and the effect of that on a typical individual within the community," and he further notes, "Having lived during the period when the United States still recognized, however reluctantly, the fullness of Osage sovereignty, Chal internalized the maturing values that sovereignty allows."[46] Read in this way, the novel suggests an only semiconscious awareness of "Osage sovereignty," in which Chal has intimations of "values" that have been suppressed through allotment and the operation of the administrative apparatus of Indian policy more broadly. Yet the idea of "mak[ing] choices regarding... [the] future" exceeds the question of juridical sovereignty, pointing toward what I have characterized as temporal sovereignty: in this case, the ability of Osage people to have their own modes of becoming not constrained or regulated by settler interventions, interests, and imperatives.

More than "values," which might be understood as precepts or philosophies, the novel suggests the presence of a nonconscious frame of reference that guides everyday forms of perception. In suggesting how the process of "keep[ing] our place on earth" and achieving "harmony" with "this part of the earth" becomes part of Chal's ordinary experience of his body, the text suggests that he registers the legacy of collective Osage duration, which provides an orienting sensory background. With respect to Chal's life "on the prairie" as a child, the text notes, "Behind these impressions would be the silence, the tranquility of his home. Always he remembered the silence, and though he grew more loquacious as he learned to say meaningless things, he had a reverence for it as long as he lived; even when he had assumed that veneer which he believed to be civilization" (12–13). Here silence points to an unspeakable presence, or, rather, a set of connections, experiences, sensations that are too quotidian to be put into words, that provides the unarticulated background for Chal's conscious awareness. The "veneer" of "civilization" sits uneasily atop the phenomenology of "home," the former comprising activities and orientations that Chal has "learned" in the context of a postallotment Osage life and education but that do not provide the primary contextual frame for his sensation of the world. This image of layering suggests the density of Osage experience—the presence of multiple spatiotemporal formations whose coexistence and nonequivalence generate a process of negotiation between conflicting forces. However, that process, and the ways it indexes Osage experiences of time irreducible to U.S. policy, occurs at the level of sensation rather than discourse. Arguing in "Felt Theory" that "emotional knowledges" function as "community knowledges," Dian Million observes that such expressions do "not always 'translate' into any direct, political statement."[47] Mathews illustrates how particular forms of affect that do not present in "political" terms, and that are not necessarily the subject of consciousness, provide the background against which events figure as such. These modes of sensation testify to orientations that arise out of an enduring relation to Osage lands, what Glen Coulthard has called "grounded normativities."[48]

Rather than offering such feelings as the basis for a juridical framework or as leading toward a particular system of political representation, the novel traces the difficulties of representing the frame of reference in which Osage becoming occurs. In *Sundown* Osage people remain unintelligible to U.S. officials unless they speak in terms of the desired "something" of progress (championed in and through allotment). Otherwise, they are portrayed as prepolitical subjects who need to be trained into proper modes of life and governance. Instead of exploring how to craft political discourses or institutions that could overcome these conditions of (or impediments to) registering Osage political voice and agency, though, the novel's emphasis on silence points to the existence of shifting Osage ways of being that are not dependent on the effort to be heard or understood by settlers. Silence in the text is neither the refusal nor the inability to say something. In *Manifest Manners* Gerald Vizenor argues that "shadows are that silence and sense of motion in memories," adding that "the shadows are the silence in heard stories, the silence that bears a referent of tribal memories and experience."[49] Silence indexes collective, nonconscious dispositions, sensations, and trajectories that bear Osage memories and relations with each other and with the land, implicitly providing the conditions for Osage ways of reckoning with the possible. Mathews's emphasis on silence as indicative of the felt presence of Osage histories and connections to place—of distinct forms of orientation and momentum—resonates with Bergson's discussion of duration as "qualitative multiplicity," in terms of both the character of time itself as movement (rather than being divisible into countable moments or points) and the attendant potential for disparate forms and flows of temporal experience that cannot be cross-cut by a universal sense of simultaneity. In *Time and Free Will* Bergson observes, "Sometimes we fix our mind on the absolute *regularity* of... phenomena, and from the idea of regularity we pass by imperceptible steps to that of mathematical necessity, which excludes duration," and in this way the "regularity" of Osage inhabitance can appear as a static fact, potentially undone by allotment, rather than as a persistent process of "keep[ing] our place on earth"—an active and also changing relation to place that provides the background for conscious intention. Bergson notes in *Matter and Memory* that "the duration *wherein we act* is a duration wherein our states melt into each other," a movement and *melting* of sensations that provides the unnamed context for action.[50] In this way Osage duration can be thought of as less a specific amount of time residing on this particular "soil" than a quality of inhabitance and relation that gives implicit historical density to contemporary perceptions.

The Osage notion of "moving to another country" recognizes and embraces the need for periodic alterations in established patterns of social life while still affirming the unbroken persistence of Osage peoplehood.[51] The characterization of change as a shift in location suggests the significance of place as a frame of reference through which to understand the potential for action in the world, and the alteration of existing ways of being entails a kind of remapping that involves reacquiring an implicit, orienting sense of emplacement. Moreover, such transformations in Osage practice may appear abrupt but tend to operate through a subtler transition whereby old elements and dynamics merge into new ones even as they are modified in the process. For example, the most "traditional" elements in *Sundown*—the practices that seem to provide the most explicit alternatives to allotment-animated forms of sociality and affect—are themselves a product of changes in Osage life and belief occurring within the decade or so before Chal's (and Mathews's) birth. The central Osage religious practice readers witness is the Peyote ceremony, which gained prominence in the early 1890s after being brought to the Osage by a Caddo medicine man named John Wilson and which became institutionalized as the Native American Church in 1918. Similarly, the I'n-Lon-Schka, the dance in which Sun-on-His-Wings participates and that Chal yearns to join (which I will discuss further in the next section), came to the Osage in the 1880s from the Poncas and Kaws, neighboring peoples closely related to the Osage.[52] Although in adopting these practices (especially the Peyote religion) Osages called for the abandonment of the prior religious system, organized around patrilineal and clan-based priesthoods, that seeming revolution in Osage social structures can be understood as carrying forward in altered ways a range of existing Osage modes of social organization and meaning making.[53] Moreover, these developments in Osage history can be understood as related to the effects of U.S. policy, particularly the official removal from their lands in Kansas to Indian Territory and the mounting interference in Osage governance that led to the adoption of a constitution in 1881. Given the ways the I'n-Lon-Schka and the Peyote religion resonate with previous modes of Osage governance (including the fact that the I'n-Lon-Schka draws on clan associations and prominent leaders in both it and the Peyote religion came from the ranks of former chiefs), one might understand their emergence around the time of the constitution of 1881 as a process by which formal Osage governance comes to be somewhat divorced from modes of internal social order that maintain an adapted continuity with earlier formations.[54] The ability to respond to non-native displacements and impositions (the gravitational force exerted on everyday life by settler institutions), though, need not be understood as displacing a place-based experience of duration through which settler presence, discourses, and institutions are perceived—a *silent* background.

Thus, while allotment and its effects exerted pressure on ordinary Osage social formations, and that force had significant affective consequences for Osage people (as discussed in the previous section), the notion of a break in Osage time between tradition and the modernity of settler imposition cannot capture the character and continuity of the "silence" and its influence on Osage experiences of time. When attending the Peyote ceremony led by Watching Eagle, Chal "sat there for several hours, and it seemed odd to [him] that he could sit thus, silently and without moving. He was fascinated and calmed. There was a complete absence of urges" (269). The silence of the ceremony resonates with the "tranquility" generated by his earliest, and not consciously remembered, sensations of home. At this moment, the tensions produced by the disjunction between this affective complex and the allotment-animated imperatives of civilization become stilled, such that Chal feels a sense of what the text describes as "harmony" rather than the need to express an undefined "something." Mathews highlights the momentum of Osage peoplehood and the nonconscious effects of dwelling in that place, including its influence on quotidian modes of sense making—an experience of what can be characterized as temporal sovereignty. In trying to console White Deer, whose son Running Elk (Chal's childhood friend) had been murdered by whites, Watching Eagle says to him:

> Here are graves of your grandfathers. You came out of this earth here. The life of this earth here comes out of ground into your feet and flows all over your body. You are part of this earth here like trees, like rabbit, like birds. Our people built their lodges here. That which came out of the ground into their feet and over their bodies into their hands, they put into making of their lodges. They made songs of that which came out of ground into their bodies. Those lodges were good and beautiful. Those songs were good and beautiful. Thoughts which they had were good because they came out of ground here. That ground is their mother. (274)

The reference to the "graves of your grandfathers" gestures toward the temporal scope of inhabitance, suggesting continued activity over a multigenerational time span. Such invocations of the quality of duration indicate the land's exertion of influence on Native practices: the "life of this earth" as it manifests in and around the people over time affects their "making of . . . lodges," the kinds of songs they sing, and the thoughts they have.

That environment operates not as a passive stage for human sociality, or a container that holds various resources on which people might draw, but as itself a force that conditions the persistent (re)emergence of Osage peoplehood—the ongoing "flow" of the land into and through Osage people.[55] At times, though, the novel's depiction of Osage dwelling may seem to partake in modernist strands of primitivism in which Native peoples function as figures of a lost ancient wisdom, and this passage runs the risk of offering a fairly essentializing portrait of Osage identity that depends on a strict sense of filiation to past practices.[56] *Sundown*'s discussion of the Osage people as being in "harmony" with "this part of the earth" and having bodies that bear "the life of this earth here" may seem to cast historical change as loss while also emphasizing an unbroken genealogical chain of transmission that in its biological overtones may seem fairly racialized and racializing. The text's portrayal of an embodied relation to place, however, functions as less a claim about the persistence of a static tradition than the powerful effects of locational continuity—the shaping stimulation of place on human action.[57]

Being "part of" this place involves less the retrieval of something from the past than a way of occupying the present that bears within it the momentum of a much longer time frame than that posited in and materialized through federal Indian policy. Although delivered in a somewhat elegiac tone, this moment speaks to the generativity of "this earth" and the patterns of inhabitance and relation that have arisen out of Osage living and storying in this place. The narrator observes, "Most of the older Indians, those who were influenced very slightly by that which they called the Great Frenzy, lived their daily lives as the fathers had lived.... The only change being that they now lived in houses with modern conveniences; radios, telephones, bathrooms and modern furniture" (266). The text here highlights a sensation of continuity that has to do less with the maintenance of a particular kind of lifestyle (as signaled by specific forms of housing or the absence of certain kinds of infrastructure or technology) than with occupying time in ways at odds with the developmental narrative animating "the Great Frenzy" of allotment and oil production. The novel envisions ways of moving toward the future—of moving to a new country—that are not coincident with the privatizing geography of Indian policy, the model of civilizational maturation (and competence) it instantiates, or the ambition for "business" it seeks to incite.[58]

The novel suggests the effectivity of the intensifying settler pressure on quotidian Osage dispositions, affects, and interactions while still indicating the presence of forms of Osage emplacement and attendant modes of temporality that remain askew with respect to the kinds of sociospatiality materialized through U.S. policy.[59] The silence remains as a nonconscious, orienting frame of reference. During the Peyote ceremony that Chal attends close to the end of the novel, Watching Eagle recalls:

> "Long time ago there was one road and People could follow that road. They said, 'There is only one road. We can see this road. There are no other roads.' Now it seems that road is gone, and white man has brought many roads. But that road is still there. That road is still there, but there are many other roads too....
>
> "The road of our People is dim now like buffalo trail across prairie. We cannot follow this road with our feet now, but we can see this road with our eyes, and our hearts will go along this road forever." (271)

Drawing on the idiom of the road that was central to Peyotism, the passage suggests a specific connection to place that has become habituated over time but that also indicates a relation to the future, a way of moving toward tomorrow that has its own momentum.⁶⁰ While that road "seems" as if it were "gone," replaced by a disorienting settler geography that makes the landscape incoherent with respect to what it had been before, the previous road is "still there." In this way it resembles the figure of the river in Deborah Miranda's work, discussed in chapter 1: "It is a river where no gallon of water is the same gallon it was one second ago. Yet it is still the same river.... Even if the whole is in constant change. In fact, *because* of that constant change."⁶¹ More than a memory, "the road" continues as a sensory presence, "dim" yet perceivable by the "eyes" and "hearts" of those attuned to recognize it. From the perspective of this account of Osage territoriality, Chal's sensations (including his dreaming and his yearning to dance, which will be discussed in the next section) suggest the affective imprint of that collective road on him as a guide in moving toward the future. The road helps connect his personal experience of time to the collective duration of Osage peoplehood—the qualitative flow of becoming in the process of moving to another country. The text notes of this moment, "The silence that came over the lodge rang in Chal's ears. He wasn't aware of how long the silence lasted, but he was happy and contented, sitting there" (271). If silence indicates a shared set of place-based orientations, without necessarily becoming the foregrounded object of awareness, the road similarly suggests a shared process of placemaking that helps animate Osage self-understandings even as settler populations, institutions, and mappings exert forms of force that alter the material dynamics of space and time on the reservation.

In addition, the novel does not characterize the effects of place as due to a special, racialized *Indianness* that enables them to commune with the land. Instead, the text juxtaposes the intervals of Osage and settler inhabitance so as to highlight the newness of non-native presence, the formation of their life patterns in relation to another space, and their refusal to engage with the qualitative dynamics of *this* country. On his train ride from Kihekah to the university, Chal observes of the towns in between, "In the distance there was haze, and the red of the earth showing along the edges of the ravines was not out of harmony. Pleasing, until some farmhouse came into view surrounded by outhouses and wire fences; houses that looked like excrescences and tinted by the red dust; houses lonely in the midst of space" (89). The narrator adds, "Chal did not know the reason for this ugliness; this ugliness which white men seemed to produce. He did not know that these buildings were expressions of a race still influenced by an environment thousands of miles across the ocean, and that these foreign expressions were due to the fact that the race was not yet in adjustment with the new environment" (90). The "ugliness" of settler infrastructure, and by extension the alienness of settler sociality, can be traced to the ways it bears within it the influence of an environment elsewhere. Whites' buildings and practices have not yet *adjusted* to the conditions of this place, and that absence of harmony produces the sense of them as adrift, as lacking any connection to the (Osage) space they inhabit. As Watching Eagle later suggests, "White man came out of ground across the sea. His thoughts are good across that sea. His houses are beautiful across that sea, I believe.... But he did not come out of earth here. His houses are ugly here because they did not come out of this earth" (274). More than implying an inherent (racializing) connection between certain populations and landmasses, these moments underline the relative absence of settler adaptation to Native homelands—non-natives do not have "harmony" with this place and, thus, cannot experience the "silence" and "tranquility" of it as "home."[62] Here the novel gestures toward the influence exerted by the land over an extended period, which cannot be encompassed, acknowledged, or institutionalized within the foreshortened frame used by the settler state to narrate (its) history. From this perspective, non-natives might come to have a relationship with this place, but, to do so, they will need to engage with the specificities of this environment, including the presence of the Osage people—their prior inhabitance as well as their stories, memories, and experiences. This critique of settler occupation lies less in presenting a version of "authentic" Osageness carried forward unchanged from the past (or in seeking to make Osage modes of life intelligible within non-native social or political formations) than in suggesting that the territory itself provides a frame of reference for Osage being and becoming, for Indigenous duration, in ways that remain incommensurate with the imperatives realized through U.S. policy and legal geography.

Reciprocally, through its use of the color red, the novel indicates how the material life brought by non-natives remains marked by the effects of seeking to dwell in this land. In addition to observing that the "red dust" on the houses Chal passes on the train is "ubiquitous" (90), he muses about the people he sees in a town near the university that they wear "red-tinted clothes, as though the Red Beds had marked them and were claiming them as their own; coloring them so that they would not be noticeable against the things which surrounded them," adding that "they were part of the life of the country, and this fact seemed to give them assurance" (101–2). The red clay of the soil clings to settler-built infrastructure in ways that testify to the continuing force of the land, as well as highlighting the lack of harmony between this place and recently imported and imposed elements, but the land also exerts some pressure on the personal and collective expressions of non-natives (their red clothing) in their effort to make themselves feel "part of the life of the country" and to *assure* themselves of their belonging.[63] Moreover, these aspects of the novel are given further depth for readers familiar with the significance of red in Osage customs, as a way of figuring both fire and the movement of the sun.[64] The repeated mention of this color, then, indicates a consonance between Osage patterns of meaning making and inhabitance as well as signaling the potential effects of long-standing Osage ways of being on non-natives, if they would surrender themselves to the country and not seek to master it and its inhabitants through ubiquitous state-backed impositions and rearrangements. After Chal expresses his sense of the ugliness of white buildings and their dissonance with the landscape, the text notes, "He would not have dared suggest his thoughts to anyone; it would have been like a sacrilege and certainly unpatriotic. One believed in his country and his state, and accepted the heroics of the race for land in the new territory, and all the virtues of the Anglo-Saxon; the romantics and righteousness of their winning of the West, as taught by his history" (90). As with the earlier reference to the distribution of lands for allotment as history being made (47–48), history here indexes Anglo-Saxon progress in the conquest of "the West," a narrative and experience of time that contributes to a "patriotic" investment in the legitimacy of the nation and its jurisdiction—the kind of national time discussed in the previous chapter. To feel otherwise indicates an affective response due to sustained practices and processes of habitation that provide the background for everyday perception and feeling, sensations with which settlers might become familiar if they were to open themselves to the "flow" of "this earth" rather than forcibly grafting "foreign expressions" onto it and the people(s) there.

## *Feeling Queer*

Mathews's representation of the coexistence of incommensurate spatiotemporal formations casts them neither as sealed off from each other nor as internally unchanging, instead suggesting how attending to experiences of time other than government-implemented modes of liberal progress undoes the self-evidence of state mappings and jurisdiction. Discussing the ways that Native governance has tended to follow patterns imposed by the United States, Mishuana Goeman observes, "By replicating abstract space in Native nation-building, Native communities move away from imagining new possibilities beyond th[ose] mapped out for Native people in settler societies," later asking, "How do we uproot settler-colonial social and material maps that inform our everyday experiences?"[65] *Sundown* offers the potential for upsetting the maps materialized through allotment by sketching modes of relation to place that are not oriented by the administrative logics, procedures, or cartographies of U.S. law. Space appears in the novel not as a quantum carved out from a regularized, homogeneous jurisdictional grid but as a site of accreting experience in which possibilities for Osage being and becoming emerge out of a transgenerational process of dwelling that provides the frame of reference for quotidian feeling and perception. The duration of inhabitance generates rhythms and trajectories that differ from those instituted through federal Indian policy. In detailing the discrepant forces acting on Osage people (in varied ways but ones that are not reducible to a vaguely defined *mixedness* or *betweenness*), Mathews offers no mediating, neutral third term through which they could be reconciled and evaluated; he refuses to provide something like an objective or absolute sense of "now" through which either set of inclinations could be cast as merely ideas or beliefs, as contrasted with the actual facts of modernity. Rather, these formations have overlapping spheres of influence that create disjunctive effects for Native people who live their intersection. In contrast to Charles Eastman's restaging of national history in ways that emphasize Native presence, dispossession, and struggle (as discussed in chapter 2), *Sundown* highlights the quotidian feeling of a disjunction between Osage and dominant U.S. temporal framings, including the absence of a ready way of naming and explaining that sensation. *Queerness* emerges in the text as a way of naming the difficulties—the *density*—of inhabiting these discordant spatial and temporal configurations within the context of escalating settler colonial pressure and imposition.

The term queer tends to be used in the text to indicate Chal's feeling that he has failed to embody non-natives' notions of normality. The earliest use appears after his father's white cousin Ellen calls him a "little savage!" for cutting the soldiers out of a picture of Christ's crucifixion that she had given Chal; the narrator indicates, "His heart was broken. A queer world" (20). Later, when Chal notes the "ugliness" of the white buildings on the landscape and the "unpatriotic" character of this sensation, the text observes, "He kept this feeling subdued; kept it from bubbling up into the placid waters of his consciousness, so that nothing would disturb those waters to keep them from reflecting the impressions that ought to be mirrored, if one were to remain in step. He certainly didn't want anyone to know that he was queer" (90). In addition, when considering Chal's connection to one of his fraternity brothers, the narrator remarks, "Chal had always been inscrutable to Nelson, and he was ever careful in his relation with him. He thought himself queer, just as everyone else thought" (92), and before Chal goes to meet the members of one of the sororities on campus, the text notes, "At the last impression of his face in the mirror that evening, he had seen a bronze face in black-and-white. . . . He had often wished that he weren't so bronze. It set him off from other people, and he felt that he was queer anyway, without calling attention to the fact" (117). These moments all suggest that feelings of queerness emerge in relation to Chal's and others' sense of his Indianness, indicating that he cannot fit in with settler expectations due to attributes—actions, thoughts, physical appearance—taken as indexing his identity as a Native person. To be "queer" here means failing to "mirror" non-native assumptions and modes of engagement (with other non-natives and their surroundings), creating forms of "inscrutable" oddity that point back to an elliptical kind of Indian difference. In this way the term captures Chal's affective experience of being askew—out of step—with respect to the rhythms and orientations of settler sociality.

Queerness in the novel marks the presence of a boundary, a threshold of translation between social fields, and its association with sexual deviance gestures toward a broader process whereby nonnuclear modes of family and household formation and non-reproductively directed forms of desire and pleasure are understood as expressive of ingrained, racialized tendencies toward backwardness. As Michael Snyder has argued, by the time *Sundown* was written the term queer already connoted sexual abnormality, and he further explores the various forms of homoeroticism that permeate the text, noting that "the absence of discourse on Chal's sexuality [in extant analyses of the novel] indicates a problematic silence."[66] Yet, while bearing a sense of errant eroticism, the intimations of perversity surrounding Chal exceed the question of same-sex desire, drawing on extant associations of deviance with nonwhite populations as a sign of their lack of development toward civilization. Addressing the role of race in terminologies and genealogies of sexual abnormality, Marlon Ross argues, "While the perceived racial difference of an African or Asian male could be used to explain any putatively observed sexual deviance, racial sameness became ground zero for the observed split between heterosexual and homosexual Anglo-Saxon men," "such that racial difference necessarily overdetermines the capacity for sexual deviance as a bodily affair."[67] Similarly, Valerie Rohy suggests that the emergence of "straight time" as a "regular, linear, and unidirectional pattern" requires the production of "atavism": "the fantasy of a straight time assailed by racial or sexual atavism actually produces the linear temporality that it takes as primary, the order that *has been disrupted* . . . by a perverse backwardness."[68] Within late nineteenth- and early twentieth-century sexology and ethnology, the absence of nuclear family structures and lifelong, monogamous heteroconjugal desire marked people of color as less advanced. The chrononormativity of the nuclear family—its use as a means of naturalizing a particular ordering time—was further materialized through a chronobiopolitics of race. Within emergent anthropological conceptual frames, which were increasingly mobilized within U.S. federal Indian administrative and policy discourses, reprosexual marital union was conjoined with the acceptance of private property and construction of non-kinship-related governance as (racialized) markers of progress toward the achievement of civilization.[69] From this perspective, Chal's queerness signals his supposed Indian atavism with respect to allotment-instituted norms while pointing to the transposition of long-standing aspects of Osage sociality into the language of perversity.[70]

If the novel's refusal to marry Chal off by the end may indicate his homoerotic inclinations, it also may suggest a refusal to recapitulate the developmental trajectory of individual maturation, by which the spatial and temporal dynamics of federal Indian policy are normalized as simply the self-evident background for the unfolding of ordinary life. From within the logic of allotment, tribal property is to be divided up and transmitted along nuclear family lines, and centering conjugal couplehood as the atom of social life also provides the basis for defining the terms of racial genealogy (reproductive transmission from both parents of determinate quanta of Indian blood).[71] These modes of inheritance served as cornerstones of the politics of Osage allotment, and in emphasizing the noncoincidence of Chal's experience with these ways of envisioning time, his out-of-step feelings of queerness, Mathews gestures toward both the continuing presence of an alternative spatiotemporal formation and the affective implications of inhabiting it amid the "atmosphere" created by ongoing (and intensifying) settler interventions. Over the course of the text, Chal has sexual feelings for and relationships with various women, but the plot does not single out one woman with whom he will find connubial bliss. The absence of such a romance plot contributes to the sensation of a lack of narrative development, operating as part of Mathews's use of modernist techniques to complicate the reader's understanding of time (as discussed earlier). More specifically, this apparent lack of personal maturation—such as his mother's viewing him at the end of the novel as "a little boy in breech clout and moccasins" (310)—suggests a suspension of the normative process of individual development, a dynamic highlighted by the novel's structure as a bildungsroman.[72] From the perspective of the state-sponsored story of proper entry into adulthood, Chal's continued identification with the land and Osage peoplehood—often expressed in eroticized forms of dancing and dreaming (to which I will return shortly)—appears as a set of juvenile attachments to be abandoned. They are queer deviations from the trajectory of "straight time" toward privatizing (and conjugally directed) individuation. However, to the extent that queerness in the text indexes not simply aberration but translation, a movement among varied social fields (albeit in ways that overwhelmingly privilege and make paradigmatic men's experience), it gestures toward the ways the time line of personal development posited in and materialized through allotment-era policy fails fully to reshape quotidian Osage experience.[73]

More than pointing toward the survival of Osage "culture" as a residual formation that persists into a present otherwise defined in settler terms, Chal's queerness suggests a set of affective trajectories that offer the potential for other ways of reckoning the time of Osage peoplehood. In particular, his queer failure to "mirror" non-native norms (especially of progress toward marital union) suggests a different orientation to what Jean O'Brien has termed the "temporalities of race," in this case the ways Osage identity comes to be defined legally as the reproductive transmission of Indianness.[74] As discussed earlier, from 1906 onward the notion of *competency* was crucial in categorizing Osage persons and in making Osage land potentially available to non-natives through sale (even while the Osage retained a collective interest in the subsoil rights), and as of 1921, relative *bloodedness* became an additional, although not equivalent, way of determining the character of (the) Osage people. A law passed in that year removed all restrictions on allotted lands held by "adult Osage Indians of less than one-half Indian blood," and a congressional statute passed in 1929 mandated that the Secretary of the Interior give to all Osages "of less than one-half Osage Indian blood" a certificate of competency within the next decade, as well as "all of the balance appearing to his credit of accumulated funds."[75] These laws do not indicate a quantum of blood below which a person would no longer be considered Osage, restrict occupancy at the three communally held villages on the reservation based on blood, or gauge a person's ability to be a headright holder in the mineral estate based on blood. They, however, do link the capacity for civilization (as represented by full participation in the market) to a calculation of biological Indianness and conceptualize an enduring relation to the land as an Osage person as being a function of relative bloodedness.[76] Thus, while not entirely pegging access to Osage lands and benefits to amounts of reproductively transmitted Indianness, federal Indian law and policy projects a generational process of de-Indianization whereby (the) Osage people will increasingly have fewer, and more atomized, claims to the land of the Osage reservation, with the exception of the existence of the mineral estate itself.

As noted earlier, though, many scholars have read *Sundown* as chronicling the particular plight of the mixed-blood struggling to negotiate the relation between tradition and modernity. This argument recirculates the logic of allotment-era policy with respect to the Osage by positing bloodedness as the central criterion through which to assess relative change, itself plotted against a time line in which forward movement in time means leaving behind attachments to Osage collectivity. Moreover, such interpretations inaccurately characterize Chal's status. In light of the consolidation after 1921 of a significant legal distinction between Osages of one-half or more Indian blood and those of less than one-half, the terms "full-blood" and "mixed-blood" came to refer to those categories, respectively.[77] Given that the novel leads readers to believe that Chal's mother's blood is all Osage and informs us that his father is three-quarters Osage by blood (53), Chal's blood quanta would be seven-eighths, putting him within the full-blood category. Chal also likely would have a certificate of competency that would enable him to evade the restrictions on his control over property and funds to which he otherwise would be subject. Critical accounts of the novel, though, use his supposed identity as a mixed-blood as a way of naming his ostensibly intermediary position between an authentic Osage past and a settler-defined present, offering no discussion of the particular legal matrix in which these terms gain meaning and the exertion of (spatial and temporal) force through them.

If one refuses to treat such racial distinctions as self-evident, or as fully determinative of Osage perceptions, the fact that Chal's story does not tend toward procreative union suggests how his *queerness* opens onto other ways of registering the presence of Osage processes of becoming amid the intensities of settler interventions and remappings. Specifically, Chal's dancing and dreaming point to embodied experiences of a relation to place that do not fit the racially and reproductively inflected trajectory of personal development at play in allotment-era Indian policy. In this way, we might understand the novel's depiction of these actions and feelings as engaging in a form of what Elizabeth Freeman characterizes as erotohistoriography: "Erotohistoriography is distinct from the desire for a fully present past, a restoration of bygone times. Erotohistoriography does not write the lost object into the present so much as encounter it already in the present, by treating the present itself as hybrid. And it uses the body as a tool to effect, figure, or perform that encounter." She later adds that "enjoyable bodily sensations" can "produce forms of time consciousness—even historical consciousness—that can intervene into the material damage done in the name of development, civilization, and so on."[78] Chal's erotic sensations with respect to dancing and dreaming point toward his immersion in a feeling of duration that links his individual perception to implicit forms of historical experience (even when they do not explicitly rise to the level of consciousness per se), and the novel presents Chal's capacity for forms of pleasure as running counter to the modes of presentness generated within allotment's fields of force. After Chal returns to the reservation in the wake of his military service, he begins to abuse alcohol, and during this time he brings a group of white women to one of the dances at "the village." "He knew how the dancers felt about people coming out to stare at them, yet they did nothing to stop or discourage it. They danced because they felt it impossible to give up that last expression of themselves" (252), and in bringing these non-natives, Chal "felt vaguely that he was betraying his people" (253). The dances, most likely the I'n-Lon-Schka, represent an "expression" of Osage peoplehood, a materialization of it for and by Osage people, rather than a cultural practice that illustrates Indian difference.[79] Chal experiences the dancing as tied to his felt need to connect to something that defies articulation: "as usual, he had a desire to join them. He had always felt that by joining them he could express that thing which came over him at times; that something which had to be expressed, but which he couldn't possibly put into words or actions" (257). To join them would entail him physically *expressing* a sense of Osage belonging, a "something" that is not equivalent to the future "something" for which various people in the novel are waiting. This "something" appears already present, but stifled or rendered mute.[80]

Earlier in the text, before he leaves for college, Chal experiences a similar sense of constrained desire given form through his connection to the landscape of the prairie. After noting that Sun-on-His-Wings and Running Elk rode their ponies back to the village after school and that "they danced at the Roundhouse in the village every June and September" (68), the text observes that Chal "felt that there was something in him which must come out, and unable to find any other expression, he took action as a means, and raced his pony wildly as before" (70). In addition, in order to try to address "this thing within which he couldn't satisfy," "sometimes he surprised himself by breaking into an old war song" or swimming "until he could feel his legs growing weaker and weaker," and "one day he stripped off his clothes and danced in a storm and sang a war song. Sang and danced until he was almost exhausted and his body was wet with rain and sweat" (70). In these moments, "A mild fire seemed to be coursing through his veins and he felt that he wanted to sing and dance. . . . He felt that some kind of glory had descended upon him. . . . He wanted to struggle with something. . . . There seemed to be intense urges which made him deliciously unhappy" (73); "this mysterious unhappiness came to him only at times, and never except when he was alone on the prairie" (72). The sociality of the village, and the dances that are part of it, become a vehicle through which Chal perceives "urges" that are unsatisfied in other areas of his life and that emerge, without a language through which he can express them, when he is alone on the land—outside of both Kihekah and the village. When experiencing a similar feeling in college, a yearning to swim naked in a nearby river, he demurs because "if someone did see him, they'd think he was crazy" (103). Here his queerness manifests as a set of sensations that tie him to particular Osage practices (singing and dancing) that are understood as at odds with the orientations inculcated through allotment-era policy and education, as well as tying him to the place of Osage peoplehood—the silence of the land and its orientations. The urges to dance and to engage with the landscape in ways not scripted by economic development become intertwined in Chal's affective life, part of a single sensory and sensual matrix. When he seeks to dance on the prairie again close to the end of the novel, he fails to achieve a sense of release from what he cannot express, and the "sense of completeness" for which he struggles is described as "an orgasm of the spirit" (297). The novel cross-references Chal's erotic relation to his own body and his surroundings with his (unsuccessful) attempt to engage with the land and his people in a way that will resolve his sense of being out of sync—an impression of aberrance generated by the pressure of settler expectations, institutions, and mappings.

Moreover, silence marks the disjunction between Chal's sense of himself and non-native behavioral norms. The text observes of him while he is in college, "He had a feeling that the others found little interest in the things which he had to say, and naturally he became silent," adding, "He was completely at a loss to understand their attitudes and their philosophies" (139). Mathews here emphasizes a white fascination with sex as a point of distinction: "He had never been able to see anything strange or unusual about mating. He had seen it all of his life among the hills, and to him it was a part of nature" (139). In this way Chal's queerness emerges as less a particular sexual orientation than an orientation toward Osage sensations of emplacement (the silence "behind [his] impressions") drawn from accreting experiences on the reservation.

Queerness in the novel indexes the density of Osage life, particularly the ways postallotment political economy racializes and renders anomalous sensations emerging from the dynamics and duration of Osage inhabitance (being part of "the life of this earth"), and conversely, the figure of dreaming offers a way of suggesting the continued effectivity of such feelings amid the material transformations of the social landscape brought by allotment. During the process of distributing allotments, which John, as noted earlier, describes as "seein' history being made,"

> Chal moved with the crowd from place to place, but the tall white man who was making history was of no further interest to him. . . . Chal was

soon far away in a dream of his own. He was pretending that where the green of the prairie met the blue of the sky was the edge of the world.... He pretended that the road which he could see twisting over the green; parallel lines lying across the prairie, passing out of sight, then appearing again on the far hillsides; passing the tall posts which marked the gates through the barbed wire fences; would lose itself in the haze of distance as it passed through the gate in the blue wall of his fanciful edge of the world. (48–49)

As opposed to the progressive movement of a history defined by federally instituted modes of privatization, Chal slides into a "dream" in which the "prairie" constitutes the horizon of his world in ways that bypass the constricting "barbed wire fences" that demarcate ownership. While "fanciful" in the sense that it does not conform to the legal geographies in the process of being constructed through allotment, the ubiquitous atmosphere created by U.S. force, Chal's dreaming can be understood less as signaling a passive retreat from the actuality of settler modernity than as expressing in a varied form the urges that lead him toward dancing.[81] Through his dreaming, the novel suggests the persistence of feelings of relation to the landscape that do not fit within Chal's conscious commitments to "business" as the basis for assessing forward movement in time. His later dreams, particularly during college, include envisioning himself as a coyote and imagining he's among Osages chasing an enemy Pawnee such that "he wished that he were among his hills again" (111, 136–37), and while in the military "he dreamed of the blackjacks and the way they would soon be standing in the sun" (233). His dreams suggest the potential for tapping into the feelings associated with silence, the affective surround of Chal's orientation toward the Osage homeland. This impression of a presence that both is and is not realized points toward a (set of) sensation(s) that appears as perversity, passivity, anomaly, and anachronism from the perspective of a futurity ordered around privatization. Conversely, such sensation(s) can be interpreted as indicating the influence of a kind of spatiotemporal formation at odds with that instituted by U.S. policy, pressured by the latter yet not replaced by it.

Aside from the question of sexual identity, queerness in the novel highlights the kinds of atavism attributed to Chal as an Indian, his nonperformance of racially inflected reproductive couplehood, and the variety of affective inclinations (silence, dancing, dreaming) that he shares with a number of other Osages in the novel—often self-interpreted as blocks to their "doing something." Together, these dynamics point to an experience of time at odds with the rhythms, inclinations, and momentum of allotment-era policy. Chal's queer feelings suggest an ongoing negotiation shaped by the copresence of multiple modes of temporality that constitute what is putatively the "same" space, the Osage reservation, in discrepant ways. The novel juxtaposes the *force* of allotment-driven transformations—and the attendant fixation on production and commerce as gauges of meaningful movement toward the future—with the process of achieving "harmony" with the landscape. However, Mathews does not cast that tension as one between the emergent and the residual, with Chal caught in this particularly Indian time bind. Rather than posing the tension as between past and present, *Sundown* suggests that there are actually multiple spatiotemporal formations at play on the Osage reservation in ways that create complex, lived disjunctions that Osage people need to navigate. This institutionally overdetermined and ongoing process of translation is captured by Chal's sensation of queerness, which he experiences as an incoherence between these formations rather than as part of the Osage legacy of "moving to another country."

In this way the text displaces the notion of persistent Indian backwardness as well as the notion that Native people need to act in the supposedly shared "now" of modernity. Instead, Mathews points toward the difficulty of finding a rhetorical, political, or geographic position from which to negotiate among incommensurate frames of reference: one predicated on the progress of privatization defined by allotment, and the other shaped by the processes of becoming at play within Osage patterns of inhabitance and perception. As the novel suggests, oil production takes part in, and helps drive, both formations, encouraging increased non-native presence, federal intervention, and forms of settler expropriation while also enabling the Osage to defer wage work, spend more time in the three reserved villages, and maintain connections with other Osage people (such as the daily "talk" that Chal often views as an absence of action).[82] In *Sundown* Mathews casts the latter as part of "keep[ing] our place on earth." That process offers an active orientation toward the future situated within an Osage understanding and material experience of time. Thus, even as the novel illustrates increasing settler pressures on (the) Osage people, it suggests the affective and environmental influence of the land on Osage personal inclinations and social dynamics. Such Osage phenomenologies of time include the emergence of new movements, such as the Peyote religion and the I'n-Lon-Schka. Through the elaboration of Chal's conflicted experience of time—his impressions, urges, and queer anomalousness with respect to non-native norms—the narrative explores the complexity of everyday Native life under settler colonialism, particularly in the allotment period, without using settler policy as the reference point for defining the meaning of time and space.

Thinking about the copresence of divergent spatiotemporal formations and modes of becoming emphasizes the nonequivalence among ways of conceiving, perceiving, and living in the "same" place and period, and doing so engages the density of Indigenous experience without either treating Indigenous formations as surviving remnants from the past or presenting the present as singular in ways that tend (explicitly or implicitly) to privilege settler frameworks, materialities, and imperatives as the reference point for assessing contemporaneity.[83] Further, as against a notion of being caught between two cultures, *Sundown* highlights the effectivity of state policies and intervention on the Osage reservation while also gesturing toward how what might be taken as "traditional" Osage sociality remains responsive to changing conditions, such as in the development of the Peyote religion and the I'n-Lon-Schka. While deferring consideration of the institutional exercise of political sovereignty, the novel does address U.S. interference in Osage governance, offering the influence of duration as an alternative conceptual and sensory frame through which to address the character of Osage peoplehood—to address the expression of Osage temporal sovereignty.[84] Doing so does not resolve the issue of what juridical form Osage nationhood could or should take, but it does suggest the importance of understanding the effects of settler colonialism as ubiquitous yet not total. The novel suggests how the allotment regime influences Native sensations of selfhood and experiences of time and history. However, it also explores the affective momentum arising out of an experience of space and time not predicated on the terms and instantiations of allotment-era policy. In this way Mathews sketches a kind of Osage phenomenology (itself responsive to change) that provides a frame for conceptualizing and living peoplehood while also refuting the idea that settler colonialism produces a decisive rupture in Indigenous lifeways such that its force necessarily determines the shape of the present and future. The text does not offer a privileged vision for Osage governance, nor does it provide a litmus test for evaluating Osage authenticity. Instead, it highlights the uneven and unchosen process of translation among disparate frames—the density of experience—that is the effect and legacy of settler policy. The sensation of queerness that attends such translation indexes possibilities for another kind of future than that projected for Native peoples within the temporality institutionalized in the allotment era.

FOUR. **GHOST DANCING AT CENTURY'S END**

On January 1, 1889, a Yerington Paiute living in Mason Valley, Nevada, named Jack Wilson/Wovoka had a vision that he ascended to heaven and spoke with God. While the exact character of this prophecy has been the subject of much debate, it entailed reunion with the dead and the performance of a dance through which to hasten that moment, and the resulting set of beliefs and practices came to be known as the Ghost Dance.[1] There were numerous similar visions and movements throughout the period, including one among the neighboring Walker Lake Paiutes twenty years earlier, and these various movements spread over wide areas, reaching at various points over the Plains, into California and the Great Basin, and up through the Columbia River and Puget Sound regions.[2] Wovoka's revelation, though, gained greater fame, and has become a touchstone in American (Indian) history, because Lakotas who had responded to Wovoka's message were pursued by the U.S. military in late 1890, owing to the false claim that they were threatening an uprising. The most famous of the events connected to this campaign occurred at Wounded Knee Creek on December 29, when a band under the Minneconjous chief Big Foot was massacred even though they had already surrendered and were in the process of making their way to the agency on the Pine Ridge reservation. Wovoka's Ghost Dance becomes epochal, and is widely remembered, as a result of its association with an act of state violence itself often understood as signifying the end of the Indian wars.[3] If lamented as tragedy, Wounded Knee becomes a historically canonical site for marking the end of an era, that of meaningful Indigenous resistance to U.S. occupation, and, through association with it, the Ghost Dance of 1890 gains its prominent position within settler and Native accounts of U.S. national history. In this way it serves as an exceptional moment of Native visibility, only to indicate the prior and subsequent irrelevance of Indigenous presence within Euramerican history.

From one perspective, then, the Ghost Dance operates as a figure of futility, the impossibility of opposing U.S. rule over Native peoples, the official containment of Native sovereignties within government-delimited spaces, and the irreversibility of the unfolding of settler time/history. In this way it fits within the dynamics of U.S. national time and of the temporalities of the treaty system discussed in chapter 2. However, many contemporary Native writers draw on its symbolic significance (as a result of its association with Wounded Knee) in order to stage alternative visions of Native pasts and futures than those at play in narratives and enactments of settler time. These texts of a century later mobilize Wovoka's prophecy in order to envision and seek to realize forms of Native world making that are bound to neither a narrative of settler national progress nor a territorialization of indigeneity onto the sanctioned spaces of the reservation.[4] As discussed in the preceding chapters, Indian policy sought to contain Native peoples within constricted areas, consistently shrink these areas by making swaths and sections of them available for non-native use and ownership (via treaties and allotment), and reorder existing Indigenous experiences of time to bring them in line with non-native modes of jurisdiction and programs of development (personal, familial, and commercial). In *Manifest Manners* Gerald Vizenor characterizes non-native accounts as "the ruins of representations of invented Indians," further observing that these "simulations of manifest manners . . . become the real without a referent to an actual tribal remembrance."[5] Invoking the Ghost Dance, though, enables Native authors to deviate from settler historical emplotments and their associated cartographies, opening room for acknowledging Native realities in which the dynamics of settler colonialism exert force but do not define the limits of Indigenous possibility, placemaking, and perception. As Mishuana Goeman argues, "the literary . . . tenders an avenue for the 'imaginative' creation of new possibilities, which must happen through imaginative modes precisely because the 'real' of settler colonial society is built on the violent erasures of alternative modes" of Indigenous being in the world.[6]

Ghost Dance narratives, like Sherman Alexie's *Indian Killer* (1996) and Leslie Marmon Silko's *Gardens in the Dunes* (1999), story late nineteenth-century millennial visions in ways that index the persistence of possibilities for Native self-determination. Chapter 2 explored how non-native narratives of national time (particularly pivoting around the Civil War as a moment of national transcendence) position Indigenous peoples and their relations to place as forms of exception, casting Native presence as an anomaly while rendering the continued existence of the state and its assertion of jurisdictional authority over Native lands as the background and horizon for a settler sense of continuity and futurity. Chapter 3 addressed the ways that the force of U.S. Indian policy, specifically the allotment program, cannot be understood as totalizing the possible spatiotemporal formations on Native lands, instead operating as one frame of reference that occupies the "same" space as the dynamics of Indigenous duration—the long-term rhythms and immanent relations that arise from Indigenous modes of occupancy and engagement with the specificities of that place. While addressing different ways that Native peoples are forcibly incorporated into settler time, and ways they retain their own temporalities distinct from it, both of these examples depend on forms of persistence, an undivided expanse of time across which Indigenous presence stretches. However, as envisioned by Alexie and Silko, the Ghost Dance suggests less unbroken continuity than complex cross-temporal communications, impressions, and relations that exceed the unfolding of a timeline. Such cross-time proximity, the sense of direct implication across the apparent gulf of chronology, might be described as *prophetic temporality*. Unlike the silence discussed in the last chapter, prophecy emerges in these texts less as an unarticulated phenomenological background than as a catalyzing force that punctuates and animates Native frames of reference. What makes these particular texts so compelling is their exploration of the complex relation between prophecy and everyday experience: the former does not rupture the latter, instead emerging in and through ordinary conjunctures of sensation, perception, and duration; and, conversely, prophecy cannot be reduced to the mechanics of quotidian circumstances, a set of causal relations that can explain (away) the force of paranormal presence.[7]

Drawing on the paradigmatic quality of Wovoka's prophecy and movement within extant historical narratives, these texts divorce Native prophecy from the periodizing finality attributed to the massacre at Wounded Knee. Alexie and Silko cite the Ghost Dance while drawing on its iconicity to index regionally specific prophetic movements and sensibilities. The Ghost Dance, then, operates as a site for generating Native futures not bound by the presumed givenness of settler national geographies and destinies. More than a turn to the past, the novels access the potentials of prophecy as a nonsuccessive relation to time that indicates both intimacy across periods and the action of nonhuman entities as causal agents that take part in processes of becoming.[8] As Russell West-Pavlov suggests, time can be understood as the complex relations among "multiple, interwoven, immanent temporalities inhabiting entities of many types rather than providing their medium or container."[9] In this way the notion of prophecy provides a way of talking about kinds of backgrounding and storying in which the past, present, and future do not line up as an evolving, continuous causal chain but in which, rather, collective experiences of time are oriented by affects—and entities—that do not follow a developmental pattern. The spirits that animate prophecy have their own "immanent" itineraries that affect temporal relations, creating forms of juxtaposition, syncopation, fusion, and interpenetration. In both novels the Ghost Dance entails the operation of beings and forms of force and power that have no place within rationalist modes of explanation. As Scott L. Pratt notes, Wovoka's prophecy and the movements that emerge from it have usually been explained "in naturalized terms" as a derivation of something else, and doing so engages in a process of "ontological reduction" in which everything can be understood as a function of "objects in the world as experienced by the surrounding Euro-descended peoples" such that they are "fully accessible to non-Natives."[10] In invoking and mobilizing the Ghost Dance, these texts do not seek to explain it (or manifestations of prophecy more broadly) within what one might describe as a sociological framework (Native despondency, lack of resources, weakening of Indigenous governance structures, etc.). Instead, the novels suggest, in John Dewey's terms as quoted by Pratt, "a contrast, not between a Reality, and various approximations to, or phenomenal representation[s] of Reality, but between different reals of experience."[11] Via the Ghost Dance, *Indian Killer* and *Gardens in the Dunes* envision quotidian forms of Indigenous temporal experience that are galvanized and gain new meaning and momentum through collective expressions and experiences of prophecy. Such experiences further license, stimulate, and texture possibilities for living a future not determined by settler histories. Rather than understanding the quotidian and the prophetic as opposed or contradistinguished, the novels present prophecy as taking part in ordinary life in various ways: as an extension and intensification of everyday affect, as a mode of ritual with its own regular periodicities, as a practice embedded in geographic and historical relations encountered in everyday ways (in terms of prior sites of prophecy or areas referred to in prophecies), and as a set of principles, guides, and frames for everyday perception and decision making.[12]

These novels do not so much supplement history, something like filling in what has been missing in available narratives, as reconceptualize *historicity*, including the principles by which to understand the relation between the past and the present and the possibilities for Indigenous futurity/ies. In this way the texts might be thought of as seeking to theorize temporal sovereignty in their insistence on the limits of linear accounts of Indian history and their emphasis on how prophetic movements express nonchronological modes of Indigenous experience and collectivity. The performance of the Ghost Dance in both novels undoes the self-evidence of settler becoming, the ways the creation, extension, and consolidation of the settler state and its social and legal geographies serve as the implicit milieu within which meaningful change over time occurs. Highlighting the potential nonequivalence between Native and non-native phenomenologies, these novels trace how settler stories become realized through their accumulation over time and the violences they enact while also indicating the existence of other temporal rhythms and forces that bear alternative possibilities for Indigenous being in the world. *Indian Killer* highlights the accretion of non-native representations of Indianness and the way they provide the context for ordinary non-native perceptions that affect not only non-natives' engagements with Natives but Native people's own conceptions of Indian authenticity, especially inasmuch as it becomes bound to the space of the reservation. Alexie positions Native prophecy as a means of breaking with this regime of imposed settler perceptions and the particular brutalities they enable, engaging with the continuing legacy of settler expropriation in ways that resonate with prophetic legacies among Coast Salish and Columbia Plateau peoples. The characters in the text continually draw on nineteenth-century Indian history as a means of naming current struggles, and, in doing so, they suggest both how the present-day pursuit of Indigenous self-determination gets dismissed as an anachronism and how contemporary events replay supposedly superseded dynamics from the past. Within this double-sided set of citations, the text's invocation of the Ghost Dance gives material shape to ongoing histories of Native grievance. Neither separate from the history of settler colonial violence nor simply derivative of it, the prophetic dancing in the text bodies forth new relations among the following: the legacies of the Indian wars, the continuing assertion of expansive Native claims to place that exceed modes of state recognition, Indigenous ontologies of spirit, and the potential for a transformed Indigenous horizon of possibility.

While in Alexie's novel the Ghost Dance produces a singular figure who is the condensation of everyday forms of Native anger at being routinely rendered unreal and unhistorical, for *Gardens in the Dunes* prophecy coalesces the potential for regeneration. Silko envisions forms of Native survival that exceed the chronobiopolitics of Indian policy and the annihilating tendencies immanent within settler institutions and economies. The novel displaces the questions of authenticity that lie at the center of *Indian Killer*, instead foregrounding the presence of complex networks whose indigeneity is not dependent on the generational transmission of quanta of blood Indianness. As against notions of the Indian real as a form of racial lineage maintained through proper sexual order and contained within governmentally regulated spaces, Silko offers a portrait of indigeneity as a dynamic and expansive matrix of transtemporal connections, for which the Ghost Dance (as it manifested on the Arizona-California border) provides the paradigmatic example. The text suggests that the state seeks to realize a certain vision of Indianness through the imposition of reservation geographies and the insistence on the need to train Native peoples in bourgeois familial and gender norms. In contrast to this set of heteronormative temporal scripts and their dependence on a view of Native place as fetishistically bounded, Silko's storying focuses on desire as a means of indicating the variability of Native social ties, their ability to engender new formations without decimating the old ones, and the presence of modes of development that are multivectored and not the linear unfolding of progress or civilization. If the realization of prophecy in Alexie's novel arises out of grief and rage, in *Gardens* prophecy bears hope through its ability to promote everyday forms of creation that weave together varied and apparently incongruous histories and future potentials, providing an alternative to the conception of modernity as newness which the text casts as inherently organized around destruction.

## At the Limits of Indian Realness

Set in Seattle, *Indian Killer* centers on the life of John Smith, a Native man adopted by a white couple from birth, without any knowledge of the tribe from which he comes. The novel leads readers through John's everyday tribulations, which suggest significant mental illness modeled on schizophrenia.[13] John's story, then, suggests a series of questions about what constitutes reality, both his own as a Native person (in the absence of connection to a specific people or meaningful information about his genealogy) and his impressions of the world (given the ways they are punctuated by voices, hallucinations, and persistent paranoid episodes). The questions circulating around how to conceptualize John's Indianness expand to include the other Native characters in the novel, each of whom fails in some way to fulfill the criteria for authenticity articulated by the others. Moreover, the novel suggests how the self-subjection of Native people to such regimes of inspection emerges within the context of accumulating non-native representations of Native people(s) that come to serve as the basis for settler-Indigenous relations. As John's difficulties intensify and the reader is introduced to multiplying conflicts over the contours and character of "real" Indianness, a rash of murders are committed by a figure referred to only as "the killer," who over the course of the novel is increasingly associated with the Ghost Dance and who is described in other-than-human terms (usually through bird imagery). The text draws on the Ghost Dance to indicate the presence of forms of Native feeling and becoming that, while responsive to continuing modes of non-native occupation, exceed the representational capacity of conventional conceptions of the real.[14] In doing so, Alexie sketches the cumulative material force exerted by such narratives over time while simultaneously gesturing toward the existence and counterforce of Native experiences of temporality—modes of prophetic duration—that exceed the parameters of settler historicity.

The novel highlights the ways that non-native accounts of Native pasts and presents come to serve as the baseline against which to define the truth of Indianness, which itself can then be claimed by non-natives as their own possession.[15] As scholars have noted, the novel repeatedly invokes the proliferation of settler stories about Indians, stories that then are treated as the basis for settlers' understandings of what constitutes the reality of Native life and history. In *Muting White Noise* James Cox argues that "the correlation that Alexie draws between storytelling and violence is so explicit that the storytelling *is* the violence that leads to John Smith's death," adding that the text "emphasizes the power of these stories to define worldviews that encourage acts of violence against Native peoples."[16] Discussing the various kinds of collection at play in the novel, Janet Dean explores the "intangible forms of violence in the text," specifically "the ways institutional and private archives designed to authenticate Native American identity threaten the very cultures they would define and purportedly preserve."[17] Examples of this pattern include the narrator's observation that John's adoptive mother, Olivia Smith, has learned all she knows about Indians "from books, Western movies, documentaries" (12), as well as the text's comment in discussing Jack Wilson, the otherwise white writer who claims Shilshomish descent with no proof, that "he knew about real Indians. He'd read the books" (178).[18] Furthermore, Olivia says to Wilson, upon meeting him at John's apartment, "I like your books. You really get it right" (355). More than illustrating the circulation of non-native portrayals of Native people as (if they were) the real of Indigenous experience, these moments suggest the role of temporality in the process of materializing such images as the basis for non-native perception: they gather and reinforce each other over time. Each builds on the previous one, such that non-natives come to treat their own past experience of such representations as a phenomenological, conceptual, and evidentiary basis for engaging with dynamics and accounts in the present—Olivia's confidence that Wilson "get[s] it right." Through this implicit process of adjudication, the archive of non-native representations functions as a cumulative repertoire of narratives through which to assess Indianness, giving a historical density to non-native versions of Indianness and providing the intuitive basis through which Native people become actual within settler sensations of the world.[19] The narrator informs readers that the book on which Wilson starts working in the wake of the murders is also called *Indian Killer* (227), and later the text notes, "Wilson knew that he was writing more than a novel. He would write the book that would finally reveal to the world what it truly meant to be Indian," adding, "He wanted the world to know about the real Indian Killer, and not just somebody else's invention" (338–39). This work of nonfiction, in which he presents John Smith as the killer (415), appears as the apotheosis of the questions of authenticity, narrative authority, and documentation around which the novel has been building, offering itself as the "truth" of not only the killer's identity and actions but Indigenous being.[20]

Moreover, the construction of Indian realness around settler stories facilitates a temporality of *becoming Indian*, whereby non-natives can cast themselves as not only possessors of the knowledge about Native people supposedly contained in such representations but also, in consuming and circulating these stories, as coparticipants in Indianness. The text observes that "Jack Wilson grew up white and orphaned in Seattle. Dreaming of being Indian, he'd read every book he could find about the First Americans," and he "recreated himself in the image he found inside those books" (157). Additionally, the mystery novels he writes feature Aristotle Little Hawk, "the very last Shilshomish Indian" (162). In terms of his sense of himself and his creative work, Wilson inhabits Indianness with a sense of authority and authenticity, an orientation, feeling, and momentum animated by his lifetime of reading (non-native) texts about Indians. Similarly, in a Native American literature class taken by Marie Polatkin (one of the novel's central characters, a Spokane woman who meets John at a powwow on campus), the professor, Clarence Mather, assigns a series of as-told-to autobiographies rather than texts produced by Native authors. In justifying his choices in the face of Marie's repeated critiques, Mather asserts, "One would hope that we can all benefit from a close reading of the assigned texts, and recognize the validity of a Native American literature that is shaped by both Indian and white hands. In order to see that this premise is verifiable, we need only acknowledge that the imagination has no limits. That, in fact, to paraphrase Whitman, 'Every good story that belongs to Indians belongs to non-Indians, too'" (60–61). From this perspective, "Native American literature" testifies to a history of collaboration between whites and Indians, one in which Mather putatively takes part through his recirculation of these books.[21] In "Writing Off Treaties" Aileen Moreton-Robinson suggests that within settler accounts indigeneity functions as "a white epistemological possession" and that, through the process of recognition, "tribes become constituted as an epistemological possession of the nation state."[22] Indian stories "belong" to everyone because they have no ontological status outside of their ability to be *imagined*, narrated, and experienced by non-natives.

Beyond displacing Native people's representations and understandings of their own histories, both Wilson and Mather contribute to a temporal dynamic whereby the accreting legacy of non-native depictions of past and present Indians provides the horizon for imagining the potential for a Native future, or the absence of one. In *Queer Phenomenology* Sara Ahmed suggests that the perception of an object as "having qualities" is less "a perception of what is proper to the object" than a reflection of the object's ability "to enable the action with which it is identified," and objects gain such qualities, imagined as inherent, from being repeatedly taken up in similar ways.[23] The Indian emerges in Alexie's text as such an object for whites, as a habituated potential for non-natives to orient themselves as belonging to this place through the imagination of a shared history. When Native people(s) do not illustrate such qualities by capacitating white presence, they are construed as, in Ahmed's terms, "bad object[s]" or "the cause of the failure," as having failed to be properly Indian and thus being in need of disciplining.[24] In this sense the novel suggests that the amassing of settler portrayals of Indianness does not simply defer Indigenous accounts but actively works to replace them in ways that affect the material possibilities available to Native people. With respect to Mather's syllabus, Marie notes, "It's like his books are killing Indian books" (68), and at a demonstration during one of Wilson's readings at a local bookstore, in which protestors hold signs such as "ONLY INDIANS SHOULD TELL INDIAN STORIES," Marie says in response to a reporter's question, "Books like Wilson's actually commit violence against Indians" (263–64). More than indicating something like a proprietary relation to Native histories, the insistence that "Indians should tell Indian stories" arises out of the sense that the momentum of non-native narrations directly functions as a form of settler colonial force, affecting possibilities for Native action in the world (in ways that might be read as similar to the field of force exerted by allotment, discussed in the previous chapter). The effects of the intensifying proliferation of, in Vizenor's words, the "simulations of manifest manners," whose "invented Indians" replace "actual tribal remembrance," extend to bodily wounding, going beyond the discursive and ideological to the corporeal.[25] Any attempt by readers to cast such claims by Marie as hyperbole is undermined by the novel's repeated insistence on the causal relationship between accumulating white stories about Native histories (the temporality of simulation) and assaults on Indigenous persons and communities in the present.[26] Characterizing non-native accounts as "killing" Native ones suggests that the former enact a decimating aggression against the latter (on a spectrum with physical assault and murder), cutting off the potential for storying *by* Indians—their ability to offer alternative visions of the past and present that could eventuate in a different future for Indigenous people(s) than that projected in settler stories (especially given the strategy of *lasting* that pervades such texts—as in Wilson's description of Little Hawk).[27]

While the text critiques non-natives' circulation of "Indian stories," Alexie also raises questions about the effects of such narratives on Native people's sense of their own Indianness.[28] Marie articulates a number of criteria for Indigenous authenticity, including speaking a Native language and participating in ceremonies (33), tribes' having clear "records of membership" and the attendant ability to find "documentation" of one's Indian identity (67, 264), and having "lived on a reservation" (246). Many of the Native characters, including Marie herself, though, do not fit these criteria. Marie was not taught to speak Spokane by her parents "because they felt it would be of no use to her in the world outside the reservation"; instead, "they bought her books" (33). Neither Reggie (Marie's cousin) nor John ever lived on a reservation, and John and Carlotta Lot, a Duwamish woman living on the street, raise further problems with the issue of documentation. John was taken from his mother and adopted by whites with literally no way of accessing any records that indicate the people from whom he descends, and Carlotta's people, the Duwamish, do not have federal recognition or prior reservation lands.[29] As she says when John meets her, "I ain't homeless. I'm Duwamish Indian. You see all this land around here. . . . All of this, the city, the water, the mountains, it's all Duwamish land. Has been for thousands of years" (251). If John provides an example of nonbelonging due to one kind of white intervention—extratribal and cross-racial adoption—Carlotta speaks to the ways histories of treaty making, settler invasion, and bureaucratic (in)visibility can contribute to peoples' lacking legally recognized membership lists or land bases, undermining their ability to *document* their indigeneity.[30] These dynamics indicate the problems involved in utilizing the kinds of temporal recognition for which Marie sometimes advocates—a notion of unbroken succession in which Indianness remains visible as such in ways intelligible to settler institutions (especially as maintained within the space of the reservation).[31]

Such problems of recognition and "tribal" identity are particularly pointed with respect to both the Puget Sound region (in which the novel is set) and the area of the Columbia Plateau (in which the Spokane reservation is located and from which Alexie himself comes).[32] Numerous peoples in both regions were not recognized through treaties, surrendered land via treaties in exchange for promised reservations that were never created, or refused to come into treaty relations because they did not want to be confined to a reservation or to be under the stifling superintendence of an Indian agent. With respect to the lands recognized via formal treaties, often they were in areas from which subsistence could not be sustained via Euramerican farming, did not contain the hunting and fishing areas still vital to Native peoples' seasonal self-provision (rights of access that themselves were guaranteed within treaties), were not considered by Native people as their exclusive site of residence, and/or were inhabited by what previously had been multiple peoples whose relations with each other might be complicated and strained. Given the official end to U.S. treaty making by congressional statute in 1871, agreements with respect to land cessions and the creation of reservations after that point (which occurred a great deal in the Columbia Plateau) had a much more insecure legal status than those reached through treaties, such that many were not fulfilled or were not fulfilled until years later (sometimes a decade or more). In other cases, agreements were substantially and unilaterally altered after they were adopted (which also happened with treaty lands, although doing so had less legal sanction) and/or were revoked entirely (sometimes with land provided elsewhere, sometimes not). Given the delays and constrictions in land recognition for Indigenous peoples in these regions, as well the limits of reservation lands for self-provision and trade and the desire to gain at least some distance from the authority of Indian agents, many Natives filed claims under the Indian Homestead Acts of 1875 and 1884 to get plots of their own, which, while not officially requiring a severing of "tribal relations," after 1884 sometimes were treated as such. Also, many Indigenous people sought seasonal wage labor, especially in the hop fields. In addition, the emergence of the hop economy brought Native people(s) to the area from much farther north, leading to greater mixture among Indigenous populations as well as a legacy of increased presence of members of such groups in the Puget Sound region. With respect to Native collective identities, the tribe was not a precontact structure, instead being generated as part of the treaty process (with non-native leaders, such as Washington's first governor, Isaac Stevens, appointing "chiefs" who could speak for larger groups so as to centralize consent and facilitate land cessions). Further, the centrality of villages made of extended kinship networks and the tendency to marry outside one's kin group (in order to secure access to resources elsewhere) positioned individual Natives within an elaborate matrix of relations on which they could call, including potentially moving to different villages and accessing different resource sites.

While the novel does not itself highlight these historical dynamics in particular, Alexie alludes to the varied histories of settler-induced dispossession, erasure, and impoverishment that contribute to such complicated diffractions of Native identity (individual and collective). The novel implicitly raises a question posed by Malea Powell in "The X-Blood Files": "Whose complicated histories and messy relationships to conquest and colonization simply become unimportant, unheard, absent?"[33] Moreover, Alexie's choice to set the novel in an urban center and continually to cast Seattle as a legitimate—if embattled—site of Native placemaking reflects the ongoing complexities of Native mobilities, networks, and processes of becoming in the region.[34] This portrayal refuses the implicit depiction of the reservation as the proper container for Indianness—one that facilitates forms of settler anachronization (as noted in the discussion of Charles Eastman's work in chapter 2). In *Mark My Words* Goeman argues that "the debate about authenticity, or who is actually a 'real' Indian, is often motivated by the spatial politics" of the reservation—"supposedly progressives leave the rez and traditionalists stay at home"—such that the image of the reservation as the proper site of Indian *realness* simultaneously contains the latter both spatially and temporally.[35]

All the Native characters in *Indian Killer* suffer from insecurities about their authenticity, a dynamic that can be understood as resulting from the effects of non-native discourses and institutions on Native perception. In discussing Marie's sense of her own identity as Native, the narrator notes, "Indians were always placing one another on an identity spectrum, with the more traditional to the left and the less traditional Indians to the right. Marie knew she belonged somewhere in the middle of that spectrum and that her happiness depended on placing more Indians to her right. She wondered where John belonged" (39). In Vizenor's terms, a settler *simulation* replaces *the tribal real*, or, put in the language of the preceding chapters, the gravitational force of settler formations can reorient Native frames of reference in ways that work to align them with dominant non-native accounts and experiences of time (individual, juridical, and national), leading to efforts to adjudicate others' place on a (singular) "identity spectrum."[36] If Eastman and Mathews offer portraits of how Native people might reorient U.S. nationalist discourses or live with the friction generated by incommensurate settler and Indigenous spatiotemporal formations, *Indian Killer* takes up the question of the effects in the *longue durée* of accreting forms of settler imposition. What becomes of Indigenous frames of reference? Are such experiences of duration simply overwhelmed by non-native presence, imperatives, and simulations? At what point does everyday indigeneity become no more than (to recall the discussion of Weheliye in chapter 2) being reduced to Indian flesh?[37]

Yet the absence of particular markers of proper Indianness does not serve for the text as a basis to disqualify the characters' representations and understandings of themselves as Indigenous, nor do the presence of settler simulations and their effects on Native self-understandings simply displace other ways of experiencing being and becoming. At one point Carlotta observes, "There's even a bigger difference between what Indians think about each other, and what you and I know about ourselves" (252). The novel takes up the Ghost Dance, and prophecy more broadly, as an expression of Indigenous temporalities—what Native people "know about ourselves"—that do not take official geographies and "records" as their frame of reference.[38] Alexie stages two confrontations over the dance's meaning, one between Mather and Marie and an earlier one between Wilson and Reggie Polatkin (Marie's cousin). About halfway through the novel, Wilson (who has the same name as the prophet of the Ghost Dance of 1890) has a conversation with Reggie in a local bar. After Wilson shares some details of the first of the killer's murders that he learned from the local police, Reggie asks, "You know about Bigfoot?," to which Wilson replies, "He died at the Wounded Knee massacre in 1890. He was Minneconjou Sioux, I think. He was killed because he was leading the Ghost Dance.... [It] was supposed to destroy the white men and bring back the buffalo. Ghost Dancing was thought to be an act of warfare against white people." Reggie presses Wilson, "What color was the man who killed Bigfoot?," and after Wilson observes, "He would've been white," Reggie insists, "Exactly, Casper. Think about that" (185). The association of the Ghost Dance with both the Wounded Knee massacre and "warfare against white people" highlights a history of conflict that the novel suggests extends into the present, thwarting Wilson's sense of an easy slippage between his history of whiteness (in terms of being perceived and living as such) and his appropriation of Indianness (in his claims to be Shilshomish).[39] In refusing Wilson's project of Indianization, and the sense of unconstrained possibilities for settler development on which it draws, Reggie implicitly figures the dance as something other than a failure, other than a story of Indian disappearance. While Reggie's comments do not themselves indicate a persistence of the dance into the present, the fact that he raises the matter in association with the actions of the killer implies such a relation, insinuating that the killer is a contemporary incarnation of similar forces as in the Ghost Dance. The narrator heightens this sense by noting that after Reggie spread word of the scalping, "Most Indians believed it was all just racist paranoia, but a few felt a strange combination of relief and fear, as if an apocalyptic prophecy was just beginning to come true" (185). The text here gestures toward the idea that the killer may, in fact, be the fulfillment of the earlier prophecy. The killer appears as an expression of enduring Indigenous sentiments and sensations that do not count as something from the past. Instead, linking the killer to the Ghost Dance suggests the potential of "Indian stories" to remake relations among the past, the present, and what's to come.

Marie's argument with Mather extends and further concretizes these impressions, detailing the Ghost Dance's role in anticolonial struggle while also more fully sketching the ways that the killer may be a manifestation of continuing histories of Native grievance.[40] During a meeting in the office of the chair of the Anthropology Department, to which Marie has been summoned in order to address a complaint filed by Mather against her, she challenges Mather's claims to expertise in ways that she routes through the invocation of the dance: "You really think you know about Indians, don't you?... You think you know about the Indian Killer, huh? Well, do you know about the Ghost Dance?" After she reiterates that it would cleanse the Americas of Europeans, Mather observes, "Yes, it was a beautiful and desperate act," later noting, "The Ghost Dance was not about violence or murder. It was about peace and beauty."[41] In response, Marie asserts, "Yeah, you don't believe in the Ghost Dance, do you? Oh, you like its symbolism. You admire its metaphorical beauty, enit? You just love Indians so much. You love Indians so much you think you're excluded from our hatred. Don't you see? If the Ghost Dance had worked, you wouldn't be here. You'd be dust" (313). The text suggests that, for Mather, the dance functions as a means of identifying with Native people(s); he claims the Indigenous past as a symbol through which to express feelings of potentially peaceful relation, albeit on settler terms. Making Native history into a vehicle for this kind of imagined reconciliation, though, requires an assertion as to the *real* dynamics of the Ghost Dance, what it was ("peace and beauty") and what it was not ("violence or murder"). Doing so works to sustain control over history as a means of shaping the possibilities for the future, one in which non-native "love" for Indians serves as an alibi for ongoing settler occupation, and Native (counter)articulations remain merely "metaphorical," in the sense of something solely figurative with no direct impact on the world.

For Marie, as for Reggie, more than expressing the irresolvable antagonisms produced by the ongoing compulsions and imperatives of settlement, the Ghost Dance of 1890 indexes the historical density of Native experience, which exceeds the colonial temporality of unending settler succession and futurity ("If the Ghost Dance had worked, you wouldn't be here"). The Indian Killer operates as the physical condensation of Indigenous perspectives, affects, and knowledges. As Marie indicates, "so maybe this Indian Killer is a product of the Ghost Dance. Maybe ten Indians are Ghost Dancing. Maybe a hundred. It's just a theory. How many Indians would have to dance to create the Indian Killer? A thousand? Ten thousand? Maybe this is how the Ghost Dance works" (313). This account insists that the Ghost Dance cannot be located at a single moment in time, which would allow it to be situated as a contained, knowable, pitiable object within a process of historical unfolding whose organizing principles axiomatically presume sustained settler presence and dominance. While John commits suicide, implying the futility of his search for authenticity, the text ends with the image of the killer itself dancing and joined by growing numbers of Native people in a ceremony "over five hundred years old" (420). The dance here indexes a continuing, collective process whereby accreting forms of everyday Indigenous imagination, sensation, awareness, and practice generate an emergent force that alters the present and seeks to realize the potential for Indigenous futurity/ies not conditioned on settler identification, narration, or possession. As Crisca Bierwert suggests of Coast Salish epistemologies in *Brushed by Cedar*, the dance can be seen as based "on ways of knowing that are *imbricated* with colonial transformations" but are not "in the grasp of state power" or merely a reaction to non-native formations.[42] After Mather declares to Marie that "we ... are on the same side of this battle" and reacts to her repudiation of this gesture by slamming his office door in her face, the narrator notes, "She wanted every white man to disappear.... Hateful, powerful thoughts. She wondered what those hateful, powerful thoughts could create" (85), and toward the end of the novel, when John kidnaps Wilson and takes him to the top of the skyscraper that John had helped build, the text observes, "John wondered if Wilson knew the difference between dreaming and reality. How one could easily become the other" (403). Furthermore, in the penultimate chapter, the police question Marie, and after she rejects their claims about John's culpability in the murders committed by the killer, the chapter ends with her saying, "Indians are dancing now, and I don't think they're going to stop" (418). These moments all speak to the power of Native affect to shape reality, altering existing configurations of (settler-imposed) materiality through alternative processes of becoming: dancing, dreaming, creation.[43] Alexie implies that the killer arises (achieves physical presence and effect) out of pervasive yet quotidian Native feelings and perceptions that exceed the settler real—that do not figure against a background defined by non-native desires, stories, orientations, and expectations. The Ghost Dance shifts the temporal frame, less enacting a return to or of the past than stretching the earlier prophecy's trajectory and momentum across time: the dance emerges within, embodies, and coalesces present sensations in ways that alter the potentials available in the present.

In this way Alexie's novel refuses the traditional kinds of empirical narrative closure offered by mystery novels of the sort that Wilson writes.[44] Michelle Burnham reads this dynamic as the text "not even allow[ing] for the identification of this killer as anything other than *something its readers cannot know*."[45] More than a catachrestic figure for unknowability per se, though, the killer consistently is linked with the Ghost Dance, suggesting the presence and power of forms of relation (in the present and across time) that do not obey the terms of Euro-American realism or historicism—opening onto other potential modes of being-in-time. Seeing the killer as simply a symbol of incomprehensibility itself engages in a form of, in Scott Pratt's words, "ontological reduction" by treating the nonrationalist aspects of the novel as figures for epistemological opacity (non-natives can't know) rather than material expressions of another mode of reality with its own dynamics and temporality/ies, including that of prophecy.[46] The novel repeatedly depicts the killer as being other than human and as possessing paranormal abilities, pointing toward an order of reality within the novel that does not obey the rules of post-Enlightenment Euro-American empiricism or historicism. In the first chapter focused on the killer, the narrator notes that during its first murder its "hands curved into talons" (54). Later, after the killer has kidnapped a white child named Mark Jones, the text characterizes its movements as "a flutter of wings" (192). In Mark's testimony to the police after the killer returns him to his house, he describes the killer as having "feathers" and "wings," and he guesses that "it could fly because it had wings" (324). In addition, the final chapter indicates, "The killer gazes skyward and screeches" (420). Cumulatively, these moments suggest a birdlike entity that has the power to transform into a human shape. The novel also indicates the killer's ability to appear as "a shadow" (71), to obscure the memory of those who have seen it (72), and to alter its features ("the killer's face, which shimmered and changed like a pond after a rock had been tossed into it" [153]). As discussed earlier, the killer is generated out of Native feelings and dancing (or at least participates in a complex causal nexus with the dancing—a point to which I'll return shortly), suggesting that it embodies accumulated forms of ordinary Native experience and action. The killer takes part in the plot as a social actor. If one refuses to see its paranormal or nonrationalist being as merely figurative (a symbol of unknowability), its central role in shaping the events in the novel suggests that the killer serves as a means of indexing kinds of sensation, existence, and becoming that do not conform to dominant notions of the literal and settler forms of chrononormativity. Alexie's *Indian Killer*, then, can be understood less as suggesting the impossibility of knowledge—what (presumptively non-native) readers "cannot know"—than as offering the potential for forms of Indigenous realness, "truth," and knowledge that do not fit the modes of causality and history at play in Wilson's *Indian Killer*.[47]

The dancing in the text does not really constitute a return, since the movement(s) engendered by Wovoka's prophecy did not spread to the areas Alexie addresses (the Puget Sound and the Columbia River region). Yet through the repeated citation of the Ghost Dance in relation to the "Indian Killer," the text asserts some connection between the present of its setting and the nineteenth-century context in which the dance emerged. The fulfillment of, in the text's terms, an "apocalyptic prophecy" (185) gives shape to, and is shaped by, the "hateful, powerful thoughts" (85) of contemporary Native people. The killer *incarnates* such feelings, which themselves arise out of the experiences amassed across generations of living within conditions overdetermined by the everyday materialization of settler stories. Linking the killer to the movement(s) of 1890 allows the text to stage a relation between the spiritual and the historical. In *Brushed by Cedar*, her study of Coast Salish forms of power, ceremony, and relation, Bierwert says of storytelling, "This linking of memories, this setting remembrances in motion, is not a nostalgia but an immanence," and Alexie's citation of the Ghost Dance offers less a return to what was or a bringing of the past into the present than a vehicle for indicating forms of sustained and sustaining immanence.[48] Through the dance, the text marks the potential alterity of Indigenous affects, experiences, and knowledges within non-native framings (and, thus, the violence of such framings' imposition), and it gestures toward forms of prophetic temporality that are nonsuccessive.

More than drawing on the fame of Wovoka's vision (particularly as taken up by the Lakota) to indicate a generic form of oppositional Indianness, though, *Indian Killer* implicitly references Salish histories and ontologies, and in doing so, it moves beyond the historical and geographic boundaries that conventionally circumscribe the meaning and scope of the Ghost Dance.[49] The citation of the Ghost Dance often recalls Wounded Knee and Lakota struggles more broadly (such as Reggie's mention of Chief Bigfoot, discussed earlier).[50] One might be tempted, then, to see the novel's citation of the Ghost Dance, and presentation of the killer as its realization, as merely an importation of Sioux philosophy or practice—something of a pan-Indian invocation that has little to do with Seattle or the Spokane reservation (the novel's two principal sites). However, as Gregory E. Smoak suggests in his study of nineteenth-century prophetic religion in the northern Great Basin, "Ghost Dances became part of a common process of identity formation that took place at different times and in different ways in Indian communities across the United States," adding, "The Ghost Dances were an appeal to spiritual power to overturn a world that was not of their making."[51] In this sense the text's invocation of the Ghost Dance of 1890 can be understood as figuring against the background of Salish processes of becoming and the role of prophecy within them. The novel implicitly translates between regional Indigenous frames of reference and the conventional chronology of Indian history, opening that history to forms of temporal experience that exceed the officially sanctioned expressions of authentic Indianness for which the reservation serves as the privileged site.

The terms of the vision Marie articulates, which is amplified elsewhere in the novel, resonate with various prophetic traditions in the Columbia Plateau of western Washington. In the late eighteenth and early nineteenth century, a movement of dreamers arose who had visions of traveling to heaven, speaking with a deity in the land of the dead, and being given a dance to be performed that would hasten their return.[52] In the mid-nineteenth century, a Wanapum leader named Smohalla (who had been a medicine man) died, visited the spirit world, and returned with a prophecy that, in the words of the local Indian commissioner at the time, "a new god is coming to their rescue; that all the Indians who have died heretofore, and who shall die hereafter, are to be resurrected; that as they will then be very numerous and powerful, they will be able to conquer the whites."[53] At the height of his influence, Smohalla had approximately two thousand followers from a range of neighboring peoples, and he called for a return to older ways and a renunciation of white technologies and practices, including the signing of treaties and the acceptance of government-demarcated reservation lands. Not only do these movements long precede Lakota Ghost Dancing, but such patterns of faith and vision extend beyond the time of Wounded Knee and into the twentieth century, passed down generationally to new bearers of the vision.

In addition, the text's depiction of the killer subtly invokes elements of Salish cosmology and spiritual practice. Coast Salish stories include the presence of a Transformer figure, linked to the moon, who while taking something of a human shape has the power to alter the form of other beings, including changing humans into animals and features of the landscape such as rocks. Transformer's ability to do so suggests not only the entity's power (understood as enacting a kind of moral authority to punish wrong conduct) but more ambient possibilities for alteration in the world and the existence of the capacity for humans to tap into such modes of power.[54] The killer in Alexie's novel only ever appears by moonlight (the narrator notes that the killer's knife "would soak up all the moonlight" [53]), taking a human-analogous form such that it can hold and engage with objects as a person would and can walk among people without being viewed as something not human. However, as noted earlier, the killer is cast consistently as having extrahuman characteristics, including birdlike aspects (a movement between human and animal often attributed to Transformer) and a face whose appearance "change[s] like a pond after a rock had been tossed into it" (153). The narrator observes in the first description of the killer that "the killer felt powerful, invincible, as if the world could be changed with a single gesture" (49), a characterization that matches the moments in Salish stories when the Transformer punishes people for their misbehavior by changing them into animals or things with a gesture. Further, the fact that the final chapter is called "A Creation Story" and features the killer, whose songs and dancing animate a gathering of hundreds of Indians who themselves start dancing, can be read as alluding to the primal generative potential of Transformer and its role in the process of creation. The Lushootseed (the language of peoples in the Puget Sound region) word for Transformer carries the same prefix, *duk*, that serves as "the root for a host of concepts including worry, dissatisfaction, anger, infirmity, and ferocity" while also serving as "the root of the words for 'yesterday' and 'tomorrow.'"[55] The figure of the killer could be read as being drawn from this matrix, in which movement and shifts over time gain expression as "anger" in ways that also carry the sense of a fundamental and persistent possibility of new creation. This linkage to Coast Salish stories helps explain the multiplication of causality in the final chapter, which suggests that the killer's dancing brings others to dance in ways different than in Marie's account, where the killer arises from Indians dancing (313). If Transformer bears a power of alteration that could become active at any time (especially given the linguistic connection to temporality itself), then this potential remains ambient, awaiting a conjuncture for its emergence. From this vantage, causality appears more polyvalent than in dominant narratives of the real, sharing something with the asynchronous rhythm of prophecy.

The killer, and its complex generation of and by the Ghost Dance, suggests a particular kind of vision and experience of time and its movement that cannot be captured within Euramerican historicism. In Maurice Merleau-Ponty's terms, perception entails "*reckon[ing] with* the possible," and *Indian Killer* suggests a more expansive sense of what constitutes the possible.[56] In this way the killer in the novel also resonates with extant ceremonial practice among Salish people in the Longhouse religion (otherwise known as Washat).[57] In it, people are visited by their personal spirit (or *syowen*), who enters them as a result of their current psychological or emotional distress, resulting in what is referenced in English as "hollering" in ways that have been known to sound like the expression of an animal.[58] The killer's "screech" at the end of the novel may recall this particular ceremonial exchange, invoking the ability of spirit to enter the realm of human action and thereby to make history. Its modes of participation in personal and collective dynamics of becoming are not readily amenable to the forms of *documentation* through which Indianness is produced and authenticated, but such forms of engagement still depend on what might be termed historical circumstances—the specific alignment of persons, affects, events, and practices that provides the ecology for the entry of this form of power. From this perspective, the nexus created in the novel between the killer and the Ghost Dance gains its momentum from the specific forms of distress faced by Native people(s), modes of Indigenous knowledge and feeling not registered within accreting settler stories. In addition, within Salish languages on the Columbia River the generic term for "bird" links to the term for "guardian spirit," connecting the depiction of the killer's birdlike characteristics to its potential status as a spirit presence within Washat and other ceremonial systems.[59] Yet the generative forces (and temporalities) at play around those experiences and associations are not themselves explicable—do not count as real (efficacious, existent, and/or causative)—within the terms of settler-sanctioned history.

The killer does not exist outside of time but illustrates a process of causality that does not comport with the explanatory procedures and modes of verification at play in rationalist historicism. To say that prophecy and subsequent spiritual movements like the Ghost Dance emerge out of settler colonial conditions is not to say that they should be understood within dominant Euramerican frames—as psychological fantasy or merely a response to the extremity of desperation, dislocation, hunger, and so on produced by Indian policy.[60] Rather, the potential for prophecy, for a syncopated relation to time that may include the impress of varied nonhuman entities, is affected by the extant social and physical environment, in which settler presence and violence play a significant role. In citing the Ghost Dance of 1890, then, Alexie draws on a prominent (set of) event(s) within Euramerican historiography to gesture toward and provide a means of naming the kinds of forces, sensations, understandings, knowledges, and beings that occur within a social landscape heavily influenced by the ongoing materialization of dominant settler stories as the supposedly given basis for reality. In this way Alexie's *Indian Killer* does make a bid for the real that is quite distinct from the "truth" posited by Wilson's *Indian Killer*. The novel sketches the potential for Indigenous temporalities that may be affected by non-native occupation but that need not be understood as reducible to the settler real, including its accumulating simulations of Indianness.

### *Histories of Rage*

In *Indian Killer* time is out of joint. Non-natives in the text often experience Indigenous people(s) as uncanny—as ghostly remainders or eruptions from a previous era. Not unlike the nineteenth-century officials discussed in chapter 2, settler perception in the novel remains oriented by a sense of national futurity in which Native sovereignties can be apprehended only as an aberrant, backward pull against the momentum of progress. To be a *real* Indian, then, entails being construed as spatially and temporally constrained.[61] To the extent that the present is not singular, that there might be multiple and intersecting temporal formations at play in ways that exceed settler historicism and rationalism, the appearance or invocation of things from the past might signal less a form of performative recycling than the coexistence of disparate temporalities (in ways reminiscent of the overlapping spatiotemporal formations discussed in chapter 3). The killer less emerges out of the past than expresses an Indigenous sense of time in which the continuities of settler violence and the attendant Native anger and grief provide the background for prophecy, engendering conditions ripe for its articulation, arrival, and realization. More than staging a return of the repressed, *Indian Killer*'s portrayal of the Ghost Dance suggests that the dance expresses, condenses, and catalyzes forms of Native feeling. The killer arises in response to such ordinary affects and orientations, and that process suggests how prophecy serves as part of and emerges through the historical density of everyday sensation. In this way Alexie differentiates between the search for origins as a ground for contemporary action, such as John's futile attempts to authenticate his past, and the ways prophetic time arises out of and transforms everyday circumstances while remaining turned toward the future.

A crucial part of the accounts of Indianness offered by non-natives in the novel lies in their projection of it into the past, treating it as itself somehow inherently anachronistic. To be Indian in the present is to be out of time, in the sense of both being in the wrong era and having no future. Readers learn that Reggie's father, Bird, served as area director for the Bureau of Indian Affairs during the early 1970s, and speaking from within Bird's perspective, the narrator notes that the region at that time

> was under siege by the American Indian Movement. All over the country, hostile AIM members had been attacking peaceful BIA Indians and non-Indians. Bird had known that the murder rate in Pine Ridge, South Dakota, was the highest in the country. All because of the hostiles.... It had been happening since Europeans first arrived. In the nineteenth century, while a peaceful and intelligent chief like Red Cloud had been trying to help his people, a hostile Indian like Crazy Horse had been making it worse for everybody. But Bird had always believed that Crazy Horse got what he deserved, a bayonet to his belly, while Red Cloud had lived a long life. (92)

The text here alludes to the significant forms of activism in and around Seattle in the late 1960s and throughout the 1970s, which included the extensive campaign of fish-ins that eventually resulted in the Boldt decision of 1974 (allocating half the annual fishing catch to Native peoples based on mid-nineteenth-century treaties), the organization of an international conference for urban Indians from the United States and Canada in 1968, the takeover of Fort Lawton in 1970, and the transformation of the fort into a cultural center in 1976.[62] Notably, these events speak to enduring Indigenous relations to place (including Native connections to urban spaces) but do not necessarily accept the institutionalized geographies of the reservation system as the paradigmatic way of marking such relations. Bird understands these events through the prism of nineteenth-century Lakota history. Not only does he invoke Pine Ridge, the site of the Wounded Knee massacre in 1890 and the standoff in 1973 between Native activists and federal officers, but he casts the contemporary struggle by Natives as an extension of that of the "hostiles" who refused to stay within demarcated reservation boundaries in the late nineteenth century, and he understands the only possible future for the latter as subjection to murderous retribution.[63] Further, Bird provides childhood instruction for Reggie in his own version of Indian history in which all events end in Native death and decimation, with Bird enjoining Reggie "to know your history" (93–94). History itself entails Indian defeat, geographic containment, and disappearance, and anything that counters such an unfolding of settler time appears as an assaultive throwback to an earlier era. Similarly, the right-wing talk show host Truck Schultz offers his own historical narrative: "treaties that the tribes signed a century ago" enable them to "insist on their separation from normal society" (118), "and now comes the news that an Indian savage is killing white men. Have we somehow traveled back to the nineteenth century? Has some Godless heathen been kept on ice on the reservation for a couple hundred years?" (208–9). Within this framing, the killer is a product of temporal deviance, of non-natives having *allowed* the continued existence of the nineteenth-century anachronisms of the reservation and the treaty that preserve Indians in their ahistorical relation to "normal" forms of development.[64]

However, the elements in the text that could be cast as supernatural within a Euramerican rationalist frame are not, in fact, ghostly: there are no hauntings in the text, no revenants whose appearance indicates an untimely lingering of that which is otherwise properly past.[65] The text may allude to the history of spectral Indians in non-native writings, including the terrors of the Indian burying ground, but it does not reenact that tradition.[66] Alexie actually mocks such conventions through reference to Chief Seattle's skeleton. When discussing a collection of tapes of Spokane elders telling traditional stories stored in the basement of the Anthropology building at the university, the narrator notes,

"Some rooms had not been opened since the early part of the century.... The basement even had its own mythology. Chief Seattle's bones were supposedly lost somewhere in the labyrinth. And the bones of dozens of other Indians were said to be stored in a hidden room" (139). At one point when Mather is down there listening to the tapes, the lights go out, and while walking in the darkness, Mather hears noises and thinks of "the forgotten bones and fragments of clothing. Chief Seattle's bones." After running to avoid that ghostly presence and slamming into an overhang, he discovers the "rattle" he heard was only the janitor (140–41). This moment presents such fears of Indigenous haunting as one of the many "Indian stories" told by non-natives while it also distinguishes such tales from the materiality of the killer and the complex causal matrix of its relation to the Ghost Dance.

In invoking Chief Seattle, Alexie also calls forth the long legacy of non-natives circulating stories about dead Indians as a means of negotiating their relation to the Puget Sound region and its Indigenous histories. A Duwamish headman, Seattle (or Seeathl) developed an alliance in the early 1850s with settlers in what would become the city of Seattle, and a speech he supposedly gave has become (in)famous owing to its ability to signify the passing away of Native peoples, the continuation of Native wisdom (despite the disappearance of actual Natives), and the attitude of melancholic nostalgia with which non-natives can experience the past of settlement.[67] Not only does the speech circulate as a sign of the quasi-ghostly presence of the otherwise definitively departed, but it closes with the statement, "The white man will never be alone. Let him be just and deal kindly with my people, for the dead are not altogether powerless." The speech itself is a kind of ghost story in which the vague allusion to a haunting substitutes for the actual presence of Duwamish and other peoples, who will dissipate as "the changing mists on the mountain side ... before the blazing morning sun."[68] Moreover, Seeathl and the speech attributed to him serve as only one of the Indian ghost tales the city of Seattle tells of and to itself. As Coll Thrush observes, "Every American city is built on Indian land, but few advertise it like Seattle. Go walking in the city, and you will see Native American images everywhere in the urban landscape"; "the city's Native American imagery ... define[s] Seattle as a city with an indigenous pedigree." Thrush further characterizes the city's self-presentation as "Seattle's Indian ghost stories," "historical creations" that "spring out of the city's past" and that work "to make sense of that past."[69] Such stories, though, are marked precisely by their *pastness*, the sense of existing across a chronological gulf that can be traversed only in ethereal ways in which Native peoples are cast as lacking substance in the present.[70]

As against such apparitional remainders, invoked through Chief Seattle's bones and even John himself, the novel presents the killer as a realization of the vision of the Ghost Dance, one made possible by the accumulating violences of settler policy.[71] Setting the final scene of the novel in a graveyard, Alexie transforms the phantasmagoria of the ghost stories circling around (Chief) Seattle into a commentary on the social and spiritual ecologies engendered by continuing colonialism. The location is a "cemetery on an Indian reservation. On this reservation or that reservation. Any reservation, a particular reservation" (419), repeating the language used to open the first chapter and speaking to the genericizing of Indians within non-native accounts as well as the shared circumstances of settler occupation that transect extant tribal distinctions. The narrator adds, "There are many graves, rows of graves, rows of rows" (419). The image is suggestive of mass death, due to disease, warfare, and/or starvation, and while not referring to Seattle in particular, the image of the cemetery subtly alludes to the repeated need in the region to move Native gravesites because of the shrinking of Indigenous land bases, multiple displacements of Native communities, and wholesale rerouting of waterways to facilitate non-native housing and trade patterns.[72] This image of the killer among the graves actually echoes a moment in Chief Seattle's speech, when he insists "that we will not be denied the privilege, without molestation, of visiting at will the graves of our ancestors and friends."[73] The novel can be read, then, as ironically citing that particular specter, and the meanings that attach to Seeathl and his supposed speech, in order to differentiate the melancholic anachronizing such stories perform from the present corporeality of the killer and the powers of prophetic time that flow through and around it.

The novel implicitly contrasts the account of the killer as a historically stunted residue having no relation to the present or future—a ghost or a "hostile" out of the nineteenth century—with the movement of prophecy, in which contemporary events call forth a spirit or vision that was also active in the past but that offers the potential for transformation in the present. When John's father, Daniel, is searching for him on the streets of Seattle, he runs into a homeless Native man who insists that he knows "who did it" with respect to the recent murders and kidnappings: "It was Crazy Horse.... This Indian Killer, you see, he's got Crazy Horse's magic. He's got Chief Joseph's brains. He's got Geronimo's heart. He's got Wovoka's vision. He's all those badass Indians rolled up into one" (219). Except Wovoka, all of these figures took part in warfare against the United States in ways that directly followed from government efforts to contain their peoples within strictly regulated boundaries, and, as such, they are the very kind of "hostiles" to which Bird alludes. Describing the killer as composed out of aspects of these well-known warriors implies that the killer is also taking part in a struggle against forms of settler constriction and management. The man adds, "Every Indian is keeping score. What? This Killer's got himself two white guys? And that little white boy, enit? That makes the score about ten million to three, in favor of the white guys, enit? This Killer's got a long ways to go. Man, he's the underdog" (220). The killer newly incarnates a long trajectory of Indigenous resistance to settler geographies and imperatives, one that defies the archaism attributed to such opposition (as in Wilson and Mather speaking of the Ghost Dance in the past tense). While Wovoka's prophecy was not itself directly articulated against non-native expansionism, prophet-led movements can be understood as responsible for animating much of the struggle against settler relocations in the Columbia Plateau in the late nineteenth century.[74] Such resistance includes the war in 1858 to which Alexie refers both in his use of the name Polatkin (the name of a Spokane chief whose daughter was married to a Yakima chief named Qualchan, often cited as the leader of the Native opposition) and in Reggie's reference to the Battle of Steptoe Butte (352).[75] Conceiving of the killer as an expression of "Wovoka's vision," or as possessing a similar kind of vision, casts the killer as a vehicle for carrying forward a prophetic movement whose power lies in its capacity to realize greater possibilities for Indigenous self-determination.

Forms of prophetic citation and sensation, then, add temporal depth, form, and force to feelings in the present. Addressing the importance of historical imagination as part of contemporary Native self-conception, Craig Womack observes, "Until we imagine these stories for ourselves, however, they mean little more than facts and dates.... I am talking about more than developing a capacity to empathize with people from our pasts. This has to do with placing ourselves inside their stories, becoming participants in history," adding, "History is a vision quest, the quintessential religious experience. How else, if not through vision, can we access these experiences from the past so we may also experience them?"[76] *Indian Killer* enacts the converse: rather than illustrating how to use one's imagination in order to live inside stories from the past, the Ghost Dance in the text intimates how (previous) prophecy comes to live inside the corporeal experience of the present. Addressing contemporary citations of earlier prophetic visions among Yukon women, Julie Cruikshank notes, "These prophecies are evaluated by contemporary narrators not in terms of whether they altered social circumstances, but in terms of their ability to forge legitimate links between knowledge experienced by past prophets and events experienced by present tellers."[77] The putative failure of the Ghost Dance of 1890 to do what, in Wilson's terms quoted earlier, it "was supposed to" indicates a particular kind of closing off by which such an event, vision, movement becomes locked into a completed moment, and from this perspective its citation later must necessarily be anachronistic, the return of a thing that had disappeared. As Veena Das suggests in ways discussed in chapter 1, though, "regions of the past" can become "actualized and come to define the affective qualities of the present moment" by rotating into present perception: "It was as if the past had turned this face toward them—not that they had translated this past story into a present tactic of resistance."[78] The novel suggests that various knowledges from the mid- to late nineteenth century, gathered around and through the Ghost Dance, gain material form in the contemporary moment because they resonate with current experiences—in Das's terms they operate as "unfinished stories."[79] Thus, they are not so much a kind of reversion as a projection or animation, less an uncanny reincarnation of what's gone than the syncopated enfleshment of prophetic vision within current frames of reference.

If prophecy requires conditions in which that spirit or vision can manifest, part of what provides that possibility within the novel is the scope and intensity of Native anger. Early in the novel, the narrator notes of John, "He didn't want to be angry. He wanted to be a real person. He wanted to control his emotions, so he would often swallow his anger" (19), and toward the end of the text, after she fends off an attack on John by a gang of white men, Marie "was shocked by her anger, and how much she had wanted to hurt those white boys. Nearly blind with her own rage, she had wanted to tear out their blue eyes and blind them" (375), echoing the killer's tearing out the eyes of his first victim (54).[80] The fact that, for John, his anger makes him less "real" hearkens back to the ways that Indianness circulates as a function of non-native stories in which settlers are welcomed as shared participants, such that negative and antagonistic Native affect has no place within the social geographies realized through those stories. As Dian Million illustrates in *Therapeutic Nations*, bad Native feeling is acceptable to non-natives when presented as arising from the pain of past trauma and as redressable through depoliticized healing and national reconciliation (a feeling in which all can share), but not when it engenders critiques of ongoing structural violence and points toward the desire to realize forms of Indigenous self-determination.[81] John's *swallowing* his feeling, directing it back into himself and away from the world, parallels his retreat to a fantasy of the reservation as origin, but Marie's direction of that feeling outward toward the world via aggression suggests the potential for materializing a different kind of real that is oriented by Native agency, interests, and understandings.[82]

Some critics have characterized the killer's violence and the affects out of which it arises as necessarily destructive, as failing to build or resolve anything and, thus, indicative of the novel's ambivalence toward the killer or repudiation of its actions.[83] However, as Ahmed suggests in *The Cultural Politics of Emotion*, "crucially, anger is not simply defined in relationship to a past, but as opening up the future. In other words, being against something does not end with 'that which one is against.' Anger does not necessarily become 'stuck' on its object.... Being against something is also being for something, but something that has yet to be articulated or is not yet." Earlier Ahmed observes of settler responses to Indigenous expressions of pain, "The impossibility of 'fellow feeling' is itself the confirmation of injury. The call of such pain, as a pain that cannot be shared through empathy, is a call not just for an attentive hearing, but for a different kind of inhabitance. It is a call for action, and a demand for collective politics, as a politics based not on the possibility that we might be reconciled, but on learning to live with the impossibility of reconciliation."[84] From this perspective, the concentration of Native anger in the novel, and the momentum it generates for the killer, suggests neither a pathological response nor a historical or political dead end. Rather, such feelings point to the pain generated by a background of accreting and ongoing histories of settler violence while turning such quotidian sensations into a collective force that can enable a different future, one in which the kind of reconciliation projected by Wilson or Mather (on settler terms) need not function as the trajectory of Indigenous aspiration. When John maims Wilson before jumping off the skyscraper he previously had been employed to help build, he says, "Let me, let us have our own pain" (411), and this moment can be read as less an expression of pathos than itself an articulation of self-determination.

Indigenous pain and attendant anger index the everyday accumulation of the effects of persistent and intensifying settler colonial displacements—a particular experience of duration that provides the context for prophetic emergence (as opposed to ghostly return). This relation to the past marks not a backward-looking identification but the potential for a changed orientation toward the parts of the present that continue to engender such negative affects. The killer, then, arises as the condensation of that possibility, as a materialization of active histories of rage in ways inflected by regional Indigenous ontologies of spirit. At the end of one of Truck Schultz's broadcasts in which he presents Native people(s) as willfully and murderously refusing the gift of civilized instruction, he asserts, "This Indian Killer is merely the distillation of their rage. He is pure evil, pure violence, pure rage. He has come to kill us because we have tried to help him. He has come to kill us because his children have moved beyond him" (346). Here the killer's status as a manifestation of collective feeling appears as indicative of its fixation in the past, a "rage" against change that bespeaks a longing for stasis as against the forward movement of progress. Yet, given the novel's consistent portrayal of Schultz's explanations as racist nonsense, his formulation here might be inverted, suggesting that Native rage itself produces effects in the world, of which the killer is the most palpable one. The killer indicates the presence of other modes of being and becoming in which everyday feeling provides a conjuncture in which (prophetic) force gathers in ways that can transform existing circumstances, opening up alternative futures than those taken by Schultz to be inevitable. The killer's violence, then, can be interpreted as marking the (re)appearance of a vision or spirit that gains material form within and because of the historicity of everyday forms of Native feeling and perception, the accreting contexts and dynamics of Indigenous anger. In this way Alexie does not offer a clear, teleological path forward based on an easily defined sense of Indianness, instead suggesting that in contemporary indigeneity there remains, in Ahmed's terms, "something that has yet to be articulated" that can be expressed only through invocations of what appears as the past.[85] The everyday is less haunted by the ghostly than saturated by non-native aggressions and Native anger in ways that call forth memories of the *longue durée* of settler colonial violence while providing the occasion for the prophetic materialization of the spirit of the killer as a force for forging a different future.

*Networks of Pleasure*

The citation of Native history in *Indian Killer* positions the Ghost Dance and the killer as neither anachronisms nor ghostly remainders. Instead, in the text spirit gives material form to pervasive Native affects that themselves arise owing to the dynamics of ongoing settler occupation, including its making unreal of continued Native presence except inasmuch as it can be inserted into accreting narratives of Indianness. In *Gardens in the Dunes*, though, the realization of prophecy coalesces less around rage than around longing—for contact with ancestors and the (re)generation of the conditions for continued Native life. The novel is set in 1900, a decade after the movement and forms of resistance to white dominance borne by the Ghost Dance supposedly ended in the carnage of Wounded Knee, and its ceremonial dancing for the coming of the Messiah appears belated.[86] However, in its simultaneous embodying of what has been and what could be, the dance breaks with settler emplotments of history, in which it would seem exceptional or residual (in similar ways to the dynamics discussed in chapter 2). The past and the future appear as virtually present in the current moment, and prophecy enables ways of accessing and materializing those potentials. Moreover, *Gardens* suggests how such prophetic experience gains momentum from defying the heteronormative temporality of Indian policy, particularly in its efforts to define Indianness by reference to racial genealogy as well as to subject Indians to a pattern of civilized maturation organized around hetero-monogamous union (as discussed in chapter 3). In this way the novel offers an account of Native experiences of time as a heterogeneous network that cannot be reduced to a reified Indian lineage or descent from a properly Indian origin (both of which are retroactively projected from, and regulated through, the reservation system).[87] Prophecy eschews a vision of unidirectional, linear development in favor of a sense of multiplicity that does not dichotomize continuity and change and that connects chronologically disparate sites.

The novel begins with a Ghost Dance occurring in 1900 in Needles, California that is interrupted by Indian police and soldiers. The central characters are a pair of sisters, Indigo and Sister Salt, from a fictional group from the lower Colorado River valley named the Sand Lizard clan, and they are in Needles with their mother and grandmother, Grandma Fleet.[88] Their mother is lost in the raid that sends them running south, back to the dunes of the title in which their people had lived for centuries, especially when fleeing from the "bloodshed and cruelty" brought by "aliens" "long, long ago" (15).[89] When their grandmother passes away of natural causes, they head north in search of their mother, only to be caught by the police. Indigo is sent to an Indian boarding school, the Sherman Institute, from which she eventually escapes; she meets a white couple (Hattie and Edward Palmer) who live nearby and ends up traveling across the United States and to Europe with them before being reunited with her sister. Sister Salt ends up doing laundry and sex work at a construction site near the Colorado River Indian reservation, getting pregnant, having a child (whom they call little grandfather), and eventually returning with her sister to the dunes. The dancing in Needles bookends the novel, and authorities find it threatening because of the various kinds of inappropriate associations occurring there. The narrator observes, "The United States government was afraid of the Messiah's dance" (14), later noting, "Federal officials feared the dancers were a secret army in disguise, ready to attack" (45). The hundreds of people who come to dance for the coming of the Messiah, following Wovoka's prophecy (22), violate the principles at play in Indian policy in several ways: Indians themselves should be living on reservations (17–18); the children are not in federally run Indian boarding schools, which "was the law" (21); they are engaged in forms of Native worship, the kind of dance celebrations that previously had been outlawed and that were particularly suspect in the wake of Wounded Knee (22); and the dances included whites, particularly

Mormons, who themselves had been targeted for federal assault owing to their polygamy (44–45). Wovoka's vision and the movements that arise from it, then, appear in the novel as generating forms of combination, connection, and inhabitance that thwart the regulatory imperatives and mappings guiding U.S. policy. In this way prophecy both marks and animates Native peoples' ongoing deviations from the sociospatial order instituted through settler governance.

Much of the commentary on the novel, though, has framed Indigenous displacements of non-native expectations, including the imposition of criteria for defining proper Indianness, as forms of *hybridization* or syncretism.[90] In "Ghost Dancing through History in Silko's *Gardens in the Dunes* and *Almanac of the Dead*," David L. Moore argues that this way of characterizing sociospiritual life relies on a "lamentable theoretical qua horticultural discourse." He argues, "For all the mixing and blending seemingly implied in hybridization, it has a reverse effect of separation and alienation precisely because of its dualistic limits," further pointing to the role of such ways of envisioning "mixture" in calculations of blood quantum which gathered much greater force within postallotment Indian policy (as discussed in chapter 3).[91] The concept of hybridity, as Moore notes, draws on the notion of mixing species, and it was a prominent way of figuring interracial sexual relations and procreation in the nineteenth century and beyond. In discussing how earlier notions of "amalgamation" as anomaly exceed the sense of reproductive doubleness at play in the Civil War–era emergence of the term *miscegenation*, Tavia Nyong'o asks, "Is it not possible to unyoke racial hybridity from its association with progressive, heterosexual time? Into what alternate temporalities might it then fall?," and this question further develops a challenge he poses in the introduction: "A critical approach to race should encompass both the history of racial ideas and the forms of historicity and temporality embedded in those ideas and practices."[92] To speak in terms of hybridity offers a reproductively inflected sense of union among things taken to be opposed or incommensurate that then enables a rapproachement between them, facilitating a more harmonious future. The event of ostensible hybridization projects two prior lines of self-contained (racial) unfolding as the background against which to register its transformative effects.[93] The prophetic temporality at play in Wovoka's vision and in the experience of the dancing in the novel remains askew with respect to such straightening of collective tradition and development. Ahmed suggests in *Queer Phenomenology*, "Things seem 'straight' ... when they are 'in line,' which means when they are aligned with other lines. Rather than presuming the vertical line is simply given, we would see the vertical line as an effect of this process of alignment," and the straight line of history, proper relations of cause and effect, gains coherence through its congruence and overlay with other lines, such as conceptions of familial descent, racial lineage, and cultural inheritance.[94] Ahmed further notes that in "the requirement to follow a straight line,... straightness gets attached to other values including decent, conventional, direct, and honest," and, reciprocally, "any nonalignment produces a queer effect."[95] The imposed temporal, spatial, and sexual orders work in and through each other, functioning as mutually reinforcing alignments through which history—the connection between the past and the present on the way toward the future—becomes imbued with a lineal immanence that parallels the genealogical transmission of racial identity and property. The failure to fit that pattern generates a sense of *queerness*, of deviation from the proper unfolding of time.

Violating the terms of dominant conceptions of causation (the straight line of history), the Messiah and the dance he inspires indicate less a developmental understanding of indigeneity—some point at which it becomes *hybridized* through engagement with Christianity—than the interanimating copresence of what might otherwise function as distinct periods.[96] In "Basin Religion and Theology," Jay Miller discusses how understandings of power (or *puha*) within the Great Basin, which includes the Colorado River area in which the novel largely takes place, depend on sensations of "flux, action, and process": "It is not static or concrete, but rather kinetic, always moving and flowing throughout the cosmos, underpinning all facets of the universe" in ways that are "rhythmical." Power, he suggests, functions like a web that "is pulsating and multidimensional," and since power accrues "wherever life gathers for however long," apparently incommensurate belief systems can be practiced simultaneously "because they all lead to the same center."[97] This kind of rhythmic, multidimensional movement characterizes the temporal dynamics of the Messiah dance.[98] A Paiute woman who came to Needles to escape the cold farther north tells Grandma Fleet of Wovoka's vision, that she "had seen Jesus surrounded by hundreds of Paiutes and Shoshones and other Indians": "Jesus talked to them, and talked all day. He told them all Indians must dance, everywhere, and keep on dancing. If they danced the dance, then they would be able to visit their dear ones and beloved ancestors. The ancestors were there to help them. They must keep dancing.... The clear running water and the trees and the grassy plains filled with buffalo and elk would return" (23). The narrator adds, "Wovoka wanted them to dance because dancing moves the dead. Only by dancing could they hope to bring the Messiah, the Christ, who would bring with him all their beloved family members and friends who had moved on to the spirit world after the hunger and the sadness got to be too much for them" (26).[99] As opposed to the unidirectionality of inheritance, the dance realizes a transgenerational connection that allows a reunion with the ancestors and materializes the conditions for further life (as opposed to those that generate "the hunger and the sadness"). The prophecy promises to gather chronologically disparate potentials and to actualize them in the present so as to enhance the possibilities for survival, and in doing so, it creates something like a flow of time around a common center, a weblike network, rather than a succession that follows the straight line of genealogical order or conventional historicism.[100]

This process of prophetic realization entails the potential for forms of relation that do not fit Euramerican alignments, and the seeming oddity, or queerness, of such conjunctures reveals the violences involved in securing state-sanctioned orientations and trajectories.[101] The novel suggests that a sense of Indianness as a form of lineal unfolding emerges as a back-formation from the terms and dynamics of late nineteenth-century Indian policy. *Gardens* makes clear that Native people's and peoples' residence outside of reservations, such as in the old gardens among the dunes, was increasingly unacceptable to U.S. officials and that part of the role of the military and Indian police was to round up those who previously had chosen to remain apart from such sanctioned spaces. The text notes that Grandma Fleet refused to move to the reservation because "there was nothing to eat" there: "Reservation Indians sat in one place and did not move; they ate white food—white bread and white sugar and white lard" (17). Later the narrator observes that "farming was easy along the river but getting along with the authorities was not" and that the "Sand Lizards preferred to rely on the rain clouds and avoid confinement on a reservation" (48), and readers are told that a "barbed-wire fence marked the entrance to the reservation" at Parker (394). The novel's Messiah dances in Needles are modeled on ones that took place among the Hualapai outside of Kingman.[102] As Jeffrey P. Shepherd observes, after Hualapais had escaped internment on the Colorado River Indian reservation eight years before, the creation of the Hualapai reservation in 1883 (just east of where much of the novel takes place) "signaled the possible preservation of a piece of their aboriginal homelands."[103] Even as reservations could provide legal recognition for some forms of Indigenous territoriality (as discussed in chapters 2 and 3), they increasingly functioned from the 1850s onward (and even more so after the formal end of treaty making in 1871) as spaces of confinement through which Native peoples were largely segregated from those on other reservations—often forbidden from leaving without passes from the agent—and subjected to supposedly civilizing regimes that reorganized extant modes of social life (particularly in the wake of the implementation of allotment starting in the late 1880s).[104] Offering a somewhat representative portrait of the reservation system as it emerged in the second half of the nineteenth century, Secretary of the Interior Jacob D. Cox in his annual report of 1869 described it as having "two objects": "First, the location of the Indians upon fixed reservations, so that the pioneers and settlers may be freed from the terrors of wandering hostile tribes; and second, an earnest effort at their civilization, so that they may themselves be elevated in the scale of humanity."[105] As Goeman suggests of this legacy, "Rather than construct a healthy relationship to land and place, colonial spatial structures inhibit it by constricting Native mobilities and pathologizing mobile Native bodies."[106] The production of Indian realness, including its role in the assessment of bodies and lineages, relies on the reservation as a privileged space through which to determine the boundaries of belonging and authenticity.[107]

The novel addresses such dynamics with respect to both the Chemehuevis and the Sand Lizards. A pair of sisters Sister Salt befriends at the construction site—Maytha and Vedna—return to the Chemehuevi reservation and are greeted with suspicion by their Christianized neighbors. When lands on the reservation are flooded as a result of the backup from the dam being built on the Colorado, the local Chemehuevi minister lays the responsibility on the sisters: "they heard a man's voice call out behind them: the wantonness and drunkenness of them and others had angered God so much he sent this flood!"; "While he mopped at his forehead and caught his breath he glared at them; they were not really Chemehuevis but Lagunas and didn't belong there. They were damned, contaminated—a risk to all others" (435). The sisters' entry into the space appears as a change that results in destruction, as a result of both their "wantonness" and their failure to fit a particular reproductive line—not being full-blood Chemehuevi and thus "contaminated."[108] Despite extensive, ongoing histories of kinship among groups in the region, especially given pre-reservation patterns of decentralized leadership and flexible matrices of relation, the minister and his supporters insist on proper lineage as a condition of belonging to the space of the reservation.[109] The institutionalized terms of authenticity produced through Indian policy engender, in Vizenor's terms, "simulations of manifest manners . . . [that] become the real without a referent to an actual tribal remembrance," and as Deborah Miranda suggests in *Bad Indians*, "my tribe must reinvent ourselves—rather than try to copy what isn't there in the first place."[110] The sisters' ostensibly mixed parentage and putative licentiousness become construed as improper divergences from what is taken to be the straight line of Chemehuevi descent and inheritance, as forms of sexual impropriety (including marrying outside the "tribe") that deviate from a (retrospectively constructed) Chemehuevi history imagined as pure.[111] Similarly, "at Parker [the Colorado River Indian reservation], if some poor person had even one parent who was Chemehuevi or Mohave, the others might jeer and tell them to go back to their own reservation." Moreover, those of Sand Lizard heritage on the reservation "were married to people of other tribes[,] . . . went to church every Sunday[,] and spoke English," and while they "did not turn Sister Salt away, . . . they shook their heads and whispered . . . about the young Sand Lizard woman . . . [who] lived out in the hills too long" (205). Those who, like Grandma Fleet's family, refused to reside on the reservation or to assent to the civilizing imperatives implemented there appear as out of sync with respect to the narrative of Sand Lizard development that arises in the wake of the creation of the reservation.

Not only do other Chemehuevis and Sand Lizards adopt forms of supposedly enlightened home and family life that direct them away from "wanton" dissipation, but those *straightened* modes of living inculcated on-reservation enact a temporal framing—a retroactive sense of lineage—that pathologizes and deauthenticates kinds of Indigenous experience that do not fit the social, spatial, and spiritual coordinates of the identities realized through Indian policy. Within the terms of the civilization program, those practices and lifeways cast as tradition appear as anachronisms to be superseded, but from that same perspective tradition provides a sense of developmental movement in being narrated as an inheritance from a prior time, tracing a line of Indianness heading toward modernization. In this way the reservation system generates lines that demarcate the boundaries of Native space, which then provide the frame for a particular genealogical and chronological model of *tribal* identity. Discussing the dismissal of claims to be Native by particular peoples (especially those on the east coast), Daniel Justice observes in "'Go Away, Water!'" that "the line of logic used by many anti-Native forces, namely, that blood quantum and phenotypically 'Indian' features are the fullest measure of cultural authenticity[,] . . . [means] that those who are lacking in these qualities are, by definition, no longer Indian—if they ever were."[112] Through its depiction of conflicts over occupancy of reservation space, the novel indicates how the historical sense of progress along a timeline (which provides the context for notions of hybridization) emerges in connection with a naturalizing image of heterofamilial reproduction.

Setting the Ghost Dance just outside Needles allows Silko implicitly to draw on that town's characterization in the late nineteenth century as an extra-reservation space of moral depravity. In their annual reports, agents for the Colorado River Indian reservation repeatedly note the scope of Mohave inhabitance in Needles, indicating that it regularly equaled or exceeded the number of residents on the reservation proper. The very fact of not being contained within the reservation is understood as contributing to this population's general disorder, including their sexual perversity and participation in practices that supposedly have been eliminated on the reservation. The characterizations of those living in the vicinity of Needles include the following: "Year by year the Mohaves at the Needles and the Hualapais situated along the line of the railroad are degenerating and growing worse morally. They are not only spreading syphilis among themselves but among the whites as well"; "The Mohaves living in the vicinity of Needles and Fort Mohave are in deplorable condition as to morals and progress toward civilization. . . . They retain all their old-time superstition and barbarous customs and have added to them the vices of a border railroad and mining town"; "These are the same tribe and kindred; in many cases brother, sister and near kindred, yet at Needles one sees savage life with all its horrors, its crime, its disease, its superstitions, its barbarisms, its utter hopelessness."[113] Needles, then, stands as the limit of reservation governance, indicating forms of occupancy and sociality that exceed the reach of control by the Indian agent and that therefore appear as lacking proper (familial) order. As portrayed in the novel, the version of the Ghost Dance that arises in this region defies the sense of lineal generational succession and the limiting conception of biological and property inheritance around which such succession is organized. Instead, the Messiah ceremony offers a temporal assemblage that suggests the presence of a multiplicity of relationships that remain at hand and actively in process. It facilitates a potentially vast network in which each person participates and on which he or she might draw in the present, like the extrareservation social geographies that make possible the Native community surrounding Needles.

This vision and experience of expansive relation runs counter to the U.S. government's efforts to normalize its jurisdictional and property mappings through official citation of monogamous marital union as the model of national life and time. *Gardens* explores how understandings of desire, generationality, and the organization of the family are enmeshed with legal orderings of space. Silko addresses the ways the chronobiopolitics of heteroreproductive lineage (and the generational transmission of racial Indianness) abet and are animated by the chronogeopolitics of assertions of U.S. sovereignty. In other words, the dominant timeline of Indian development (including when read as leading toward hybridization) depends on treating the contours of U.S. jurisdiction as, in Merleau-Ponty's terms, the "fixed points" or background against which to perceive temporal movement.[114]

One of the most significant intimations of that nexus in the novel, albeit a subtle one, lies in Silko's invocation of the sustained U.S. assault on Mormonism. In addition to noting that Mormons participated in the Messiah dance in Needles (14, 29), the text indicates that the husband and sister wives of a

Mormon friend of Grandma Fleet's, Mrs. Van Wagnen, had been arrested: "The old church had been brushed aside by demons, she said. But Grandma Fleet thought maybe the other Mormons got tired of resisting the U.S. government. The government said only one wife, and now the new church said one wife, so the old Mormons moved to remote locations. For years and years, the U.S. soldiers chased Mormons when they weren't chasing Indians" (38).[115] After we hear that Mrs. Van Wagnen's husband has been murdered by members of the reformed Mormon Church, the narrator observes, "The U.S. government had been after the old Mormons for a long time, killing their men and burning their farms wherever they went until they escaped to the west," adding, "Old Mormons believed they were related to the Indians" (44). Through these moments, the novel alludes to the government's decades-long attack on the Mormon Church for its support of polygamy and for its perceived attempt to form a separate government on U.S. territory not amenable to federal jurisdiction, which culminated in the church's official renunciation of polygamy in 1890 and the admission of Utah to statehood in 1896.[116] In the twenty-five years before statehood, well over two thousand criminal cases were filed, almost entirely for crimes related to sexual and marital relationships.[117] The text links the persecution and self-understandings of Mormons to Native people(s), and this connection is affirmed by a statement by John Randolph Tucker, one of the sponsors of a crucial piece of anti-Mormon legislation passed in the same year as the General Allotment Act (1887): "We dissolve tribal relations of the Indians in order to make the Indian a good citizen; so we shatter the fabric of this church organization in order to make each member a free citizen of the Territory of Utah."[118] The wrongness of Indian and Mormon sociosexual life indicates an eruption of barbarism, of an anachronistic communalism, within the space of the nation. That threat to retard or reverse the forward movement of the country requires an overwhelming show of force in order to "shatter" those modes of life so that they can be reassembled in ways that fit the terms and temporalities of U.S. citizenship, themselves consistent with the legal geographies of the state.[119] While the mention of Mormons in the novel suggests that they suffer because of their resemblance to Indians, such references also highlight how the supposed challenge that Indians pose to national futurity—the straight line of national development—lies in their putative failure to conform to proper forms of reproductive generationality, ordered around the procreative line of the nuclear family.[120]

Silko explores the potential for less rigidly lineal, privatizing, and property-oriented ways of conceptualizing and experiencing time—ones more consonant with the text's account of the Ghost Dance—through discussion of Sand Lizard notions of sexual relation. The narrator observes that, in response to Sister Salt's ribald humor, "Maytha and Vedna giggled...; she was like the old-time people their mother talked about—before the missionaries came. In those days, the Chemehuevis really knew how to enjoy one another; only Sand Lizards knew how to enjoy sex more, Maytha joked, and Sister Salt nodded proudly. It was true: Sand Lizards practiced sex the way they all used to, before the missionaries came" (206).[121] Rather than being a forbidden topic, eroticism is a source of joy, one that indicates possibilities for continuity and camaraderie among varied peoples.[122] Such feelings about the expression of desire indicate notions of interpersonal and intergroup relation that do not fit the heteronormative framework of genealogical transmission institutionalized within Indian policy. Not only is sensual pleasure not inherently limited to, in Dana Luciano's terms, *reprosexual* ends, but it engenders nondiscrete modes of imbrication that facilitate the construction of extended social networks.[123] In explaining what Sister Salt describes as "Sand Lizards' wild sexual practices," the text notes, "Sand Lizard mothers gave birth to Sand Lizard babies no matter which man they lay with; the Sand Lizard mother's body changed everything to Sand Lizard inside her. Little Sand Lizards had different markings, and some were lighter or darker, but they were all Sand Lizards. Sex with strangers was valued for alliances and friendships that might be made" (202).[124] This vision of Sand Lizard identity offers, in Justice's terms, "an understanding of a common social interdependence within the community, the tribal web of kinship rights and responsibilities that link the People, the land, and the cosmos together in an ongoing and dynamic system of mutually affecting relationships."[125] Such relations do not depend on a sense of inheritance, in the sense of something in particular passed from generation to generation that makes a person Sand Lizard.

Rather, there are various kinds of relationships instituted through intercourse, all significant whether procreative or not, and Sand Lizard–ness exists within this multiplicity of (sorts of) connections. With respect to Sister Salt's participation in sex work at the construction site on the Colorado, the text notes, "Preachers condemned the sale of sex, but Sister always felt happy after her walks with the men.... Naked on the river sand she always felt as free and joyous as that River Girl character in the old stories the twins heard at Laguna" about how River Girl's relationships with various entities allowed her people to receive vital goods like venison and buffalo (400).[126] In addition, the text earlier remarks, "The old-time Sand Lizard people believed sex with strangers was advantageous because it created a happy atmosphere to benefit commerce and exchange with strangers. Grandma said it was simply good manners" (219). The resulting expansive web of relations certainly exceeds the geographies of the reservation, the heteronormative parameters of reservation residence and the official calculus of Indianness, and the idea of the nuclear family as the explicit or implicit model for the transmission of indigeneity. Notions of power in the region can themselves be characterized as weblike, including what have been termed the modes of syncretism at play in the Ghost Dance, and when one thinks back to the attempts to generate intimacy through the Messiah dance with those who have passed on, this nonnuclear assemblage greatly extends the potential reach of a reference like "ancestors."[127] Also, during the Messiah dance itself, as noted earlier, the text indicates participants' desire to commune with "their dear ones and beloved ancestors" and "their beloved family members and friends" (23), refusing to understand the matrix of cross-temporal connection as one defined by even the most capacious conception of family. The prophetic power of the Messiah dance, then, lies in its capacity to express, condense, and extend the "whole perceptual context" of Sand Lizard sensation, which includes expansive and flexible notions of relation and belonging.[128]

What grounds a Sand Lizard sense of self is, literally, the ground—connection to the old dunes. Those of Sand Lizard descent who moved to the reservation and acquired standing there through marriage have ceased to be in contact with those in the dunes, a break enforced by settler authorities (49). That severing is what constitutes a rupture in Sand Lizard time—or, more precisely, what institutes lineal inheritance as the paradigmatic way of experiencing temporality. By contrast, the dunes orient cross-time connection for Sand Lizard people, with guidance about how to live responsibly in that place being transmitted between generations of women.[129] That knowledge, though, can be understood as an enduring intimacy with place, as the center around which possibilities for belonging and becoming pivot. The porousness of Sand Lizard identity does not indicate the absence of a notion of indigeneity. Miranda suggests in "A Gynostemic Revolution," particularly of the text's portrayal of Hattie's connection with Indigo, "Let me be clear: Silko is not advocating a cavalier philosophy in which who we are and where we come from is not important, or worse, that anyone can make up an 'Indian name,' help out a few needy Indians, and 'become' instantly indigenous."[130] Rather, to be Sand Lizard entails a continuing connection with the dunes themselves:

> Grandma Fleet told them the old gardens had always been there. The old-time people found the gardens already growing, planted by the Sand Liz-

ard, a relative of Grandfather Snake, who invited his niece to settle there and cultivate her seeds. Sand Lizard wanted her children to share.... The first ripe fruit of each harvest belongs to the spirits of our beloved ancestors, who come to us as rain; the second ripe fruit should go to the birds and wild animals, in gratitude for their restraint in sparing the seeds and sprouts earlier in the season. (14–15)

This gratitude to the Sand Lizard for creating the gardens and to various other beings for the ongoing modes of interspecies sharing that make possible continued life there offers something less like familial legacy than like a profound sense of the multiplicity always already at play within this site and a sense that the present remains permeated by such chronologically extended and disjointed forms of mutuality. Inhabiting the gardens over time, then, functions less like a generationally iterated property claim for a particular "Indian tribe" than as an indication of the extent of a people's enmeshment in the shifting networks through which this place is (re)created. Their collective sense of duration emerges out of the shared background of the garden itself, and the webs of connection through which the place and their relationship to it are sustained serve as an expression of what I have been characterizing as temporal sovereignty (in contrast to the forms of Native governance officially recognized as such by the United States). As the narrator observes, "the people called themselves Sand Lizard's children; they lived there for a long time" (15).

As in *Indian Killer* the Ghost Dance in *Gardens* can be seen as further catalyzing and materializing existing, everyday kinds of feeling. It prophetically intensifies ordinary forms of cross-temporal sensation, an experience of the present as actively permeated by other times. The novel suggests that such temporal sensations remain at odds with the forms of Indian identity engendered within the reservation-era context, where a group's past, present, and future came to be understood as organized around properly directed modes of (racialized) procreation—the (re)production of Indian flesh. Within Indian policy's frame of reference, dominant models of lineal genealogy become the implicit basis for conceptualizing movement through time. The novel's portrayal of the Messiah dance emphasizes a set of temporal principles and experiences that do not fit the generational unfolding of heterofamilial lineage, thereby producing a queer effect. The Messiah dance suggests the operation of a complex and expansive matrix of relation that neither aligns with the vision of Indian realness projected through the reservation system nor obeys a temporality of inheritance. *Gardens* offers a vision in which nonnuclear modes of pleasure and connection (to persons and nonhuman entities), as well as nonpropertied modes of inhabitance, enable experiences of time as an immanent multiplicity (versus a chronological evolution)—a set of affects oriented by a commitment to connection to the land and most powerfully accessed and conveyed through prophecy.

*Beyond Settler Death-Worlds*

Silko presents the Ghost Dance in its manifestation in the Arizona-California border region as assembling a heterogeneous network that crosses seemingly disparate time periods and does not obey the forms of genealogical straightness and inheritance that characterize the nuclear family. Instead, prophecy opens the present to both the past and the future (and the transits among them), thwarting the sense of a developmental trajectory in which the past moves toward the future as a linear, or historicist, sequence. As contrasted with the popular association of the Ghost Dance with death, specifically the massacre at Wounded Knee, *Gardens* presents Wovoka's vision as expressing and amplifying the conditions of possibility for further life. More than enabling contact with those who have died, Wovoka's prophecy promises a remaking of the earth such that it can continue to sustain human survival, as against what the novel casts as the necropolitics of Euramerican modernity.[131] Uncoupling Native identity from the procreative transmission of Indianness, as organized around the forms of tribal alignment institutionalized through the reservation system, the text opens up the range of possibilities for envisioning what it means to sustain indigeneity as an ongoing form of incipiency. In the version represented in the novel, the Messiah dance illustrates and enhances a broader everyday tendency toward connection, recontextualization, and creation that gains orientation not through reference to lineal familial inheritance but through an enduring, capacious connection to and (re)creation of place. This continuum of birth and becoming, expressed perhaps most directly through the novel's account of the movement and meaning of seeds, differs both from heteronormative genealogy as well as Euro-American notions of newness, which the text suggests are predicated on the destruction of what is.

The novel emphasizes the Ghost Dance's powers of rejuvenation by differentiating it from the forms of decimation that characterize Euro-American development. Several critics have noted the distinction in the text between the search for commercially viable forms of rareness—through seizures of plants from "exotic" locations and processes of grafting them onto more familiar ones—and the movement of people and seeds represented by Indigo (to which I will turn shortly).[132] Before Edward's introduction to Indigo (who is found in the garden of their house in Riverside, California, by his wife, Hattie, after Indigo ran away from the boarding school to which she had been sent), he took part in a failed expedition to Brazil in search of rare orchids, during which he received a wound that left him impotent. Edward's journey illustrates the ways that the pursuit of newness within European and American economies entails the production of devastation.[133] He was accompanied on the trip by two men, one of whom, Mr. Vicks, was sent by the English Department of Agriculture to steal samples of rubber plants to combat the blight that "was destroying Britain's great Far Eastern rubber plantations" (129). The other, Mr. Eliot, ends up setting fire to an entire valley in order to prevent anyone from a rival company from gaining access to a particular species of orchid: "Rival hybridizers would be stymied when they sent out their plant collectors now that this Pará River site was destroyed" (142). The incineration of this habitat parallels earlier forms of murderous violence against Indigenous peoples of the region: "Now, Indians knew the value of wild orchids, but frequently white brokers came upriver and demanded their entire stock of a species to corner the market. Indians who did not cooperate were flogged or tortured, much as they were at the Brazilian and Colombia rubber stations," themselves often built on the sites of villages that had been burned and cleared of their former inhabitants (133). The creation of novelty as a commodity, as a salable experience of newness, requires a kind of exclusivity that necessitates practices of elimination. In order to manufacture the rareness that engenders a sense of something as unique, other examples of that item must be obliterated. Moreover, in this vein, newness indicates the extraction of something from one context and its distribution and appreciation as anomaly within other sites (as opposed to integrating a once-alien object into the shifting series of relations that constitute the place of its arrival). The appearance of an exotic addition from elsewhere, or the aura of alienation that leads one to experience something as exotic, then, enables the feeling of futurity as rupture, a disjunction between what was and what will be that retains the continuity of dominant frames of reference while generating the sensation of the unexpected. The novel indicates, though, that such an impression of change for the privileged consumers of these fetishized objects depends on the actual shattering of environments and lifeways for those from whose lands newness is extracted as an exploitable resource.

In this sense Silko presents the Euro-American movement toward the future as necropolitical. Achille Mbembe characterizes *necropolitics* as "the ultimate expression of sovereignty" through "the power and capacity to dictate who may live and who must die," and one of the chief expressions of such power is through "the creation of *death-worlds*, new and unique forms of social existence in which vast populations are subjected to conditions of life conferring upon them the status of *living dead*."[134] In addition to noting the de facto seizures of Native land in the name of industrial progress, *Gardens* further depicts the maelstrom of violence unleashed against those who do not conform to normative lifeways.[135] When Indigo and Sister Salt go in search of their mother after Grandma Fleet's death, they reach the home of Mrs. Van Wagnen, their grandmother's Mormon friend who had participated in the Ghost Dance, only to find the house burned to the ground, the food thrown on the ground to spoil, and the fruit orchards chopped down. Sister Salt thinks of the pointedness of this assault: "If this was what the white people did to one another, then truly she and the Sand Lizard people and all other Indians were lucky to survive at all. These destroyers were out to kill every living being, even the Messiah and his dancers" (61). The intensity of this assault brands whites as "destroyers" fully prepared to annihilate all modes of life that deviate from sanctioned patterns of intimacy, occupancy, and association. Not only does Mrs. Van Wagnen represent the broader Mormon challenge to national monogamy and sovereignty discussed earlier, but her connections with Native people via the Ghost Dance indicate a potentially dangerous set of alliances that must be utterly foreclosed. Similarly, the text later emphasizes the brutality of Hattie's assault and rape by local white men, apparently motivated by her attempt to bring provisions to people on the Chemehuevi reservation, an attack that nearly kills her (456–58), and just before the end of the novel, Sister Salt and Indigo return to the dunes from Needles to find that "terrible things had been done at the spring" by "strangers," including "hack[ing] to death" both the snake that lived there and the apricot trees Grandma Fleet had planted (476). While not characterized as official acts by agents of the state, these last two scenes combine with the previous instances to suggest a broader pattern in which the assertion, imposition, and maintenance of settler sovereignty entail the attempted elimination of the conditions of life for countervailing social formations, particularly those of Indigenous peoples.[136] That system(at)ic exertion of the capacity to make die becomes justified as part of producing progress, creating the conditions of (settler) futurity through the eradication of obstructive impasses. Against such a background, reservations function as spaces of anomaly to which Indianness can be consigned as a temporal oddity, aberration, and/or vestigial artifact even as it is straightened into modes of tribal lineage.

The novel suggests that the Messiah and his family serve as objects of necropolitical state violence (being pursued by the soldiers, with the dancers subject to arrest and removal), but also as an alternative to it, offering an experience of temporality not organized around ostensibly progress- and novelty-generating modes of devastation. Through dancing, "the used-up lands would be made whole again and the elk and the herds of buffalo killed off would return"; "great storms would purify the Earth of her destroyers," while "the winds would dry up all the white people and all the Indians who followed the white man's ways, and they would blow away with the dust" (23). The need for such transformation arises from the fact that "the invaders made the Earth get old and want to die" (26). From this perspective, the Ghost Dance activates the potential for more life, engendering such possibilities through both the return of the ancestors and the extension of the conditions for the survival of humans and nonhuman entities alike. Over the course of the novel, several characters offer versions of the sentiment that "to go on living is far more painful than death" (51), suggesting that the existing circumstances generated under U.S. rule, more than being simply oppressive, are, in fact, unlivable.[137] Silko's rejoinder to the narrative of U.S. national history as the unfolding of unending improvement, then, lies in the depiction of national time as an increasing expansion of spheres of extinction, obliteration, and bereavement. In contrast, Wovoka's vision as articulated in the novel bears within it an experience of the relation between the past, present, and future as one of interdependence, in which the movement forward in time cannot be understood as an extraction of something of value from the useless, discardable husk in which it is encased.

Moreover, the past does not so much recede as itself potentially function as a horizon for unfolding dynamics of being and becoming. As Shari Huhndorf argues with respect to Indigo's travel from Riverside, to New York, to Europe, the "journey from west to east ... reverses the direction of conventional frontier narratives (such as Frederick Jackson Turner's) of the 'settlement' of North America. It is also a journey backward in time to white America's European origins," and in this movement "Indigo believes she is following the path of the Messiah and his family on her own trek through Europe."[138] When on her transatlantic voyage, Indigo thinks, "She was crossing the same ocean that the Messiah crossed long ago on his way to Jerusalem. After they tried to kill him, he returned over the dark moving water. . . . She took heart because the Messiah and his followers visited the east and returned; she would too" (197); at several other points the narrative indicates that the Messiah and his family periodically traveled "east" (see 55, 122, 265, 277, 285). The movement of the Messiah entails a process of "return," in which what's to come appears less as the result of a unidirectional telos (such as in the Hegelian frontier narrative of history moving west) than as something of a doubling back on what already had been, which then serves as the site of further possibility.[139] The novel develops that sense of ongoing contact with what was: what from a chronological and historicist perspective would be *the past* remains immanently within the expansive and shifting network of relationships that constitutes continuing life, as a potentially generative set of forces, trajectories, presences. The Ghost Dance in *Gardens*, then, enacts a process of becoming in which the putative past is not that which is behind or which progressively withdraws. Instead, chronologically prior relations of influence, interdependence, and animation help orient actions and movements in the present, providing an active frame of reference for them.[140] Unlike a notion of inheritance or origin, an increasingly remote point from which something descends, prophecy expresses, highlights, coalesces, and intensifies a sense of cross-temporal imbrication—not as nostalgic or melancholic desire for return to what is gone but, playing on David Scott's notion of the "former future," as what might be termed the past incipient, a potential for further emergence.[141]

Indigo's trip abroad expresses this sensibility in her introduction to recently discovered ancient European places and figures of spiritual power in Bath and Corsica. These incarnations of fertility include "amulets of ivory and bronze in the shapes of breasts" (257), a stone carved in the shape of "a human vulva" (290), objects with "concentric circles" that "represented the pubic triangle" (291), and "a snake-headed figure with human arms and breasts" nursing a baby snake (296). Literally buried under later buildings or dense overgrowth, these sites and objects return as inspiring instantiations of the potential for fecundity in the present; in doing so, they suggest not so much normative heteroreproductivity as a ubiquitous capacity for creation.[142] Those encounters repeatedly remind Indigo of the Messiah and his family and of Sand Lizard relationships with Grandfather Snake at the old garden, and in this way, Silko indicates connections between geographically and chronologically disparate materializations of spirit, challenging Edward's notion of civilizational inheritance in which Europe appears as the most advanced: "It was just as well Indigo missed the serpent figures. The child was from a culture of snake worshipers and there was no sense in confusing her with the impression the old Europeans were no better than red Indians or black Africans who prayed to snakes. Hattie agreed; they must help the child adjust to the world she was in now" (302). If for Hattie and Edward "now" indicates a coherent present of Euro-American progress along a singular timeline of development in which nonwhite peoples remain backward (and in need of disciplining and tutoring in order to catch up), the *now* of the fertility figures and the Ghost Dance suggests a world permeated and animated by *then*—movements, beings, relationships, forces, and social formations from a chronologically prior period. Some critics have suggested that this similarity between European and Native expressions of the potential for creation and emergence illustrates "a form of serpentine matriarchal spirituality [that] binds peoples across nations and across time," one that "telescopes complex waves of conquest into one relationship" while "becom[ing] a way of not having to address questions of cultural translatability."[143] However, the resonance among these formations might be considered less an essential unity, one often characterized as myth (as opposed to history), than an indication of a broader potential for the prophetic emergence of the past in the present as part of the (re)generation of life—a potential that extends beyond the peoples of the Great Basin or those on lands claimed by the United States. Such linkage refuses the ontological reduction of the Ghost Dance to a set of regionally specific *beliefs* while also indicating possibilities for moving toward the future other than those at play in the necropolitics of Euro-American instantiations of time as modernity.

The novel's discussion of the movement and replanting of seeds offers a concrete example of how the Ghost Dance's spirit of incipience and its cross-time affiliations express and amplify ordinary practices and tendencies. Silko distinguishes such everyday modes of emergence from both the (commercialized) pursuit of novelty and the unfolding of (heterofamilial) lineage. Throughout her travels Indigo gathers or is given seeds from virtually every location she visits, from Hattie and Edward's garden, to Long Island, to Bath and Corsica. In doing so, she is inspired by Grandma Fleet, who "always advised the girls to collect as many new seeds as they could carry home. The more strange and unknown the plant, the more interested Grandma Fleet was; she loved to collect and trade seeds. Others did not grow a plant unless it was food or medicine, but Sand Lizards planted seeds to see what would come; Sand Lizards ate nearly everything anyway, and Grandma said they never found a plant they couldn't use for some purpose" (83–84). More than serving a specific delimited purpose or adding something unique to the landscape, seeds indicate potential. They bear within themselves the history of the place(s) from which they come, and the ecological dynamics that nurtured them there, while opening toward an unpredictable future. As with notions of power in the region discussed previously, seeds here suggest flux and action, a fundamental sense of movement that attaches to the possibility of life. Not knowing "what would come" means that seeds exceed a particular reproductive economy in which the goal is the transmission of the same—the conveyance of patrimony. Instead, they allow for the emergence of something different from what came before, less developmental trajectory or exotic newness than a contribution to the diversity immanent within growth itself. That variability, the quotidian rejuvenation of the already complex and changing organic and inorganic matrix that is the dunes, opens possibilities for action in the world ("never found a plant they couldn't use for some purpose") without necessarily enacting a radical break within extant practices and ways of being.[144] While the novel often expresses something like an ethos of balance, in that it refuses a commercialized extractive relation to land and resources (in ways I will discuss further), Silko displaces a static notion of ecology in which equilibrium is paradigmatic, instead emphasizing the constant process of alteration as itself crucial to the maintenance and extension of life.[145]

Continuity and change are not readily differentiable, in the sense that there is not a clear event of rupture that could definitively mark the onset of the new. Past and future seem to overlap and interpenetrate—like in the Ghost Dance itself. From the perspective of Wovoka's vision as presented in the novel, the land itself is used up owing to the effects of exploitation by "the invaders" who made "the Earth ... want to die," and in bringing the possibility of further life, seeds gathered from elsewhere simultaneously suggest a movement back toward a more fecund time and a movement forward toward achieving sustainable subsistence for unborn generations. In teaching the sisters to leave enough beans on the ground for "the old gardens ... [to] reseed themselves," Grandma Fleet tells Indigo, "Don't worry. Some hungry animal will eat what's left of you and off you'll go again, alive as ever, now part of the creature who ate you" (51). Introducing new seeds, then, does not so much engender innovation—uniqueness for its own sake—as amplify the potential for the conversion of death into life, a temporality organized not so much around progression as around the maintenance of an expansive matrix of relation (like the web of alliances formed by sex). As Stephanie Li argues, the novel "draw[s] explicit parallels between the act of gardening and that of mothering," with procreation suggesting less a linear unfolding than an enfolding within a multivectored network, and the planting of once-alien seeds within the garden materializes in a different key the sense of ancestors and descendants as coparticipants within the time of the Messiah dance.[146]

The novel's emphasis on the presence of seeds from elsewhere, and their transport across sometimes vast distances (at one point Indigo thinks, "Seeds must be among the greatest travelers of all" [291]), works against the account of both the Ghost Dance and Native placemaking more broadly as the desire for a return to a pristine past or to something that could count as origin. In discussing Hualapai Ghost Dancing, which provided the inspiration for the novel's account, Henry F. Dobyns and Robert C. Euler note that "the Pai [another name for the Hualapai] placed the recovery of their land at the heart of their concept of the ghost dance movement," but they then suggest that the Hualapai "like other ghost dancing Indians ... wanted to return to the good days gone by[,] ... to retrieve [their] former assets and the lifeway that went with them."[147] To the extent that the Ghost Dance both arises out of and further animates Native connections to place that exceed the terms of the reservation system, it appears here as a yearning for the return of a bygone era unsullied by settler presence. By contrast, in the novel the seeds and their movement suggest a revitalization of the land, a regeneration that is not merely repetition; they activate the capacity for further life and becoming toward which the Ghost Dance is also oriented. Discussing the geographies offered in Native women's writing, Goeman argues, "(Re)mapping is not just about regaining that which was lost and returning to an original and pure point in history, but instead understanding the processes that have defined our current spatialities in order to sustain vibrant Native futures." She later adds that "healing is not linked to an original, stagnant home ... but returning to a specific land and a community that is always in the process of creation."[148] The future appears less as novelty or an evolutionary shift away from what came before than as the continuing potential for creation itself, as a process of drawing on past and present patterns as reservoirs in whatever ways that they can enable efforts to stave off settler-induced modes of death and destruction. The novel's portrayal of seeds suggests a particular way of "reckon[ing] with an environment," such that the potential for action in the present appears not as replicating the past but as holding the current moment open to possibilities that are both emergent and residual—a prophetic sense of latency that does not run in only one direction.[149]

If Sand Lizard identity and placemaking in the text occur through the building and sustaining of an expansive web of relations with other persons and places, these practices and understandings remain oriented by a persistent commitment to the dunes as the space of home. The bringing of plants from elsewhere (whether hundreds of miles away or across an ocean) to the dunes enacts its own temporality, one that is not so much restorative or recursive as rooted. The dunes provide a meeting point for the varied itineraries of Sand Lizard people, including their encounters with other Indigenous peoples and the violences of settlement. The gardens, then, provide a nexus for connecting chronologically disparate events and movements to each other, through which they are woven together as part of Sand Lizard being and becoming. This process incorporates Sand Lizard trajectories that lead them elsewhere (such as Indigo's gathering of seeds through travel or Sister Salt's erotic alliances and the child that results), but they remain oriented by their continued turning back toward the dunes themselves, which continue to provide the background against which to figure their experiences of movement. Mobility and occupancy, then, do not appear here as antagonistic or opposed. Instead, the time of seed-gathering travel merges back into the time of residence in the dunes because the former was always shaped around making possible the latter. The future life projected through the gathering of seeds occurs within a frame of reference in which the gardens serve as the privileged site of Sand Lizard temporality.[150]

While exceeding the space of the reservation, Native identity remains centered in place. As opposed to the reproductive unfolding of lineage, a sense of descent from an origin that itself is projected backward from the *tribal* spaces produced by Indian policy, indigeneity appears in the novel as a capacious network through which various persons, practices, elements, and events are affectively interwoven in processes that remain in flux, if oriented to a particular homeland. From this perspective, time functions not as a succession of moments in a direct causal chain, a view that can lead to the notion of the new as the decimation of the old which helps drive and validate the necropolitical violence of Euro-American development. Instead, time is a multiplicity in which what is chronologically past materializes as part of and helps influence present dynamics of continual becoming. That potential reaches its most explicit form in the novel through the Messiah dance, which prophetically expresses and intensifies the quotidian capacity for intense and intimate relations across chronological time. Rather than serving as a period marker for the end of meaningful Native opposition to settler rule (via the Wounded Knee massacre), the Ghost Dance in *Gardens* indicates enduring possibilities for regeneration through everyday modes of emergence that do not obey the heteroreproductive lineality through which settler governance constitutes, regulates, and curtails Indian realness.

Both Alexie and Silko displace the inevitability of settler time. In these texts prophecy enacts forms of temporal relation that do not fit the developmental frame of post-Enlightenment historicism. The novels address the violences perpetrated by projections of futurity that take the normalization and extension of non-native occupation as an unquestionable orienting frame, as the background against which to mark the movement of time and the advent of the new. Such conceptions of historical unfolding cast Native peoples as anachronisms of one form or another. Moreover, the texts explore how these visions of development in time circulate accounts of Indianness that are largely bound to the tribally specific space of the reservation, creating kinds of realness that imagine few options for Native people(s), constrain the possibilities for Indigenous self-determination, and work to derealize (make invisible and less available) alternative means of expressing and living indigeneity. In contrast, the Ghost Dance expresses modes of temporality in which the connection between the past, present, and future need not be one of contiguous, causal sequence.

Rather, the work of prophecy lies in its ability to stimulate and coalesce a nonsuccessive relation between persistence and potential—the ability of *was*, *is*, and *will be* to enter into complex exchanges with each other that do not follow an inherent progression from one to the other. In *Indian Killer* the relation among chronologically disparate moments—the ways these conjunctures are oriented toward each other—enables a process of transformation by which the Ghost Dance gives material form to modes of Native rage. It gains momentum from while further animating quotidian experiences at odds with those realized through non-native narratives of Indianness. Less a syncopation than an overlay, prophetic temporality in *Gardens in the Dunes* intensifies quotidian Native experiences of time as a capacious network, one in which connections less succeed each other (on the model of the straight line of inheritance) than become enfolded or enmeshed in ways oriented around being and becoming in a particular homeland.

In these enmeshments, entanglements, resonances, and projections across time, the kinds of prophetic temporality circulating around and through the Ghost Dance enact a kind of becoming that is not contingent on the supersession of what's come before. What from a chronological assessment belongs to the past here appears as actively influencing the future (and vice versa), and thus, what would usually be considered residual serves as a vehicle for emergence and becoming, such that the horizon for the future can entail an orientation toward the past. Prophecy in these texts involves the potential for a given moment in time to be permeated by noncontiguous moments and presences, whether understood as a kind of periodicity, an actualization of potentials that have been foretold, the intermittent presence of ancestors and other beings, and/or a renewable and flexible matrix of relationships. In this way the present neither replaces the past nor extracts something from it that provides a kernel that can be transformed in order to generate the future, whose very status as future in this formulation (as the prospect of the new) depends on its difference and separation from the past.

The conception or experience of indigeneity as other than lineal unfolding does not fit within the terms of a model of Native identity as that which persists from the past, as that which always must be fundamentally oriented around its priorness to the settler state.[151] That very insistence on priorness—an existence as landed political entities whose claims precede those of the state formed on top of them and through their domestication—operates as a key part of arguments for recognizing Native sovereignty and self-determination. Yet the emphasis on such a relation to the past as the mode of legitimizing contemporary articulations of peoplehood can perpetuate the sense of Indigenous peoples as inherently belated, while also casting any break in the self-same continuity of Native identity, collectivity, and territoriality as indicative of peoples having ceased to exist as such. In this vein, tradition serves as the sign that guarantees Native authenticity, but as Alexie's novel indicates, such narratives of Indian realness usually operate as a retrospective projection—a form of settler simulation. Insisting on the modernness of Native people(s), their inhabitance of a shared present with non-natives, though, does not remedy this problem. As discussed in chapter 1, to the extent that modernity (or modernities) functions as the frame in which to recognize Native people's and peoples' existence in time, they remain subjected to the violences and vicissitudes of settler recognition. They remain affectable others whose contemporaneity must always be in question and consistently deferred, a frame of reference in which settlement implicitly functions as the background against which to register presentness.[152] By contrast, in Alexie's and Silko's texts, what could be characterized as tradition appears not as a vestige but as chronologically discontinuous forms of knowledge, experience, memory, extrahuman force, and relationship that can become realized in the now in spectacular and quotidian ways that are potentially transformative, individually and collectively. From this perspective self-determination appears less as a particular and properly modern mode of performing peoplehood than as the expression of the multiplicity of Indigenous peoples' ways of being and becoming.

CODA. **DEFERRING JURIDICAL TIME**

*Beyond Settler Time* has sought to explore the friction between two modes of Indigenous critique: the insistence that Native people be recognized as contemporary or modern (rather than seen as anachronistic, stunted, or vestigial) and the refusal to pursue non-native recognition on the basis that it is part of the colonizing interpellation of Indigenous peoples into settler social forms and dynamics. The notion of *temporal sovereignty* emerges from the effort to track the force exerted through processes of temporal recognition (including the insertion into dominant periodization schemes, treatment of Native opposition as an eruptive aberration, reorganization and privatization of personal development, reorientation toward the market economy, and subjection to anachronizing models of Indian realness) while envisioning Native being and becoming as nonidentical to these imposed frames of reference, even as Indigenous temporalities are affected and shifted by such colonial imperatives. One prominent question that arises out of my use of this phrase, though, is, how do we address the relation between temporal sovereignty and the kinds of juridical structures and fields that tend to provide the de facto referent for sovereignty (such as the constitutional division of powers, lawmaking, administrative regulations and enforcement, court systems, determinations of jurisdiction, etc.)? While I've discussed at length the ways U.S. policy form(ul)ations work to warp Native experiences of time, I have been hesitant to address institutions of Native governance as such. Partially, this choice is due to the fact that the Native materials I engage are nongovernmental in character. However, more substantively, there is the question of the degree to which modes of governance officially recognized by the United States as sovereignty can express forms of temporality that differ from dominant Euramerican frames of reference. Rather than offering a conclusion per se, then, this coda is meant as an extended meditation on that question and its implications for pursuing certain kinds of work in Native studies. I want to argue for deferring juridical time—or, more specifically, to argue for the potential value of provisionally suspending the question of how temporal sovereignty, as I've described it, could or should be operationalized as part of juridical apparatuses and processes.

Invoking "sovereignty" indicates the presence of a distinct polity by articulating Indigenous peoples' pursuit of self-determination in a political language intelligible to the state while at the same time seeking to mark the limits of settler-state authority. In that double-sided dynamic, the relative balance between rendering legible through translation and asserting relative autonomy remains continually in process. Moreover, that negotiation further entails addressing the issue of what constitutes a polity or people as such, and the form(s) of Indigenous peoplehood recognized by the state may not be the same as the form(s) of collectivity either actively desired by a given people or implicitly at play in everyday modes of experience. To what extent, then, in the context of ongoing settler colonialism, should the juridical apparatus of Native governance be taken as expressive of the contours and dynamics of peoplehood (and its temporalities)? Jean Dennison powerfully asserts that "the academic compulsion to deconstruct sovereignty threatens to aid settler colonial efforts to discredit indigenous authority," and in "For Whom Sovereignty Matters" Joanne Barker invokes "the work of contemporary indigenous scholars and activists" in order to underline the ways they have argued that "sovereignty emanates from the unique identity and culture of peoples and is therefore an inherent and inalienable right of peoples to the qualities customarily associated with nations."[1] However, Dennison also foregrounds the forceful presence of modes of "entanglement," "the inherent power dynamics within the ongoing colonial context," and Barker notes, "Of course, translating indigenous epistemologies about law, governance, and culture through the discursive rubric of sovereignty was and is problematic."[2] Does the designation of nonjuridical aspects of Native social life as expressions of sovereignty necessarily make them congruent with official institutions of governance? Put another way, does the use of the term *sovereignty* imply a definite relation between institutions of governance and those other areas, textures, or quotidian dimensions of social life? These questions open up conceptual and political space for exploring the potential ambiguities, ambivalences, and indeterminacies in the way the term sovereignty circulates and the productive possibilities generated by suspending the effort to collate the work of sovereignty into the institutional dynamics of Native national governments. Scholarship in Native studies over the last decade has stressed both the importance of political sovereignty (as a means of characterizing Native polities' existence as such) and its limits (the role it can play in inserting and translating Indigenous modes of governance and social life into settler forms). This tension generates possibilities for approaching experiences of peoplehood, including temporal frames of reference, as both related to and distinct from formal institutions of governance.

The insistence that Indigenous peoples properly possess and exercise sovereignty works as a way of challenging their treatment as bearers of depoliticized *cultural* identity. In a context in which questions about the contours and character of Native governance are so often recast as about cultural difference within the (settler) nation, Audra Simpson indicates that "sovereignty" can mark "something beyond difference," pointing to the fact that what's at stake in Indigenous-settler relations is "the grounds of governance" and also contesting the notion that Native political authority can be understood as endowed by the settler state.[3] Yet, as Simpson asks, "how does one assert sovereignty and independence when some of the power to define that sovereignty is bestowed by a foreign power?" She further argues that such claims that the state has "bestowed" power on Indigenous peoples "force us to ask how one is to define a citizenship for one's people, according to one's political traditions while operating in the teeth of Empire, in the face of state aggression."[4] The difficulty to which Simpson points lies in the need for a way of signifying and defending the existence of forms of Native political life not inherently superintended by the settler state while at the same recognizing the ways the state does, in fact, regulate Native political institutions, determining their proper boundaries and casting them as sitting atop foundational non-native geopolitical claims and/or as arising through non-native beneficence. In this vein *juridical time* can be understood as referring to the temporal dynamics at play in performing modes of governance understood by settler institutions as such, including the following: adopting conventional periodicities of liberal governance (such as those of elections, court sessions, and annual budgets); using a fairly foreshortened timescale in regular decision making (not invoking events decades or centuries prior as a basis for action); pegging legislation, policy making, and judicial findings to extant developments in federal Indian law and policy; separating kinship and spiritual relations (and their temporalities) from the sphere of governance; discounting nonhuman entities as causal factors in explaining phenomena or developing policy; and treating dominant notions of proper individual development, such as participation in nuclear family homemaking, as the metric for measuring the health and well-being of the population (and, thus, the basis for determining what constitutes good or effective policy). These implicit dynamics of institutional time can come to serve as the orienting background for governance as such. Other ways of envisioning and experiencing time can end up consigned to the sphere of culture, as signs of unique Indigenous "difference" divorced from Native peoples' right to determine for themselves how they could or should operate as polities—what will count as the contours and content of *politics* and *governance*.

To be recognized by the state as possessing sovereignty means having the potential for political autonomy and self-determination acknowledged *and* having that potential constricted, (mis)translated, and redirected in ways largely driven by non-native interests and trajectories. Juridical instantiations of Native peoplehood recognized by the settler state are pressured to express political life—via sovereignty—in ways oriented by the momentum of institutionalized non-native frames and forces. As against the *recognition* of "difference," Simpson offers *refusal* as "a political alternative," noting, "Refusal comes with the requirement of having one's *political* sovereignty acknowledged and upheld."[5] If asserting sovereignty acts as a bulwark against depoliticizing attributions of cultural difference, how is the sovereignty acknowledged by the state distinct from the sovereignty regulated by the state (as a condition of recognition)? If sovereignty is the means of asserting and signifying an Indigenous people's existence as a self-determining polity to non-natives *and* is the means by which the settler state defines and regulates the terms of Indigenous political authority, how can the forms of the one be distinguished from the impositions of the other? To what degree can Indigenous political understandings that are not consonant with settler frames be realized as sovereignty within such institutions when Native nations remain "in the teeth of Empire"? Dennison suggests that in the face of these dynamics, "The key is making something out of this structure that does not mirror the oppression of the colonizer," but she also notes, with respect to the ways settler governance constricts the scope and character of Native governance, "Such circumstances do not mean . . . that we should think of indigenous sovereignty as somehow inherently limited but must instead find or make other spaces in which these sovereignties can be realized."[6] In the face of ongoing settler impositions, regulations, and displacements, expressions of peoplehood may lie in a proliferation of sovereignties that are not reducible to formal governmental institutions and processes. Refusing state recognition can be understood as opening up room for political maneuvering—for engaging with varied possibilities for what might constitute politics or legitimate forms of collective life—by suspending the moment when such a politics needs to be realized as governmental policy, when it needs to become intelligible to the settler state as a structure of governance. The notion of temporal sovereignty occupies this space of potentiality and difficulty, partaking of the need to signify Native being-in-time while also attending to how becoming temporally intelligible to settlers may be the vehicle for enacting forms of "state aggression" and interpellation. In this way sovereignty indicates both the need to engage nonnative discourses and expectations (such as the anachronizing images of static Indianness) and the importance of acknowledging modes of temporal experience that do not conform to settler orientations, backgrounds, and frames.

What, though, are the limits of refusal as a political strategy? Repudiating state recognition entirely can produce conditions in which Native peoples play no substantive role in the ways they are narrated within settler institutions. Dale Turner suggests, "*As a matter of survival*, Aboriginal intellectuals must engage the non-Aboriginal intellectual landscapes from which their political rights and sovereignty are articulated and put to use in Aboriginal communities," and in doing so, an "asymmetry arises because indigenous peoples must use the normative language of the dominant culture to ultimately defend world views that are embedded in completely different normative frameworks."[7] Engaging with the state may involve translating Native frameworks into non-native terms (including with respect to time), but the absence of such an effort at translation leaves unchallenged the political and legal "landscapes" that orient settler laws and policies and that provide the background for materializing them on Native bodies and territories. From this perspective, some gambit for recognition is necessary to shift the dynamics of non-native governance such that they can be made accountable, at any level, to Native self-understandings. In this vein Turner asserts, "We need to be able to speak and write convincingly in indigenous terms *and* be able to change how these arguments are used in the institutions of the state."[8] This formulation indicates the ongoing presence of two scenes that affect each other but remain distinct—conversation and confrontation with non-natives within "the institutions of the state" *and* participation within Indigenous publics with their own "normative frameworks." These frameworks can be understood as extending beyond philosophical propositions about right conduct to include, in Dian Million's terms, "affectively informed Indigenous conceptual frames," such as those arising from collective modes of duration, the momentum of Indigenous histories, and shared forms of backgrounding and storying.[9]

The state-recognized apparatus of Native governance serves as a site for both institutionalizing Indigenous terms (such that they can be made visible to the state as a normative basis for politics) and negotiating the impositions and imperatives of settler governance (which has its own normative dynamics and modes of backgrounding).[10] In his critique of the pursuit of a politics of recognition, Glen Coulthard says of Turner's analysis that it "spends little time discussing the assimilative power" of engaging with "the legal and political discourses of the state," but Turner highlights the ongoing presence of settler colonial fields of force in which formal Native political institutions, among other sites, serve as multivalent terrains of struggle.[11] A politics of refusal opens room for acknowledging the existence of collective forms of being and becoming that give everyday shape and substance to Indigenous peoplehood but that do not mirror state frames and aims. However, such a politics supplements rather than substitutes for the overdetermined role of official Native political institutions in mediating colonial imperatives, in serving as a site of contestation over what will count as governance, and in translating alternative Native norms as policy. Dennison suggests of Osage constitutionalism, "It is only by uniting behind Osage sovereignty that we have any hope of ensuring an Osage future."[12] Viewed from this perspective, refusal gestures toward the need for a more capacious sense of the character and contours of peoplehood than what currently is institutionalized and recognized by the state. As Million asks in *Therapeutic Nations*, "What would it really take to materialize an ardent desire for self-determination into governance that could actually serve the peoples?"[13] Characterizing other possibilities for experiencing and expressing peoplehood as *sovereignty*, then, insists on a dynamic and ongoing relation between such experiences and the official mechanisms of governance without necessarily specifying the exact nature of that relation.

The various kinds of social forms, processes, and trajectories to which sovereignty can refer are therefore not necessarily equivalent to or continuous with each other. As Barker suggests, "Sovereignty—and its related histories, perspectives, and identities—is embedded within the specific social relations in which it is invoked and given meaning."[14] Designating Indigenous modes of life and experience as expressions of sovereignty asserts their political importance, especially when they are not acknowledged as political within dominant settler frameworks, but doing so also implicitly differentiates such dimensions of peoplehood from the juridical structures whose political authority also operates under the sign of sovereignty. For example, Simpson observes with respect to Mohawk experiences of belonging, "*Living, primary, feeling citizenships* may not be institutionally recognized, but are socially and politically recognized in the everyday life of the community."[15] Such "feeling citizenships" may not be the same as the official rules for band membership. While both may be designated as expressions of sovereignty, and feeling citizenships are not unaffected by official policy, such affective connections are also not reducible to the terms of policy, pointing to the existence of collective dynamics that constitute ordinary modes and expressions of peoplehood that may not be congruent with the apparatus of formal governance. Part of the power of such feelings lies in the ways they have a shared historical density that orients quotidian modes of perception and gives rise to forms of storying that help provide everyday cohesion in ways other than "institutionally recognized" criteria for belonging. Rather than providing a frame for governance, ordinary affective relations (including the forms of temporal experience I've discussed) are often displaced to a potentially anachronizing sphere of cultural authenticity, consigned to the feminized and privatized space of the family, or dismissed as inappropriate and destructive outbursts of negative feeling (like anger). Million points to "the dense amount of psychological technologies that are now in place in Indian Country to interpret our affect and emotion, to produce the speech/affect/memory we were formerly denied, to explain the feelings we weren't supposed to have, or to suggest how we 'should' feel," further noting that within such settler-sanctioned accounts "culture is good as individual/community therapeutic practice but unimaginable as relational practices that inform governments, ways of living in places."[16] To explore such potentials, including the sort of quotidian sensibilities and phenomenologies suggested by Simpson's "feeling citizenships" or Million's "felt knowledge," however, one needs to avoid taking the apparatus of governance (and the specific institutional translations and contestations in which it is enmeshed) as one's sole intellectual and political frame of reference.

The concept of temporal sovereignty provides a hermeneutic through which to name and address kinds of collective feelings and everyday experiences, specifically marking how they are subjected to colonial modes of translation not unlike those enacted through other forms of recognition. While one must bear in mind Dennison's caution quoted earlier that seeking to deconstruct the concept of sovereignty can "discredit indigenous authority," temporal sovereignty functions less as an effort to challenge juridical sovereignty than as a means of indexing the multiplicity of ways that time both operates as a vector of settler colonialism and expresses Indigenous self-determination. I seek to mark settler colonial modes of chrononormativity while also opening up the potential for treating various temporal phenomena—patterns of behavior, forms of perception, periodizations, continuities, memories, stories, prophecies—as potentially political manifestations of peoplehood. These phenomena are not inherently governmental, and it may or may not be desirable to try to incorporate them into official political processes. As Dennison indicates with respect to the question of whether aspects of Osage "culture" (including the I'n-Lon-Schka) should be incorporated into the constitution, "by insisting that the Osage government should have no part to play in 'Osage culture,' these elders were ensuring a continued space for their own authorities and practices outside of this centralized governing structure," adding that, from the perspective of many Osage people, "to incorporate any aspect of them into the constitution, or to require participation in them as part of the citizenship requirements, was seen as detrimental to the living quality of these practices."[17] Characterizing Indigenous narratives, phenomenologies, and practices of time as sovereignty, then, is not necessarily an indication that they should be governmentalized. Conversely, though, such a designation does refuse to designate Native frames of reference as cultural rather than political—or to treat them as kinds of *belief* whose significance and efficacy can be bracketed when talking about the materialities of policy, territoriality, and resource distribution. This use of sovereignty to speak of Indigenous experiences of duration seeks to hold open a sense of what peoplehood might mean while tracking the dynamics of temporal interpellation and reorientation enacted via settler form(ul)ations.

Unlike juridical assertions of sovereignty, the characterization of time in these terms does not really speak to expressions of authority or claims to jurisdiction. In fact, many of the examples I address fall outside of legally recognized Native spaces over which conventional modes of sovereignty might be exercised—the effects of the lack of federal acknowledgment on Deborah Miranda's sense of Esselen peoplehood, the displacements in the wake of the Dakota War (including Charles Eastman's growing up off-reservation), the urban Native community of *Indian Killer*, and the lives lived outside the official geographies and genealogies of late nineteenth-century Indian policy in *Gardens in the Dunes*. Rather than referencing something akin to control over those matters conventionally understood as governmental in what are officially designated as Indian spaces, the notion of temporal sovereignty operates as a negative dialectical provocation, suggesting the ways that treating time as singular, neutral, and, thus, a basis for including Native peoples within a shared modernity, may limit possibilities for envisioning and enacting Indigenous self-determination. As argued in chapters 1 and 2, the presumption of shared time can serve as an extension of shared belonging to the nation (itself treated as self-evident) in ways that occlude the potential for engaging with Native peoplehood, except as a (residual) part of national history and jurisdiction. What kinds of collective sensations, perceptions, historical accounts, engagements with nonhuman actors, and the like vanish, are subordinated, or are badly distorted when translated into dominant settler frames of reference? How does treating time as a series of synchronous simultaneities felt in common with non-natives break up lived forms of Indigenous continuity, periodicity, and historicity?

The concept of temporal sovereignty seeks to increase the possibilities for articulating and analyzing ways of experiencing time that do not depend on inclusion within settler modes of experience, backgrounding, and orientation. The scope of Indigenous self-determination encompasses what Coulthard has characterized as Native "mode[s] of life" that do not accord with the parameters and imperatives of settler governance. Such modes are part of what he earlier describes as *grounded normativity*—"the modalities of Indigenous land-connected practices and longstanding experiential knowledge that inform and structure our ethical engagements with the world and our relationships with human and nonhuman others over time."[18] Extending well beyond formal political institutions, such dynamics of being and becoming register the duration of relations with a people's homeland(s) as well as ongoing histories of displacement by non-natives, various life cycles and rhythms (ritual and not), and sustained connections to other-than-human entities and processes. The momentum of such shared experiences and stories gives rise to ways of conceptualizing, perceiving, and inhabiting time that both may be at odds with the chronobiopolitics of settler society and are largely effaced within the chronogeopolitics of settler occupation.

The political character of such feelings lies not in the ways they serve as a system of governance, or even necessarily as a model for one, but in the ways they provide a background through which Native political traditions and normative frameworks emerge as such while also serving as a target for settler aggression, including in the characterization of Native peoples as aberrant or anachronistic populations in need of discipline, aid, or enlightenment. Looking to the United Nations Declaration on the Rights of Indigenous Peoples (UNDRIP), one can see a range of terms that seek to indicate aspects of what might constitute self-determination, many of which do not seem to pertain to government as such.[19] The declaration makes mention of "social structures," "spiritual traditions," "histories," "philosophies," "knowledges," "development," "values," "ethnic identities," "religious" and "intellectual" "property," "oral traditions," and "land tenure systems," among other elements of Indigenous sociality and self-understanding. This broad spectrum of possibilities speaks to how dimensions of Indigenous lifeways "d[o] not always 'translate' into any direct, political statement," and one can understand temporal formations (which incorporate and gain their textures and trajectories from these various dynamics) as occupying a similar status—as important to the continuing sense and character of Indigenous peoplehood without necessarily coming to the fore as formal political procedures or policies.[20] Questions of temporality might bear on specifically juridical questions (such as who has claims to human remains on a particular people's lands that are hundreds or thousands of years old; how to determine laches, or how much time must pass before a set of circumstances or question can no longer be judicially assessed; to what extent and how oral tradition is incorporated into statutory law and judicial reasoning; and what the proper timeframe is for evaluating potential law and policy—years, decades, or generations). However, implementation within Native juridical institutions is not necessarily the horizon toward which an engagement with Indigenous temporalities must move. Like the provisions in UNDRIP, which seek to expand the scope of the political in order to secure Indigenous ways of being against forceful settler interventions and interpellations, the notion of temporal sovereignty seeks to open room for further engaging with how shared ways of experiencing time—continuity, change, and relations among the past, present, and future—help give substance and direction to Native peoplehood.

Such engagement with temporality as a feature of self-determination aims to forestall making settler modes of temporal recognition into the sole framework through which to register Indigenous persistence, existence, adaptation, and transformation. UNDRIP remains indeterminate as to whether all of the processes and forms of social activity toward which it gestures could or should function as part of a formal political apparatus. Rather, the definition of self-determination in these fairly broad terms works to forestall a process of culturalization through which vast swaths of Indigenous world making can be deemed not properly *political*. Like UNDRIP's capacious and institutionally open-ended account of self-determination, the concept of temporal sovereignty works to defer the normalization of settler frames of reference (in terms of experience, chronology, causality, periodization, periodicity, normative life cycle, historical memory, planning, development, etc.), and doing so requires holding open a certain ambiguity in the relation between juridical structures as such and the practices, rhythms, trajectories, and momentum that might constitute Indigenous temporalities, especially in a context in which Native governmental structures continue to be subject to intense settler scrutiny, intervention, and regulation.

While exploring the interpretive potential of more fully attending to temporality as a vector of settler intervention and Indigenous self-determination, I do not seek to present Indigenous temporal formations as utopic or as providing a clear metric for assessing authenticity. Native frames of reference are neither inherently liberatory nor transcendent. They can generate their own questionable normativities and can still be worthy of critique on various grounds. For example, in the representations of temporality addressed in *Beyond Settler Time*, one could take issue with the masculinism of Eastman's and Mathews's formulations, the apparent commitment to U.S. nationalism rather than Dakota political forms in Eastman's writing, what might be considered to be a celebration of large-scale violence in Alexie's novel, and the relative isolationism (and perhaps environmental romanticism) in Silko's text. In offering readings of these various ways of portraying Native time, my goal is less to suggest them as inherently desirable models than to draw on them to sketch possibilities for addressing Indigenous thought, feeling, remembrance, and political imagination that do not fit within dominant settler accounts of time and that may be occluded by the call to understand Native people(s) as inherently occupying a shared time with non-natives.

Reciprocally, none of the various modes of Indigenous duration addressed in this study should be treated as the exclusive basis on which to define or adjudicate *real Indianness*. The issue at play in my analysis is not so much authenticity as ethicality: is force at play here? With respect to Indigenous mappings at play in Native women's writing, Mishuana Goeman indicates that "these women's stories ... are not testaments to geographies that are apart from the dominant constructions of space and time, but instead they are explorations of geographies that sit alongside them and engage with them at every scale."[21] Similarly, the temporal dynamics I am addressing are not somehow sealed off from "dominant constructions" and orientations, yet they remain irreducible to such constructions—existing "alongside" the latter. Rather than suggesting that Indigenous temporal sovereignty indicates an insulation from settler influence or violence, a form of purity in contrast to degraded modes of temporal assimilation, *Beyond Settler Time* presents temporal sovereignty as a hermeneutic for the following: engaging with experiences and ontologies of time that appear only as belief when set against the ostensibly supervening time endorsed by Euramerican institutions, displacing the a priori insistence on an epochal break in Native time (into "the modern"), expanding what counts as historical contextualization and explanation beyond conventional historicist tendencies, addressing the varied forms an individual's life course might take and the ways they arise within collective processes and/or aspirations, engaging with the effectivity and timescales of nonhuman entities (including the land), and accounting for the continuities of settler colonial displacement and occupation. The various intellectual moves I make may or may not facilitate specific initiatives within Native governance, but they do point toward further dimensions of the politics of peoplehood, indicating challenges and alternatives to the translation of Indigenous being and becoming into settler temporal frames.

In suggesting that Indigenous modes of peoplehood might be characterized as temporal sovereignty, I also draw on a queer analytic, or sensibility, in which what constitutes political work remains productively indeterminate.[22] What will turn out to have been politically significant, and by what criteria do we gauge the boundaries and content of the political? How do we account for the roles played by feelings, memories, stories, and all manner of non–legally codified relationships in thinking about the possibilities and dynamics of collective struggles for justice? Lauren Berlant defines *cruel optimism* as occurring "when something you desire is actually an obstacle to your flourishing," "when the object/scene that ignites a sense of possibility actually makes it impossible to attain the expansive transformation for which a person or a people is striving."[23] U.S. Indian policy can be understood as seeking to engender such a relation to forms of state-sanctioned tribal sovereignty, calling on Native people to invest in modes of settler-constrained and settler-regulated governance that can actively impede efforts to pursue broader forms of self-determination. While recognizing that the actual work of Native governments can exceed the expectations of non-native policy makers, one can also inquire as to what other scenes of actual, potential, or remembered (and possibly revived) flourishing exist outside the parameters of juridical structures. As Berlant further asks, "Is th[e] refusal to go through the motions and emotions of fidelity to politics a sign of ethical failure?," and she offers the possibility of "revitaliz[ing] political action" by "valuing political action as the action of not being worn out by politics."[24] While questions of formal governance matter deeply in thinking about Native peoples' survival and well-being, including the possibilities for achieving sustained transformations of state policy, they can also direct attention away from exploring "motions and emotions" not set within the juridical apparatus—collective kinds of affect that may constitute modes of "political action" by virtue of sustaining a felt sense of peoplehood in the face of potentially being "worn out" by the translations, accommodations, and territorializations demanded by settler law and policy in the field of politics as such. Refusing such institutionalizations of Indianness does not necessarily mean repudiating official structures of Native governance, but the political project of rejecting settler efforts to manage the scope and texture of indigeneity can draw on the presence of affective formations that do not align with state imperatives and trajectories.

In this vein a queer commitment to the errant aids in formulating the potential political work performed by temporal sovereignty, which operates at an oblique angle with respect to juridical sovereignty. Elizabeth Freeman suggests that "because we can't know in advance, but only retrospectively if even then, what is queer and what is not, we gather and combine eclectically, dragging a bunch of cultural debris around us and stacking it in idiosyncratic piles," adding, "For queer scholars and activists, this debris includes our incomplete, partial, or otherwise failed transformations of the social field."[25] While the kinds of Indigenous duration I have discussed are not accretions of debris in quite the ways Freeman imagines, they seem "idiosyncratic" when viewed against extant settler orientations, and they might be understood as expressing what Freeman later describes as "the movement time of collective political fantasy."[26] Indigenous experiences of time may appear as oddities—anachronisms, aberrations, irrationalities, anomalies—when they do not line up neatly with dominant forms of chronology, historicism, and perception. As Sara Ahmed observes, "Things seem 'straight' ... when they are 'in line,' which means when they are aligned with other lines," and, reciprocally, the production of a seemingly evident timeline, trajectory, or causal sequence can be interpreted as a function of a "process of alignment" whereby temporal relations follow a "straight" path. She later notes, "Queer orientations are those that put within reach bodies that have been made unreachable by the lines of conventional genealogy," further contending that a "queer commitment" is one that does not "presume that lives have to follow certain lines in order to count as lives, rather than being a commitment to a line of deviation."[27] Relations among bodies that have been displaced, that have been subjected to escalating forms of geopolitical restriction, that occur within explicit and implicit forms of ritual and collective activity, that are linked by nonnuclear modes of desire and care, that are connected across chronologically divided periods, and that are inhabited and animated by the work of prophecy all lie outside "conventional genealogy" and the normative "lines" of settler national time and Indian policy. A queer commitment to notice and engage with such apparent failures to conform to "certain lines" of development or officially recognized Indianness opens up possibilities for acknowledging dynamics that serve as the background for collective modes of political imagination not understood as such by the state. Moreover, such modes of sovereignty are not merely a deviation from the presumptive straight line of national time or the proper lines demarcated by settler legal geographies. Rather, they express forms of Indigenous orientation that have their own immanent integrity.

To speak of Indigenous orientations suggests processes of being and becoming that emerge out of everyday life. They arise from, among other things, memory, storying, collective practices, dynamics of maturation and family formation, modes of inhabitation and connections to place, encounters with law and policy in their quotidian effects, histories of dispossession and opposition to it, and engagements with nonhuman entities of various kinds. Such processes generate durable frames of reference that guide perception while being affected, and shifted, by events in the present. These temporalities need not be understood either as antithetical to or as derived from official institutions of governance. Rather, multiple forms of sovereignty can coexist, and acknowledging this multiplicity is part of engaging with the density of Indigenous social formations. Native juridical structures continue to bear the pressure of recognition, of being intelligible to non-native institutions while also being subject to ongoing forms of state regulation, and the kinds of negotiations occurring within and through such institutions may not facilitate engagement with the felt knowledge of Native people(s), including experiences of time. To what extent do juridical modes of Indigenous sovereignty as they exist within the context of ongoing settler colonialism "presume that lives have to follow certain lines"? What kinds of political traditions, normative frameworks, forms of individual and collective development, ways of narrating and carrying forward the past, and modes of envisioning and birthing the future are effaced by the inscription of such lines or lie outside them? How might struggles for self-determination be enriched and textured, given greater scope and complexity, if such potentials were engaged more fully? These questions are not about displacing or ignoring formal governance but opening further analytical and political room for posing questions with respect to how time bears on questions of sovereignty. The aim is to make visible the presence of other potential trajectories of Indigenous flourishing. In this sense, deferring juridical time enables the exploration of temporal sovereignty as a set of possibilities that exist alongside but are not subordinate to the politics of recognition—possibilities that come into view when one ceases to take settler frames of reference as the necessary background.

# NOTES

### PREFACE

1. This preface serves as more of a sketch than a fully fleshed-out contextualization of my work within existing scholarship. For such references and engagements, see chapter 1.
2. In my exploration of this issue, I owe a particular debt to the dissertation work of Jason Cooke. While his approach is different from mine, his analysis played a crucial role in inspiring my own.
3. Walters, *Talking Indian*, 135.
4. Fabian, *Time and the Other*.

### ONE. *Indigenous Orientations*

1. Here I am alluding to Albert Einstein's theory of special relativity, and I will return to the question of frame of reference later in this chapter.
2. Cordova, *How It Is*, 108.
3. Ahmed, *Queer Phenomenology*, 15.
4. On acceleration as a mode of colonization, see Collins, *Global Palestine*, 79–108.
5. The notion of being "between two worlds" often has been used as a way of characterizing mixed-blood Native people, those who live off-reservation and those who have been educated in primarily white institutions, among other forms of "hybridity." Employed in this way, the phrase tends to focus on "cultural" difference at the expense of attending to ongoing modes of colonial power and its effects on Indigenous people(s), as well as to present Natives as if any exposure to anything non-native led to a fall from a prelapsarian Indian wholeness. However, in *Remember This!* Waziyatawin Angela Wilson observes that the Dakota phrase usually translated as "liv[ing] in two worlds" literally means "being tied to two states of being" or involving "two ways of knowing" (116, 134), and the concept might be recuperated in this sense of referring to modes of being, knowing, and becoming, in contrast to the image of sealed-off spaces of purity.
6. Miranda, *Bad Indians*, xvi. Focused as it is on the complex dynamics of peoplehood over time in relation to the violences of settler occupation, Miranda's text serves as an immensely useful touchstone in thinking through the questions about temporality posed in this chapter.

7. On thinking about Indigenous processes of becoming, rather than about what Indigenous people(s) have *been*, see C. Andersen, *"Métis"*; TallBear, *Native American DNA*.

8. I use the term *Euro-American* to indicate dominant European and U.S. formations, ideologies, practices, and the like. I use the term *Euramerican* to indicate U.S. subjects of European descent—whites.

9. In Theodor Adorno's terms, negative dialectics entails "thinking against thought" in order to address the ways "objects do not go into their concepts without leaving a remainder." *Negative Dialectics*, 5.

10. Turner, *This Is Not*, 30–31.

11. This critical orientation on my part can be understood as partially due to my personal institutional trajectory—I was trained in an English department and have always been professionally located in one. However, I also am trying to be mindful of the ways claims about the direct relevance of my work for Native political struggles can preempt Indigenous formulations and can function as a somewhat self-serving means of legitimizing intellectual work in the academy. See Morgensen, "White Settlers and Indigenous Solidarity"; Sullivan, *Revealing Whiteness*; Weigman, *Object Lessons*.

12. Bruyneel, *Third Space of Sovereignty*, 2, 171. On the ongoing history of representing Native peoples in "taxidermic" ways, see Wakeham, *Taxidermic Signs*.

13. O'Brien, *Firsting and Lasting*, xxii. On the nineteenth-century production of Indians as the past against which to define a settler modernity or futurity, see also Ben-Zvi, "Where Did Red Go?"; Carr, *Inventing the American Primitive*; Conn, *History's Shadow*; Luciano, *Arranging Grief*, 25–118.

14. Barker, *Native Acts*, 20, 28, 223. For other discussions of this problematic within Anglophone settler states, see Den Ouden and O'Brien, *Recognition*; Kauanui, *Hawaiian Blood*; Klopotek, *Recognition Odysseys*; Lawrence, *"Real" Indians and Others*; Lowery, *Lumbee Indians*; Povinelli, *Cunning of Recognition*; Powell, "X-Blood Files."

15. Barker, *Native Acts*, 221.

16. Miranda, *Bad Indians*, xiv, 136.

17. As José Rabasa suggests, "One can rush to demand the recognition of one's history by the state, but one should also ask to what extent one is engaging in a new—perhaps even more nefarious—form of policing the past." *Without History*, 15.

18. Deloria, *Indians in Unexpected Places*, 7, 180. For an account that raises questions about such supposed rural sameness, see Blu, "'Where Do You Stay At?'"

19. Deloria, *Indians in Unexpected Places*, 231.

20. Deloria, *Indians in Unexpected Places*, 232–33.

21. Deloria, *Indians in Unexpected Places*, 232.

22. For discussion of the complex and largely unacknowledged entwinements of chronology (time as succession), periodization (division of time into units), and specific kinds of cultural content (a unit of time as itself illustrating a certain way of experiencing time) that swirl around the notion of the modern, see Osborne, *Politics of Time*.

23. O'Brien, *Firsting and Lasting*, xxiii.

24. In addressing Native people's participation in the modern, Deloria suggests that it gave rise to "*other* forms of indigenous creativity, specifically, the modern development of the idea of Indian political nationhood," adding, "In the twentieth century, the memory of indepen-

dence helped give birth to the dream that began to take shape under the name *sovereignty*." *Indians in Unexpected Places*, 234–35. Here, Indigenous sovereignty as a concept and practice emerges out of Native peoples' engagement with, and opposition to, non-native invasion and superintendence. In this way, expressions of "political nationhood" are part of Native participation in modernity. For similar arguments, see Konkle, *Writing Indian Nations*; Lyons, *X-Marks*. Yet what happens to other modes of Indigenous collectivity that do not necessarily look like liberal nation-state-hood? Are they, then, anachronisms? Must they be understood as necessarily receding before this particular modern expression of sovereignty? How are forms of juridical sovereignty mediated by other, nonmodern, dynamics? What other possibilities for Indigenous self-determination exist that are not equivalent to *modern* "political nationhood"? The equation of sovereignty/self-determination with a particular kind of statist structure and the understanding of that structure as an expression of entry into the modern can treat Native political expression as inherently dependent on participation in a temporal framework that is overdetermined by settler interests and imperatives. I seek to suggest how Native being and becoming can be understood as responsive to non-native presence and pressures while not necessarily participating in a shared formation.

25. Das, *Life and Words*, 177.
26. Deloria, *Indians in Unexpected Places*, 180.
27. Gaonkar, "On Alternative Modernities," 1, 2–9, 14–15.
28. Mignolo, *Local Histories/Global Designs*, 37.
29. Wynter, "1492," 13.
30. Rabasa, *Without History*, 136, 175. See also Escobar, *Territories of Difference*. Rabasa, however, does endorse the "undeniable fact" of "the shared temporality that makes all cultures contemporary" (160), and while challenging the anachronization of peoples encountered in the present, this assertion of contemporaneity seems to underdevelop the insights about temporal multiplicity Rabasa offers. I will develop this point further in the next section.
31. Ahmed, *Queer Phenomenology*, 31, 37–38.
32. Chakrabarty, *Provincializing Europe*, 71, 31. On the positing of such transition as the basis for continued state oversight, surveillance, and monitoring, supposedly in order to move toward eventual entry into the full personhood of freedom, see Nguyen, *Gift of Freedom*. On the ways the appearance of time as singular relies on processes of translation, see also Lim, *Translating Time*.
33. Lyons, *X-Marks*, 3, 9–10, 32–33.
34. Throughout Lyons's text, *modernization* serves as a term for marking the effort to adapt to changing circumstances, but Native movements or social formations that do not fit within a particular conception of the modern are understood as resisting change per se (as "cultural" rather than "political," or as the *ethnie* rather than the "nation"). For example, he argues, "Tecumseh and Tenskwatawa (to mention only two of a long and esteemed line of cultural resisters) said No to the imposition of cultural and political dominance, but they didn't say Yes to modernity. Much to the contrary, they and others like them rejected modernity precisely to keep their existing way of life intact." *X-Marks*, 121.
35. Mehta, *Liberalism and Empire*, 49, 63.
36. Denise Ferreira da Silva (*Toward a Global Idea of Race*) argues that processes of racialization position people of color as "affectable" others who continually must demonstrate their

humanity and capacity for being self-determining, where the implicit model of such selfhood is whiteness. Similarly, one might consider the way Native peoples are positioned as temporally affectable others whose capacity to be seen as self-determining depends on demonstrating their presentness, where the implicit model for such is dominant non-native modes of sociality.

37. O'Brien, *Firsting and Lasting*, 123, 118.

38. Coulthard, *Red Skin, White Masks*, 3, 14.

39. Simpson, *Mohawk Interruptus*, 11, 22. For further discussion of the ways contemporary modes of recognition by settler states can work as a means of intensifying superintendence and fortifying the presumption of settler governments' right to oversee Indigenous peoples as domestic subjects, see Barker, *Native Acts*; Engle, *Elusive Promise*; Million, *Therapeutic Nations*; Povinelli, *Cunning of Recognition*. Such recognition, especially when indigeneity is cast as a cultural or racial residue, can be understood as rendering Indianness as a kind of disability. See Samuels, *Fantasies of Identification*; Senier and Barker, "Introduction."

40. Simpson, *Mohawk Interruptus*, 158. Simpson says of contemporary Native "feeling citizenships" that "part of their citizenship and political consciousness stems from another time, a past that is very much alive in the present and a past that gets pushed forward into the present" (187). This relation between the past, present, and future clearly entails ways of experiencing and existing in time that are not equivalent to those of the non-natives who seek to impose modes of recognition.

41. Miranda, *Bad Indians*, 77.

42. On the Esselens' lack of federal acknowledgment, see Laverty, "Ohlone/Costanona-Esselen Nation."

43. Mehta, *Liberalism and Empire*, 49.

44. Chakrabarty, *Provincializing Europe*. In *The Erotic Life of Racism*, Sharon Holland argues against the tendency to attach certain histories to particular kinds of bodies, specifically appending slavery and its legacies to black persons as opposed to understanding those legacies as shaping conditions of life and feeling for everyone. Bearing in mind this caution, I'm not suggesting that settler colonialism can be understood as affecting Natives rather than non-natives. Rather, I'm suggesting that experiences of time within settler colonial formations may be discrepant: the idea that all are affected by settler colonialism need not require understanding everyone as participating in a shared time.

45. Lyons, *X-Marks*, 13.

46. Lim, *Translating Time*, 84.

47. As Peter Osborne (*Politics of Time*) argues, the idea of participation in a periodizing formation like "modernity" (itself understood as unevenly geographically distributed around the globe, with certain populations entering "later" than others) is not the same as the "cosmic time" or "natural time" of chronology—the idea that everyone occupies the same now. However, as I have been suggesting, the notion of mutuality in the present (a shared now) is not separate from positioning such supposed natural jointness against a presumptively shared background (such as the modern) that makes temporal unity and movement meaningful.

48. In thinking about temporal multiplicity as varied trajectories, I am drawing on Doreen Massey's reformulation of space "as the dimension of multiple trajectories, a simultaneity of stories-so-far." *For Space*, 24. She suggests that the same space might be criss-crossed by itin-

eraries and becomings (of persons, communities, nonhuman entities, institutions, etc.) that intersect and influence each other. By virtue of doing so, however, they do not become a single story, in which all events belong to "the stately progress towards modernity/modernisation/development on the Euro-Western model," which denies "the possibility of other trajectories." *For Space*, 70. However, as I will suggest, the idea of simultaneity is not itself a meaningful concept outside those trajectories. For a differently configured critique of Massey that is also processual in character and concerned with what often is characterized as cultural transmission, see Ingold, *Being Alive*, 145–55.

49. Cordova, *How It Is*, 70.

50. West-Pavlov, *Temporalities*, 3. It should be noted that West-Pavlov's analysis of time itself emerges out of consideration of Indigenous temporalities in Australia.

51. West-Pavlov, *Temporalities*, 51, 176.

52. Such ways of understanding time can be characterized as "relational," as opposed to "substantivalist" theories that present time as having an existence separate from its contents. See Dainton, *Time and Space*, 1–12. On recent efforts within physics to conceptualize completely relationalist accounts of space and time, see Smolin, *Three Roads to Quantum Gravity*.

53. For examples of such taxonomies, see V. Deloria, *God Is Red*; Fixico, *American Indian Mind*. For critiques of such accounts, see Lyons, *X-Marks*, 13–21; Turner, *This Is Not*, 101–4.

54. Fabian, *Time and the Other*, 42. In *Taxidermic Signs* Pauline Wakeham observes that "the denial of coevalness is reinvented and reinscribed through various forms of time-lagging and time-warping that continue to deny the 'active occupation' of shared time between Euro-North American society and the figure of the aboriginal" (17–18).

55. Fabian addresses what he calls the "conditions of intersubjective knowledge," asserting, "*Somehow we must be able to share each other's past in order to be knowingly in each other's present.*" *Time and the Other*, 92. However, people may engage with each other in meaningful ways while contextualizing and interpreting each other's actions within divergent systems of meaning, so that one can recognize that someone with whom one is engaging cannot be understood as indicative of one's past (of a less developed social stage, for example) without claiming that everyone occupies the *same* present. As Lim asks, "Does shared time—that time of coevalness, encounter, and intersubjective exchange—presuppose only one time? Can multiple temporalities be shared?" *Translating Time*, 248.

56. As Mimi Thi Nguyen observes, "the invitation to coevality also imposes violence—including a politics of comparison, homogenous time, and other commensurabilities—through the intervention (a war, or development) that rescues history for those peoples stalled or suspended in time." *Gift of Freedom*, 17.

57. Miranda, *Bad Indians*, xiv.

58. Chakrabarty, *Provincializing Europe*, 73–74. On "homogeneous empty time," see also Osborne, *Politics of Time*, 138–59.

59. The idea of a shared future also serves as a means for settlers to stage a renunciation and transcendence of past wrongs in ways that engage fully with neither Native sovereignties nor Indigenous peoples' continuing orientation by those histories. See Ahmed, *Cultural Politics of Emotion*; Coulthard, *Red Skin, White Masks*; Kauanui, "A Sorry State"; Million, *Therapeutic Nations*; Povinelli, *Cunning of Recognition*; Reilly, "Sovereign Apologies."

60. While Fabian (*Time and the Other*) observes that "the substance of a theory of coevalness ... will have to be the result of actual confrontation with the Time of the Other" (153), the "Time of the Other" here does not shift the existence of a clearly shared "now" formulated in terms of the linear progression of "natural" time. In *The Anthropology of Time* Alfred Gell seeks to dispel the idea of "unfamiliar, exotic, temporal worlds" (4), emphasizing the existence of forms of "non-transcendental relativism" that account for "inter-cultural differences in beliefs, attitudes, and values within one encompassing reality" (69). However, he also notes the existence of "socially established periodicities"—"a set of collective representation[s] of social and natural processes" (35–36)—and later observes, "Homogeneous duration, outside the technical or laboratory context, is a myth": "Practical time is non-homogeneous because any given stretch of duration is cognitively salient only in conjunction with socially relevant processes, governed by a scheme of expectations" (108). If periodicity, "practical time," cognition, perception, and social relevance are all dependent on extant "cultural" structures, meaning they are affected by collective patterns and are not necessarily consistent across groups, what precisely is the "one encompassing reality" of which Gell speaks?

61. My turn here to Albert Einstein and Henri Bergson is due to the specific value of their ways of rethinking time for opening up possibilities for conceptualizing temporal multiplicity, but also to the ways these figures are quintessentially *modern*. I draw on them as part of developing an immanent critique through which to unthink the supposed simultaneity of modernity. On immanent critique, see Adorno, *Negative Dialectics*.

62. On the problems for simultaneity posed by special relativity, see also Canales, *Physicist and Philosopher*; Dainton, *Time and Space*, 311–42; Greene, *Fabric of the Cosmos*, 39–76; Kern, *Culture of Time*, 10–35; Mermin, *It's About Time*; Savitt, "Time in the Special Theory of Relativity"; Smolin, *Time Reborn*, 54–75.

63. Galison, *Einstein's Clocks, Poincaré's Maps*, 13, 254–55.

64. However, as N. David Mermin observes, "how do you know that a frame of reference is inertial? This is just another way of posing the deep question of how you know motion is uniform. It would appear that you have to be given at least one inertial frame of reference to begin with, since otherwise you can ask[,] 'Moving uniformly with respect to what?'" *It's About Time*, 4.

65. Dainton, "Time, Passage, and Immediate Experience," 322.

66. Savitt, "Time in the Special Theory of Relativity," 561.

67. Mermin, *It's About Time*, 63.

68. The constancy of the speed of light plays a central role in displacing the Newtonian conception of absolute space and absolute time. If absolute space exists as an actual part of the physical universe, it should play some role in affecting movement. Its presence is marked by the use of the term *ether* (or *aether*). However, if the ether has no effect on physical processes, in what way can it be said to exist? Following from this question, if the velocity of light is constant and is unaffected by the relative motion of its source (light from a flashlight on a moving train does not move any faster than light from a flashlight at the station), than the movement of the source with respect to the ether (its supposed shift of position in relation to absolute space) is irrelevant to determining the speed of light. Thus, the constancy of light suggests that there is no ether, since it has no material effect on light. Returning to the example of the train and the station, if there is no absolute space to provide a universal

framework against which to assess movement, then there is no objective standard through which to adjudicate the differing accounts of simultaneity offered by those people on the train and those at the station. Without some basis for adjudicating between them, or for measuring their distance from some third objective standard, both accounts of simultaneity are equally valid, such that there can be no absolute time, either.

69. Green, *Fabric of the Cosmos*, 49; Dainton, "Time, Passage, and Immediate Experience," 323.

70. As Bruno Latour argues in "A Relativistic Account of Einstein's Relativity," the triangulation of different perspectives through reference to the constancy of the speed of light allows all the "reports" from varied frames of reference to be "superimposable" into a coherent account (14), creating a "transversal link that allows all frames, no matter how unstable and pliable, to be *aligned*" (17). Hermann Minkowski is responsible for transforming Einstein's insights about light and simultaneity into the notion of a four-dimensional manifold of spacetime, and this conception of spacetime served as the basis for Einstein's development of general relativity, which includes the effects of acceleration/gravitation (the warping of spacetime by matter) not present in special relativity.

71. Bergson, *Time and Free Will*, 64, 76, 105. On the debate between Einstein and Bergson, see Canales, *Physicist and Philosopher*. On Bergson's broader cultural significance, see also Deleuze, *Bergsonism*; Guerlac, *Thinking in Time*; Lim, *Translating Time*; Massumi, *Parables for the Virtual*. For recent discussions of the implications of a block conception of spacetime for considerations of human perceptions of time, see Dainton, "Time, Passage, and Immediate Experience"; Ismael, "Temporal Experience"; Savitt, "Time in the Special Theory of Relativity."

72. In *Time and Free Will* Bergson tends to treat space as homogeneous, in contrast to the heterogeneity of time. Massey (*For Space*, 20–48) critiques him on this score, instead arguing for the multiplicity and dynamism of space. However, Bergson also acknowledges that the abstraction of space as a divisible quantity may not be the actuality of the physical world, especially in human experience ("We shall not lay too much stress on the question of the absolute reality of space: perhaps we might as well ask whether space is or is not in space" [91]). In *Matter and Memory* (246–91), he extends the dynamics of qualitative relation and multiplicity raised previously with respect to human consciousness and psychology to the external world, or what commonly is referred to as space. See also Guerlac, *Thinking in Time*, 106–72.

73. Some accounts seek to avoid this problem by simply insisting that the geometry of Minkowskian spacetime is the shape of physical reality and that the coordinate systems and commensurations among them used in mathematically negotiating the Minkowskian grid directly reflect the real. For an example, see Maudlin, *Philosophy of Physics*.

74. Bergson, *Time and Free Will*, 115, 120. As Lee Smolin suggests in *Time Reborn*, "It's assumed that some method can give the whole set of possible configurations ahead of time—that is, before we watch the actual evolution of the system. The possible configurations do not evolve, they simply are. A second assumption is that the forces, and hence the laws the system is subject to, are timeless. They don't change in time, and they also presumably can be specified ahead of the actual study of the system" (44).

75. Bergson, *Matter and Memory*, 250, 290. Smolin notes in *Time Reborn*, "The mathematical representation of a motion as a curve does not imply that the motion is in any way identical to the representation" (34). On this problem, see also Massumi, *Parables for the Virtual*.

76. Miranda, *Bad Indians*, xvi.

77. For other philosophical, largely phenomenological, critiques of the idea of the "punctal" present (of an isolatable point in time that can serve as the universal reference point for *now*) see H. Andersen, "Development of the 'Specious Present'"; Arista and Lloyd, "Subjective Time"; Gallagher, "Time in Action"; Ismael, "Temporal Experience"; Phillips, "Temporal Structure of Experience"; Trigg, *Memory of Place*. One can also understand the notion of quantified space (and the possibility of an isolated geometric point) as equally an abstraction. In addition to the previous sources, see Casey, "How to Get from Space to Place"; Lefebvre, *Production of Space*; Marratto, *Intercorporeal Self*.

78. Smolin, *Time Reborn*, 103–4, 116. Smolin, though, also offers an account of time as "a succession of moments" (xiv), and in his efforts in the epilogue to think about the need to address the impending catastrophe of climate change, his singularization of time results in deeply conventional and disturbing claims about Native peoples in the Americas as well as the character of human progress (253–58).

79. One could argue that if one wants to use relativity (special and general) to challenge notions of simultaneity, one has to accept the substantivalist account of space and time—as a unified, four-dimensional spacetime—that relativity offers, and that relational ideas of time as immanently emerging through motion and change (like those offered by Cordova [*How It Is*] and West-Pavlov [*Temporalities*]) need to be suspended. However, beyond the challenges of Bergson and others to the mathematization of space and time within physics, there are possibilities within theorizations of spacetime through which one might reconcile these two approaches. One way is to argue that time is essentially relational and emergent at a fundamental level since it arises out of interactions among infinitesimal subatomic particles (whose dynamics along with the lines of force that connect them also create space) and that quality of relation also pertains at larger scales. Another approach is to suggest that there are forms of, in Savitt's terms, "local becoming" that occur along an entity's worldline, or the path it traces within the four-dimensional spacetime manifold. See Dainton, *Time and Space*, 368–86; Savitt, "Time in the Special Theory of Relativity"; Smolin, *Three Roads to Quantum Gravity*.

80. On the ways dominant temporal frames of reference efface the experiences of those oppressed by them, see Nixon, *Slow Violence*; Povinelli, *Economies of Abandonment*.

81. Cordova, *How It Is*, 49, 62. As Simpson (*Mohawk Interruptus*), Barker (*Native Acts*), Sandy Grande (*Red Pedagogy*), Karen Engle (*Elusive Promise*), and others have noted, "culture" can serve as a way of deferring discussion of Native sovereignty and self-determination by presenting Indigenous peoples as not really political, as subnational entities whose cultural identity makes them internal to the (settler) state. However, Cordova does not use "culture" in these terms, instead seeking to mark the coherence of peoplehood itself in ways that allow for talking about variation among lifeworlds.

82. Chakrabarty, *Provincializing Europe*, 83.

83. Turner, *This Is Not*, 81.

84. As Darieck Scott suggests of forms of black abjection, one can "examine [the] deleterious effects [of racism] not only for the purpose of demonstrating their injurious outcomes but to see how the effects, indeed the injuries themselves, may themselves be tools that can be used either to model or to serve as a means of political transformation." *Extravagant Abjection*, 9.

Here, Scott seeks to suggest how experiences of subjection become part of reenvisioning possibilities for being in the world, rather than simply marking exclusions from a presumptive norm of proper, generic personhood. See also Allewaert, *Ariel's Ecology*; Cervenak, *Wandering*; Holland, *Erotic Life of Racism*; Keeling, *Witch's Flight*; Weheliye, *Habeas Viscus*.

85. Miranda, *Bad Indians*, 77, 34.

86. Miranda, *Bad Indians*, xiv.

87. Das, *Life and Words*, 215.

88. Such invention also can include the emergence of new forms of Indigenous peoplehood in the wake of Euro-American contact. For examples, see C. Andersen, *"Métis"*; G. Anderson, *Indian Southwest, 1580–1830*; Lowery, *Lumbee Indians*; Shepherd, *We Are*. Those processes of becoming also need not be understood as categorically different from the ways in which new forms of peoplehood emerged before such contact.

89. Merleau-Ponty, *Phenomenology of Perception*, 9, 125. Merleau-Ponty's conceptualization of perception as a holistic engagement with a horizon of possibility based on the potential for action profoundly resonates with Bergson's analysis in *Matter and Memory*. On Merleau-Ponty's sustained engagement with Bergson's philosophy, see Guerlac, *Thinking in Time*, 5, 173.

90. Merleau-Ponty, *Phenomenology of Perception*, 159, 483, 509, 249, 277. Chris Andersen's account of "density" in "From Difference to Density" colors my thinking about the meaning and implications of Merleau-Ponty's notion of "historical density."

91. Das, *Life and Words*, 97, 100.

92. Trigg, *Memory of Place*, 11, 17, 106.

93. Merleau-Ponty, *Phenomenology of Perception*, 278. On the role of the future as a basic part of perception that exerts a causal pressure on the present, see Gallagher, "Time in Action"; Gallagher and Zahavi, "Primal Impression and Enactive Perception"; Kohn, *How Forests Think*.

94. On how collective affective tendencies emerge through the entraining of individual sensory capacities, see Feld and Basso, *Senses of Place*; Gallagher, *How the Body Shapes the Mind*; Keeling, *Witch's Flight*; Seremetakis, "Memory of the Senses"; Protevi, *Political Affect*.

95. On the role of spirit, noncorporeal entities, and ostensibly nonanimate entities as acting in the world, and therefore helping produce history and causality, see Boyd and Thrush, "Introduction"; Nabokov, *Forest of Time*; Povinelli, *Cunning of Recognition*; Stevenson, *Life beside Itself*; Thornton, *Being and Place*; Wilson, *Remember This!* On the linguistics and politics of attributions of animacy, see Chen, *Animacies*. On the importance of an expansive understanding of what can constitute an agent in sociohistorical modes of explanation, see Latour, *Reassembling the Social*.

96. Merleau-Ponty, *Phenomenology of Perception*, 460.

97. Ahmed, *Queer Phenomenology*, 38.

98. Turner, *This Is Not*, 98.

99. Mehta, *Liberalism and Empire*, 49.

100. Smith, *Everything You Know*, 85.

101. While I'm focusing here on Indigenous temporalities, I also should note that the violent claiming of the land by settlers and the justifications offered for doing so orient settler experiences of time, in ways that direct attention away from enduring Indigenous presence

and sovereignties. In earlier work I've sought to conceptualize the regularity of such orientations as "settler common sense." See Rifkin, *Settler Common Sense*.

102. For a discussion of the phenomenological implications of land loss for Native people(s), see Palmer, "Devil in the Details." I should be clear, though, that I am not seeking to present mobility as the opposite of an authentic Indigenous rootedness. On forms of mobility, chosen and not, as modes of Indigenous being and becoming, see Clifford, *Returns*; Goeman, *Mark My Words*; Povinelli, *Labor's Lot*; Ramirez, *Native Hubs*; Vizenor, *Fugitive Poses*.

103. Miranda, *Bad Indians*, 202.

104. Simpson, *Mohawk Interruptus*, 22.

105. This approach also goes beyond the notion of culture, given the ways that concept gets employed to signal static ways of being and subpolitical forms of *difference* that can be managed within the overarching multicultural governmentality of the settler state.

106. Cordova, *How It Is*, 70.

107. On the notion of "resurgence" as a way of naming that process, see L. Simpson, *Dancing on Our Turtle's Back*. Highlighting the (re)generation of stability works to reorient the emphasis on the new within affect studies. For discussion of this tendency, see B. Anderson, "Modulating the Excess of Affect"; T. Dean, "Bareback Time." For examples, see Bennett, *Enchantment of Modern Life*; Grosz, *Time Travels*; Seigworth and Gregg, "Inventory of Shimmers."

108. Miranda, *Bad Indians*, 135.

109. On the emergence of patterns of regularity within complex systems that then constrain and shape the functioning of those systems, see Ingold, *Being Alive*; Mitchell, *Complexity*; Protevi, *Political Affect*.

110. Mehta, *Liberalism and Empire*, 63; Chakrabarty, *Provincializing Europe*, 71.

111. My way of approaching the effects of experience across generations differs from notions of "prosthetic memory" or "postmemory." In the former the emphasis lies on interaction with mass-mediated representations as a way of experiencing the past, and the latter addresses feelings and memories as they are transmitted from one generation (those who lived through the events) to another generation. See Hirsch, *Generation of Postmemory*; Landsberg, *Prosthetic Memory*. Rather than marking a clear distinction between actual memory of events and that which has been mediated in some fashion, I want to suggest that historical processes and patterns provide forms of collective orientation that help shape ordinary perception.

112. Barker, *Native Acts*, 223; Lyons, *X-Marks*.

113. Miranda, *Bad Indians*, xi.

114. On story as a verb, see Basso, *Wisdom Sits in Places*; Cruikshank, *Social Life of Stories*; Doerfler, Sinclair, and Stark, *Centering Anishinaabeg Studies*; Goeman, *Mark My Words*; King, *Truth about Stories*; B. Miller, *Oral History on Trial*; Sarris, *Keeping Slug Woman Alive*; Shorter, *We Will Dance*; L. Simpson, *Dancing on Our Turtle's Back*; Wilson, *Remember This!* In this vein Tim Ingold suggests in his discussion of "tools" that "things *are* their stories" and "the functions of things are not attributes but narratives": "the functions of tools, like the meanings of stories, are recognised through the alignment of present circumstances with the conjunctions of the past." *Being Alive*, 56–57. He later argues with respect to story that it is "the retracing of a path through the terrain of lived experience," that "it is not passed on as a compendium of information from one generation to the next, but rather subsists in the current of life and consciousness" as part of "a continuous process of becoming" (161).

115. Cruikshank, *Social Life of Stories*, xv, 41.

116. Wilson, *Remember This!*, 27.

117. Merleau-Ponty, *Phenomenology of Perception*, 249.

118. Wilson, *Remember This!*, 95–96. While Wilson largely addresses Dakota stories as a specific body of knowledge held by those trained to recall and interpret them, her conclusions about the significance and work of story for quotidian modes of Indigenous peoplehood can be extended to include less specialized kinds of recollection, circulation, and deployment. On Native oral historians as themselves specialized knowledge producers with expertise that should be acknowledged as such by non-natives (including in courts), see also B. Miller, *Oral History on Trial*.

119. Million, "There Is a River in Me," 31–32.

120. Miranda, *Bad Indians*, 203.

121. Das, *Life and Words*, 104.

122. Miranda, *Bad Indians*, 74. In addition to addressing nonnuclear forms of relation and transmission of stories, knowledge, and belonging, Miranda explores the historical role of people who, from a Euramerican perspective, would be understood as gender nonconforming (31–32). See also Miranda, "Extermination of the *Joyas*."

123. Halberstam, *In a Queer Time*, 1–4.

124. The desire to fit into exactly such a model of maturation and personal development, except with same-sex partners rather than different-sex ones, has been characterized as "homonormativity." See Duggan, "New Homonormativity"; Manalansan, "Race, Violence"; Puar, *Terrorist Assemblages*.

125. Freeman, *Time Binds*, 3, 5.

126. Luciano, *Arranging Grief*, 9.

127. See Ahmed, *Queer Phenomenology*, 65–108.

128. For "straight time," see Rohy, *Anachronism and Its Others*. In a different vein, see also Cervenak, *Wandering*.

129. See Barker, *Native Acts*; Brant, *Writing as Witness*; Denetdale, "Chairmen, Presidents, and Princesses"; Miranda "Extermination of the *Joyas*"; Piatote, *Domestic Subjects*; Rifkin, *When Did Indians*. On the effort to cast Native identities and genealogies as the biological transmission of DNA, see TallBear, *Native American DNA*.

130. Dinshaw et al., "Theorizing," 185.

131. Rohy, *Anachronism and Its Others*, 129; Goldberg and Menon, "Queering History," 1609. On this notion of a transcendence of or break with the past as central to narratives of modernity, see Cheah, *Inhuman Conditions*; Fritzsche, *Stranded in the Present*; Latour, *We Have Never Been Modern*; Osborne, *Politics of Time*.

132. Goldberg and Menon, "Queering History," 1616; Rohy, *Anachronism and Its Others*, 130.

133. Luciano, *Arranging Grief*, 47, 82. On this point, see also Ben-Zvi, "Where Did Red Go?"; Conn, *History's Shadow*; O'Brien, *Firsting and Lasting*.

134. Cordova, *How It Is*, 70.

135. Miranda, *Bad Indians*, 177.

136. Freeman, *Time Binds*, xxi, 63, xiii. For a differently configured but complementary effort to think about emotional investments in such futures that might have been, and might still be, see Muñoz, *Cruising Utopia*. In another context, David Scott (*Omens of Adversity*)

has referred to such unrealized potentials as "former futures," although he offers a far less optimistic sense of their significance for contemporary political imaginings and movements.

137. Freccero, *Queer/Early/Modern*, 80.

138. Christopher Nealon (*Foundlings*) illustrates how white sexually nonnormative people's "overwhelming desire to *feel historical*" often has taken the form of "analogies that equate an inarticulate homosexuality with a distant racial, ethnic, tribal, or national form" (8, 11), including seeing sexual nonnormativity as a kind of primitivism that aligns queers with Native people(s). See also Morgensen, *Spaces between Us*; Roscoe, *Changing Ones*; Ross, "Beyond the Closet." On the ways (queer) desire cannot be understood as exterior to histories of racism, see Holland, *Erotic Life of Racism*.

139. Brant, *Writing as Witness*, 45.

140. On the potential disjunction between queer political and intellectual projects and the concerns of Indigenous peoples, see Driskill, "Doubleweaving Two-Spirit Critiques"; Driskill et al., *Queer Indigenous Studies*; Morgensen, *Spaces between Us*; Rifkin, *When Did Indians*. On the relation between Indigenous sexuality and sovereignty, in addition to the previous sources, see Akiwenzie-Damm, "Erotic, Indigenous Style"; Denetdale, "Chairmen, Presidents, and Princesses"; Gould, "Disobedience"; Justice, "Notes toward a Theory of Anomaly"; Miranda, "Dildos, Hummingbirds"; Rifkin, *Erotics of Sovereignty*; Tatonetti, *Queerness of Native American Literature*.

141. Ross, "Beyond the Closet," 175, 173. See also Cohen, "Punks, Bulldaggers, and Welfare Queens"; R. Ferguson, *Aberrations in Black*; Roscoe, *Changing Ones*; Somerville, *Queering the Color Line*.

142. See Hartman and Wilderson, "Position of the Unthought"; Weheliye, *Habeas Viscus*; Wilderson, *Red, White, and Black*. For challenges to such an account of black history and experience, see Holland, *Erotic Life of Racism*; Warren, *What Was African American Literature?*; Wright, *Physics of Blackness*. For a discussion of notions of "Afro-modernity," see Hanchard, "Afro-Modernity."

143. Byrd, *Transit of Empire*.

144. On the phenomenology of whiteness, see Sullivan, *Revealing Whiteness*.

145. Allen, *Republic in Time*, 11.

146. Pratt, *Archives of American Time*, 3, 156.

147. Pratt insists on the "hybrid" character of experiences of time in the United States, but what engenders relation among these experiences is their mutual participation in "the heterochronic time of modernity," with whatever modernity might designate providing a regularity and orientation amid these other differences. *Archives of American Time*, 195–96, 199.

148. Miranda, *Bad Indians*, 132.

149. Miranda, *Bad Indians*, 203.

150. Blaeser, "Wild Rice Rights," 241.

151. While scholars sometimes distinguish Indigenous "history" from "myth," noting that there can be a distinction between these two modes of storying in Indigenous languages, they are not necessarily distinct from each other in perceiving the landscape and its potentials. See Basso, *Wisdom Sits in Places*; Carlson, *Power of Place*; Doerfler, Sinclair, and Stark, *Centering Anishinaabeg Studies*; Nabokov, *Forest of Time*; Ortiz, *Tewa World*; Povinelli, *Cunning of Recognition*; Rappaport, *Politics of Memory*; Shorter, *We Will Dance*; Thornton, *Being and Place*.

152. Goeman, *Mark My Words*, 34.

153. However, as Evelyn Peters and Chris Andersen caution in their introduction to *Indigenous in the City*, "privileging a connection to ancestral homelands as a marker of Indigenous identity reinforces dominant visions of Indigenous people as authentic only if they live in remote areas and engage in 'traditional' lifestyles or, conversely, only if we assume that these homelands are located exclusively in such areas": "the preoccupation with an Indigenous relationship to land has deflected attention away from an understanding of the ways that many contemporary Indigenous people express their identities in contemporary urban settlements" (8–9). Yet landed storying can also involve shifting locations, histories of migration (chosen and coerced), and remembered relations to prior places of inhabitance that do not invalidate one's relation to where one is now.

154. Basso, *Wisdom Sits in Places*, 32, 101. On the relation between Native narrations of time and particular relations to place, see also B. Miller, *Oral History on Trial*; Ortiz, *Tewa World*; Rappaport, *Politics of Memory*; Shorter, *We Will Dance*; Thornton, *Being and Place*.

155. Cruikshank, *Social Life of Stories*, 41.

156. Doerfler, Sinclair, and Stark, "*Bagijige*," xxiii. See also Garroutte and Westcott, "Story Is a Living Being."

157. Goeman, *Mark My Words*, 24, 34, 207.

158. In this way, such texts move beyond the role of serving as a version of the *native informant*, instead performing a refusal (in Audra Simpson's terms) of settler temporal intelligibility. On the native informant, see Spivak, *Critique of Postcolonial Reason*.

159. Sedgwick, *Touching Feeling*.

TWO. *The Silence of Ely S. Parker*

1. While incited by my thoughts after viewing *Lincoln*, this chapter was also inspired by C. Joseph Genetin-Pilawa's reimagining of Parker's career in *Crooked Paths to Allotment*.

2. See Armstrong, *Warrior in Two Camps*.

3. Kushner, *Lincoln*, xvi. For quotations from the film, I refer to the published screenplay by Tony Kushner.

4. Kushner, *Lincoln*, 91, 140, 157.

5. My discussion of the film's storying of U.S. history can be differentiated from what Alison Landsberg has termed "prosthetic memory," which she defines as the ways "cinema and other mass cultural technologies have the capacity to create shared social frameworks for people who inhabit ... different social spaces, practices, and beliefs": "With prosthetic memory, ... people are invited to take on memories of a past through which they did not live." *Prosthetic Memory*, 8. Instead, I seek to suggest how mass-mediated narratives emerge from and gain meaning within everyday forms of sense making in which dominant impressions of the inevitable, encompassing presence of the nation-state shape experiences of time and history. The film expresses in mass-mediated form a much more mundane dynamic in which the continual renarration of certain events as central to the nation's history and becoming helps (re)generate quotidian forms of temporal orientation—situating one's experience within a particular historical trajectory that influences what appears possible now and in the future.

6. Merleau-Ponty, *Phenomenology of Perception*.

7. Thanks to one of the anonymous readers of the essay version of this chapter for highlighting this point. On the work of multicultural forms of racial inclusion, and the ways they continue to animate modes of erasure and destitution, see Cacho, *Social Death*; Chow, *Protestant Ethnic*; R. Ferguson, *Reorder of Things*; J. Lee, *Urban Triage*; Melamed, *Represent and Destroy*; Nguyen, *Gift of Freedom*; Nyong'o, *Amalgamation Waltz*; Reddy, *Freedom with Violence*. On the ways the Civil War serves as a normative frame for narrating U.S. history, see Blight, *American Oracle*; Hogan, *Lincoln, Inc.*; Kelman, *Misplaced Massacre*; Schwartz, *Abraham Lincoln*. For recent scholarship that performs this framing, see Finkelman, "Lincoln"; Guelzo, *Fateful Lightning*; Holzer, *Emancipating Lincoln*; Neely, *Lincoln*; Witt, *Lincoln's Code*.

8. Agamben, *State of Exception*, 1. He characterizes the transformation of such subjects into legitimate objects of state aggression as their reduction to "bare life." On the limits of Agamben's notion of bare life as a way of framing forms of institutionally generated abjection, see Povinelli, *Economies of Abandonment*; Weheliye, *Habeas Viscus*.

9. In this sense the film, and this way of envisioning the Civil War and national temporality, does not allow for the possibility of "tragedy" in David Scott's sense. *Lincoln* illustrates a "confidence in history's promises" that cannot engage with "the nature of human action in time ... [in] its boundlessness, its unpredictability, its uncertainty, its unreliability," such that time itself appears to inevitably follow a certain narrative trajectory that confirms a particular moral order. See Scott, *Omens of Adversity*, 44, 62.

10. On the failures and inaccuracies of such a narrative with respect to African American freedom, see Blight, *Race and Reunion*; Dayan, *Law Is a White Dog*; Hartman, *Scenes of Subjection*; McCurry, "War, Gender, and Emancipation"; Sexton, *Amalgamation Schemes*; Stanley, *From Bondage to Contract*; Weheliye, *Habeas Viscus*; Wilderson, *Red, White, and Black*.

11. Ahmed, *Queer Phenomenology*, 38.

12. Byrd, *Transit of Empire*, xxiii, xx.

13. As Michelle H. Raheja argues in *Reservation Reelism*, Indian images in popular media can reinforce existing settler narratives, but "these cinematic and televisual experiences also enable Indigenous [and, I would add, non-native] spectators to engage critically with the artifacts of imagined cultural knowledge produced by the films and their long political, narrative, and historical context" (xi).

14. Povinelli, *Economies of Abandonment*, 4, 118.

15. I use Dakota rather than Sioux since the former conforms more to the self-designation of the peoples in question (Mdewakantons, Wahpekutes, Sissetons, and Wahpetons) and the latter has a wider frame of reference. Sometimes the term Dakota is used to refer to all seven peoples commonly grouped together as Sioux, but I am drawing on the term's common use to refer to the four eastern peoples of those seven (not including the Yankton, Yanktonai, and Lakota or Teton).

16. Million, *Therapeutic Nations*, 46, 57.

17. Goodwin, "Foreword," ix. For alternative accounts that emphasize the ways the ideal of equality articulated in the Revolution and in its wake gains meaning within the context of extant racialized systems of value, see Kazanjian, *Colonizing Trick*; Silva, *Toward a Global Idea of Race*; Sweet, *Bodies Politic*.

18. Scott, "A President Engaged."

19. Kushner, *Lincoln*, 8.

20. Kushner, *Lincoln*, 99, 115.

21. On this specific legal aporia for African Americans and its legacy after the war, see Crenshaw, "Race, Reform, and Retrenchment"; Harris, "Whiteness as Property"; Hartman, *Scenes of Subjection*; Wagner, *Disturbing the Peace*.

22. Kushner, *Lincoln*, 127.

23. Kushner, *Lincoln*, 154.

24. Reddy, *Freedom with Violence*, 39, 203.

25. In an article for *USA Today* on the 150th anniversary of the Emancipation Proclamation that was published while *Lincoln* was still in theaters, Rick Hampson observes that, in light of the presidency of a black man, "This... [anniversary of the proclamation] is the first major one when black people can fairly be called free," while citing Martin Luther King's "I Have a Dream" speech at the Lincoln Memorial in 1963, in which King insisted, "A century after Emancipation, 'the Negro is still not free.'" Hampson then later notes that "the [Civil Rights] movement led by King, [John] Lewis and others allowed the nation to realize the ideals of emancipation taught in its schools," and closes by quoting Lewis's statement that "as a part of that tradition, you had to free yourself." The actual moment of the realization of "emancipation" keeps shifting, as does the agency that emancipates. What remains consistent is the presumption that true freedom is always already on the way, for which state action (such as the Emancipation Proclamation) serves as the avatar, if not always the most effective implementer. See Hampson, "Lincoln's Bold Move."

26. Kushner, *Lincoln*, 35–36. On the constitutional controversies created by Lincoln's assertions of various kinds of executive authority, see Neely, *Lincoln*.

27. Scott, "A President Engaged"; Ferguson, *Reorder of Things*, 170.

28. Quotes from Rhea, "As Lincoln, Daniel Day-Lewis Stands Very Tall," and Chan, "Lincoln," respectively.

29. Weheliye, *Habeas Viscus*, 35, 51.

30. Weheliye, *Habeas Viscus*, 113; Wynter, "1492." In *Economies of Abandonment*, Povinelli also explores the "alternative social projects" that emerge in spaces of abjection and precarity, although focusing more on indigeneity (primarily in the Australian context) than blackness. However, she also raises a series of significant ethical cautions about treating such projects as inspirations for radical imagination given the ways they depend on conditions of immiseration.

31. On this dynamic more broadly, see Byrd, *Transit of Empire*; S. Jackson, *Creole Indigeneity*; Moreton-Robinson, "Writing Off Treaties"; Rifkin, *Manifesting America*; Wolfe, "Settler Colonialism."

32. Nichols, "Contract and Usurpation," 102, 110.

33. Moreton-Robinson, "Writing Off Treaties," 84. Moreover, the Civil War can be understood as about what modes of settler inhabitance and production would dominate in the occupation of Indigenous lands. While noting the increasing importance of questions about differentiating "slave" and "free" territory in the decades leading up to the war, scholars do not necessarily address how that crisis depends on holding constant non-native rights to lands incorporated into U.S. domestic space. See Guelzo, *Fateful Lightning*, 54–93. On this dynamic, see Kelman, *Misplaced Massacre*, 1–43. Such a reinterpretation of the Civil War

would align with work in the last twenty years that has reframed the American Revolution as a conflict among non-natives over the possibilities of settlement. See Calloway, *Scratch of a Pen*; Jones, *License for Empire*; White, *Backcountry and the City*; Yirush, *Settlers, Liberty, and Empire*.

34. Ahmed, *Queer Phenomenology*, 37.

35. On counterinsurgent historiography, see Guha, "On Some Aspects."

36. Waziyatawin, *What Does Justice Look Like?*, 30.

37. On Dakota history leading up to the war, the conflict itself, and its immediate aftermath, see Board of Commissioners, *Minnesota*, 162–326; G. Anderson, *Little Crow*, 89–179; G. Anderson and Woolworth, *Through Dakota Eyes*; Berg, *38 Nooses*; Chomsky, "The United States-Dakota War Trials"; Gilman, *Henry Hastings Sibley*, 162–209; Herbert, "Explaining"; Hyman, *Dakota Women's Work*, 93–142; Martínez, "Remembering the Thirty-Eight"; Meyer, *History of the Santee Sioux*, 109–54; D. Nichols, *Lincoln and the Indians*, 61–128; Peacock, "Account of the Dakota-US War"; Pexa, "Transgressive Adoptions"; Wilson, "Decolonizing the 1862 Death Marches"; Wingerd, *North Country*.

38. Little Crow had been removed as speaker for the Mdewakanton quite recently because he was perceived as not committed enough to the project of "civilization" privileged by the Indian agent. See G. Anderson, *Little Crow*, 119–21.

39. Reinforcements of about twelve hundred men led by Sibley arrived at Fort Ridgely on August 27, and relatively soon thereafter this increase in U.S. forces in the vicinity of the fort brought an end to active conflict in the south. During this time Dakotas had been gathering in Little Crow's village, which came to serve as their base of operations through August 25, at which point they decided to move northward to try to join with Sissetons and Wahpetons at the Upper Agency. Leaving on August 26 in a train five miles long, they arrived at Yellow Medicine two days later, leading to an extended debate over the next two weeks about participation in the war and the formation of the "peace party" (which began gathering and protecting white and mixed-blood prisoners captured during the course of the war). On September 21 news reached Little Crow's camp that Sibley's forces were just south of the Upper Agency, and after a failed attack on them at dawn the next morning, Little Crow counseled those who wanted to continue the war to head toward the Plains.

40. Since there was no centralized authority for all the bands in a given people, they did not function like a state that declares war as a unit, so leaders and members of different peoples took various positions. However, the alignments tended to follow a split between those peoples connected to the Lower Agency (Mdewakanton and Wahpekute) and those connected to the Upper Agency (Sisseton and Wahpeton).

41. See "Message from the President," H.R. Exec. Doc. 68, 9, 10, 23, 25, 31, 33.

42. "Message from the President," H.R. Exec. Doc. 68, 32.

43. "Message from the President," H.R. Exec. Doc. 68, 40. Similarly, in a memorial to the president in late 1862 on the subject of what to do with the Dakotas who already had been tried, the Minnesota congressional delegation notes, "These Indians are called by some prisoners of war. There was no war about it. It was wholesale robbery, *rape, murder*. These Indians were not at war with their murdered victims," adding, "The people of Minnesota . . . have *not* violated the *law*" ("Message of the President," S. Exec. Doc. 7, 4).

44. In his dispatches to various officials and military personnel during the Dakota War, Sibley repeatedly refers to Native combatants as "guilty." See *Minnesota*, 165–282. On the

earlier appearance of a similar translation of warfare into criminal culpability with respect to the Nisqually leader Leschi in the late 1850s in Washington Territory, see Blee, *Framing Chief Leschi* (esp. 50–79).

45. For similar statements by Alex Ramsey, who in his role as the territorial governor of Minnesota also served as the superintendent of Indian affairs in the area, see *Annual Report* 1850, 76–77; 1851, 411. Galbraith attributes Dakota actions to ingrained propensities toward lawless savagery despite noting a range of sources of Dakota resentment and desperation before the war, including starvation due to the delay of the annuities, long-standing concerns about the amount of Dakota funds given over by the government to non-native traders, the nonfulfillment of treaty promises, and discontent with the agent's efforts to turn the Dakotas into Euramerican farmers.

46. "Message from the President," H.R. Exec. Doc. 68, 23–24.

47. Galbraith further asserts, "Whilst I must confine myself to the Sioux, I cannot keep out the fact that what is generally true of the Sioux is also generally true of all our other Indian tribes" (36).

48. Byrd, *Transit of Empire*, xx, 227, 125.

49. For discussion of a similar discursive and policy nexus through which Native actions in the period are understood as crime, see Rand, *Kiowa Humanity*, 58–92.

50. Governor Ramsey does describe the recent events as an "Indian war" in a letter to Lincoln on September 6, 1862, but then describes it as "being equally cruel and barbarous with those waged by that race which have preceded it in the history of our country." In a letter to the War Department that same day, Ramsey describes it as a "national war" in order to insist on the need for "the general government" to intervene, and in a letter to General Pope on September 19, Colonel Sibley also describes it as a "war" while then immediately shifting the meaning of the term by characterizing the Dakotas as "the most warlike and powerful of the tribes on this continent." See Board of Commissioners, *Minnesota*, 224–25, 234. These moments cite the name of war but qualify it by its association with Indianness, suggesting the term indicates less the political stakes of the conflict than the savagery of the scope and modes of Indian violence. Moreover, Dakota Indian agent and northern superintendency reports from the 1850s routinely use the term *war* to discuss conflicts among Native peoples, particularly the Dakota and Ojibwe, suggesting something other than, in Galbraith's terms, *regular war*. For examples, see *Annual Report* 1850, 107; 1851, 280; 1854, 273; 1856, 604; 1859, 424, 427, 449; 1860, 285.

51. Palmer, "Devil in the Details," 290. Palmer here is speaking of hagiographic accounts of the life of the Mohawk saint Tekakwitha.

52. "Message from the President," H.R. Exec. Doc. 68, 27.

53. Board of Commissioners, *Minnesota*, 257.

54. "Message of the President," S. Exec. Doc. 7, 2. For a critique of efforts to exonerate Lincoln or ameliorate his role in the executions, see Martínez, "Remembering the Thirty-Eight."

55. "Message of the President," S. Exec. Doc. 7, 7–8. Battles are included in the charges against prisoners no. 5, no. 12, no. 19, no. 24, no. 35, no. 67, and no. 70, and they are the principal charge against no. 19 and no. 35.

56. On the emergence of new understandings of the laws of war during the Civil War, see Witt, *Lincoln's Code*. However, he does not consider the role of settler-Indigenous conflict as

itself potentially formative in the construction of the laws of war from the eighteenth century onward. Thanks to Aziz Rana for highlighting this point.

57. Board of Commissioners, *Minnesota*, 256, 233, 236, 262, 270, 272, 279.

58. Board of Commissioners, *Minnesota*, 198, 288.

59. For similar sentiments in relation to Dakota lands, see *Annual Report* 1850, 77; 1851, 416–420; 1852, 350; 1859, 419.

60. "Message from the President," H.R. Exec. Doc. 68, 33.

61. Board of Commissioners, *Minnesota*, 291.

62. On the legality of the trials, see Chomsky, "The United States-Dakota War Trials"; Herbert, "Explaining"; Martínez, "Remembering the Thirty-Eight"; Witt, *Lincoln's Code*, 330–35. This period in Dakota history, during the war and in the years immediately in its wake, has been characterized as genocidal. See Hyman, *Dakota Women's Work*, 93–94; Pexa, "Transgressive Adoptions"; Waziyatawin, *What Does Justice Look Like?*, 17–70. Without minimizing the violence done to Dakota people(s) in this period, I want to highlight the continuities of political imagination in government discourses between times of "peace" and "war" in order to emphasize that U.S. settler colonialism does not rely primarily on military intervention as the tool for displacing Native peoples and securing state jurisdiction. On this point, see also Rockwell, *Indian Affairs*. On the circumstances of deprivation for prisoners and noncombatants in the wake of the war, see Canku and Simon, *Dakota Prisoner of War Letters*; Hyman, *Dakota Women's Work*, 93–142. On contemporary non-native responses to Dakota characterizations of non-native policy and actions during and after the war, particularly the forced marches and imprisonment of noncombatants, see Nunpa, "Dakota Commemorative March."

63. The treaties of 1858 provided for the allotment of Dakota lands and the patenting of such allotments by the president "as members of said bands become capable of managing their business and affairs" (Kappler, *Indian Affairs*, 2: 781); they also enabled the Secretary of the Interior to use annuity funds as he "shall deem best calculated to promote their interests, welfare, and advance in civilization" (784), giving the secretary complete power over Dakota monies so as to promote their *advancement*. See G. Anderson, *Little Crow*, 89–115; Hyman, *Dakota Women's Work*, 67–92. However, attempts to mark allotments on Mdewekanton and Wahpekute portions of the reservation were often met with outright rejection: "the process has been difficult, owing to the fact that the Indians, immediately after the surveys were made, pulled up most of the stakes and threw down the mounds" (*Annual Report* 1859, 455).

64. "Message from the President," H.R. Exec. Doc. 68, 25. For similar statements in relevant agent and superintendent reports before the Dakota War, see *Annual Report* 1850, 107; 1851, 281–83, 413–14; 1852, 350; 1858, 387; *Annual Report* 1859, 419–21; 1861, 699–700.

65. O'Brien, *Firsting and Lasting*, 105.

66. Ahmed, *Queer Phenomenology*, 121, 138.

67. On the nation as a white possession, see Moreton-Robinson, "Writing Off Treaties."

68. *Annual Report* 1851, 420; 1852, 350.

69. There were whites, mostly clergy, who did seek to highlight the devastating effects of federal policy on Dakota peoples, as well as the systemic venality and violence of non-natives as causes of hostility. However, while bemoaning Dakota suffering, such accounts still situate it within the trajectory of expanding U.S. jurisdiction and the potential for civilization. For

examples, see G. Anderson, *Little Crow*, 55, 106–8; Hyman, *Dakota Women's Work*, 89–90; Wakefield, *Six Weeks*, 100, 127. The perspectives offered by non-natives sympathetic to Dakota peoples raise serious questions about casting the social formations before the treaty of 1837 in what would become Minnesota as a "hybrid culture" rather than a complex conjunction of Native and non-native trajectories and mappings. See Wingerd, *North Country*.

70. Merleau-Ponty, *Phenomenology of Perception*, 365.

71. On the "uncanny" phenomenological experience of encountering something that does not fit one's impressions in/of the present, see Trigg, *Memory of Place*.

72. Wilson, "Introduction," 155.

73. For the text of the treaties, see Kappler, *Indian Affairs*, 2:493–94, 588–93, 781–89. For discussion of the process through which these treaties were negotiated and the history surrounding them, see G. Anderson, *Kinsmen*, 130–260; Gilman, *Henry Hastings Sibley*, 62–134; Hyman, *Dakota Women's Work*, 39–92; Meyer, *History of the Santee Sioux*, 72–108. For discussion of a similar dynamic in treaty making with respect to the Kiowa in the same period, see Rand, *Kiowa Humanity*, 33–57.

74. Kappler, *Indian Affairs*, 2: 493, 588, 591. The treaties of 1851 are almost identical in their provisions for the Mdewakanton and Wahpekute and for the Sisseton and Wahpeton, as are the treaties of 1858.

75. In 1860 the Senate finally resolved that the Dakotas did have such a right, and they set compensation at thirty cents an acre; the funds were largely claimed by traders in repayment of debts supposedly owed to them (G. Anderson, *Kinsmen*, 231).

76. Povinelli, *Economies of Abandonment*.

77. Coulthard, *Red Skin, White Masks*, 109.

78. As noted earlier, there were prominent differences in how Dakotas responded to the conditions produced by Indian policy. Put succinctly, not everyone decided to go to war, and many Dakotas chose to defend white relatives rather than participate in armed struggle. John Peacock refuses Wilson's description of those Dakotas who defended white relatives as "traitors of Dakota people" by emphasizing the ways such behavior was itself "in accordance with customary Dakota obligations." "Account of the Dakota-US War," 197–98. I am not offering an account that seeks to position one set of responses as more authentic or less assimilated, simply noting that the war can be explained as a result of mounting outrage if one acknowledges the temporality of attrition that increasingly defined Dakota experience in the decades before the war.

79. Deloria, *Indians in Unexpected Places*, 7.

80. Kappler, *Indian Affairs*, 2: 593. The article struck out by the Senate specifies that the land that "the United States do hereby set apart for the future occupancy and home of the Dakota Indians, parties to this treaty, [is] to be held by them as Indian lands are held" (Kappler, *Indian Affairs*, 2: 593).

81. Kappler, *Indian Affairs*, 2: 592.

82. *Annual Report* 1851, 417.

83. *Annual Report* 1850, 80. In that same report he also suggests that the Dakotas "live almost without law" (80). For similar sentiments in relevant agency and superintendency reports before the war, see *Annual Report* 1850, 107; 1858, 388; 1861, 703–4.

84. *Annual Report* 1855, 378–79; 1856, 606. See also Anderson, *Little Crow*, 55–88.

85. Hyman, *Dakota Women's Work*, 98; Meyer, *History of the Santee Sioux*, 140; Waziyatawin, *What Does Justice Look Like?*, 55–56.

86. Peacock, "Account of the Dakota-U.S. War," 197. On the intergenerational transmission of stories of the war and its aftermath, see Hyman, *Dakota Women's Work*, 93–117. On the commemoration of the forced removal, see Wilson, "Introduction."

87. Peacock, "Account of the Dakota-US War," 158.

88. In *What Does Justice Look Like?* Waziyatawin observes, "This is precisely why we cannot put this history behind us. The contemporary circumstances compel us to look at the root causes of our current dysfunction and pain.... When we expose the root causes of our pain and question whether those root causes still affect our lives, we see that the same oppression facing our people in the nineteenth century still exists today" (10).

89. Wilson, *Remember This!*, 95.

90. Some people at the time, though, interpreted the Dakota War as brought on by Confederate agitators. See Berg, *38 Nooses*, 99–100; Board of Commissioners, *Minnesota*, 1893, 260, 274; Gilman, *Henry Hastings Sibley*, 177; D. Nichols, *Lincoln and the Indians*, 78–79.

91. On other Civil War era violence against Indians, see Denetdale, *Reclaiming Diné History*; Kelman, *Misplaced Massacre*; D. Nichols, *Lincoln and the Indians*, 161–201.

92. Ahmed, *Queer Phenomenology*, 44.

93. Scott, *Omens of Adversity*, 125. Povinelli, *Economies of Abandonment*, 99.

94. On Parker's participation in the Tonawandas' struggle to regain their reservation, see Armstrong, *Warrior in Two Camps*, 20–70; Genetin-Pilawa, *Crooked Paths to Allotment*, 29–50; Hauptman, *Tonawanda Senecas' Heroic Battle*, 75–113; A. Parker, *Life of General Ely S. Parker*, 71–93. On the structure of the Iroquois League and its sachemships, see Fenton, *Great Law of the Longhouse*; A. Parker, *History of the Seneca Indians*. I should note that I am not trying to position Parker as paradigmatic but to use his example as a touchstone for thinking about the occlusion of Native experiences of time, specifically in relation to settler modes of governmentality. For an incisive critique of the ways Parker has served as emblematic of Iroquois "tradition," particularly given his work with Henry Lewis Morgan, see Simpson, *Mohawk Interruptus*, 67–94.

95. On the treaties of 1838 and 1842, see Conable, "Steady Enemy"; Genetin-Pilawa, *Crooked Paths to Allotment*, 29–50; Gonzales, "Cultural Colonizers," 182–220; Hauptman, *Conspiracy of Interests*, 175–212; Society of Friends, *Case of the Seneca Indians*; Valone, "William Seward, Whig Politics."

96. On the Tonawandas' struggle to regain their reservation after 1842, see Conable, "Steady Enemy," 283–324; Genetin-Pilawa, *Crooked Paths to Allotment*, 29–50; Hauptman, *Tonawanda Senecas' Heroic Battle*.

97. On the history of this preemption claim and the role of the Ogden company in Seneca politics in the first half of the nineteenth century, see Conable "Steady Enemy"; Densmore, *Red Jacket*; Ganter, *Collected Speeches of Sagoyewatha*; Hauptman, *Conspiracy of Interests*; Wilkinson, "Robert Morris."

98. Haudeonsaunee is the name used by the peoples themselves for what conventionally is known as the Iroquois League or Six Nations. This confederacy includes the Mohawks, Oneidas, Onondagas, Cayugas, Senecas, and Tuscaroras.

99. For an overview of the history of treaty making under the United States from the perspective of U.S. officials, see Prucha, *American Indian Treaties*. On the ways the treaty system

generated subjectivities conducive to U.S. legal geographies, see Cheyfitz, "Navajo"; Jones, *License for Empire*; Rifkin, *Manifesting America*; Rockwell, *Indian Affairs*.

100. My reading here runs counter to the portrait of the treaty system as a vehicle of Native collective agency. Scott Richard Lyons observes of "the x-mark," his term for the signing of treaties by those who did not write their names, "The x-mark is a contaminated and coerced sign of consent made under conditions that are not of one's making. It signifies power and a lack of power, agency and a lack of agency. It is a decision one makes when something has already been decided for you, but it is still a decision," and he argues that "there is always the prospect of slippage, indeterminacy, unforeseen consequences, or unintended results: it is always possible, that is, that an x-mark could result in something good," later noting, "An x-mark is a commitment to living in new and perhaps unfamiliar ways." *X-Marks*, 3, 169. See also Williams, *Linking Arms Together*.

101. Turner, *This Is Not*, 88; Povinelli, *Economies of Abandonment*, 61.

102. "Report of the Committee on Indian Affairs," S. Doc. 156, 61.

103. "Report of the Committee on Indian Affairs," S. Doc. 156, 65–66.

104. "Report of the Committee on Indian Affairs," S. Doc. 156, 87.

105. "Letter from the Governor of New York," June 17, 1844, Ely S. Parker Papers, American Philosophical Society, Philadelphia, PA; "Letter From the Governor of New York Stating That He Has No Power to Prevent the Execution of the Treaties," April 18, 1846, Ely S. Parker Papers, American Philosophical Society, Philadelphia, PA.

106. On Haudenosaunee treaties with the state, see Graymont, "New York State Indian Policy"; Hauptman, *Conspiracy of Interests*; Tiro, *People of the Standing Stone*.

107. "Report of the Committee on Indian Affairs," S. Doc. 156, 115.

108. "Difficulties between the Tonawanda Senecas and the Ogden Land Company," January 21, 1848, Ely S. Parker Papers, American Philosophical Society, Philadelphia, PA.

109. "Appeal for Justice Till Difficulties are Settled," February 22, 1845, Ely S. Parker Papers, American Philosophical Society, Philadelphia, PA.

110. "Appeal for Justice." The Tonawandas, though, also were able to make use of the government's official disavowal of coercion as the means of acquiring Native lands. In the case of *Fellows v. Blacksmith* (1857), the U.S. Supreme Court found that the Ogden Company had no legal right to resort to violence as a means of gaining access to the land it supposedly acquired in the Treaty of Buffalo Creek, instead indicating that the federal government had exclusive authority to manage the process of Seneca removal as it saw fit.

111. "Letter from the Secretary of War," H.R. Mis. Docs. 37, 4. In his reading of Parker's postwar career, Genetin-Pilawa emphasizes this dimension of Parker's thinking. *Crooked Paths to Allotment*, 61–72, 84–90. On Parker's term as the commissioner of Indian affairs, see Armstrong, *Warrior in Two Camps*, 137–65; A. Parker, *Life of General Ely S. Parker*, 150–61.

112. "Letter from the Secretary of War," H.R. Mis. Docs. 37, 2.

113. *Dred Scott*, 407.

114. "Letter from the Secretary of War," H.R. Mis. Docs. 37, 2.

115. The formulation "dependents and wards" dates back to the majority decision in the U.S. Supreme Court case of *Cherokee Nation v. Georgia* (1831), which invented the category "domestic dependent nation" as a way of characterizing the legal status of Native peoples, while also describing their relation to the U.S. government as like that of a "ward" to a "guardian."

For citations by agents and superintendents of such wardship with respect to the Dakota, see *Annual Report* 1851, 281, 416; 1858, 387; 1859, 418–19, 421; 1861, 698.

116. *Annual Report* 1869, 448. Genetin-Pilawa also addresses Parker's complicated attitude toward treaty making in the wake of the Civil War in "Ely S. Parker and the Paradox of Reconstruction Politics in Indian Country" (n.d.).

117. In his annual report in 1870, Parker notes, "Although I do not recommend that treaties be made with the Apaches and the several bands of Utes . . . , yet I would present for the consideration of Congress the importance of these bands being properly cared for. . . . As soon as practicable they should be placed upon a reservation, and furnished with whatever may be required to enable them to become self-sustaining" (*Annual Report* 1870, 469), suggesting some alternative to treaty making that could create a reservation but not specifying the mechanism or legal theory for doing so. Treaty making was formally ended by Congress in 1871. On this decision and the temporal narratives mobilized to legitimize it, see Bruyneel, *Third Space of Sovereignty*, 65–96.

118. "Letter from the Secretary of War," H.R. Mis. Docs. 37, 5.

119. "Letter from the Secretary of War," H.R. Mis. Docs. 37, 6.

120. In addressing the consequences of this settler tsunami, Parker's rhetoric at times seems almost boosterish of the civilization program, resembling the kinds of remarks made by Galbraith and Sibley discussed earlier. At one point, he claims, "The clouds of ignorance and superstition in which many of this people were so long enveloped have disappeared, and the light of Christian civilization seems to have dawned upon their moral darkness, and opened up a brighter future" (*Annual Report* 1869, 445), and he endorses "the policy of giving to every Indian a home that he can call his own," by which "the Indians would be more rapidly advanced in civilization than they would if the policy of allowing them to hold their land in common were continued," also indicating the need for Indians to "progress toward that healthy Christian civilization in which are embraced the elements of material wealth and intellectual and moral development" (*Annual Report* 1870, 473–74).

121. *Annual Report* 1869, 447.

122. On the continuing struggle over Seneca lands in the late nineteenth century, including during the Civil War, see Hauptman, *Iroquois in the Civil War*.

123. Parker, *Life of General Ely S. Parker*, 176, 181–88. See also Raheja, "I Leave It."

124. Vizenor, *Manifest Manners*, 51. Vizenor also observes, "His new name, education, and marriage were revolutions in his time; moreover, he was burdened with the remembrance of violence, the separation and conversion of his father, and the horror of the massacre of Wounded Knee" (47), adding, "He endured the horror of a massacre, the melancholy of racialism, and resisted federal policies on reservations and in government schools" (49).

125. Martínez, *Dakota Philosopher*, 144.

126. Lopenzina, "Good Indian," 741.

127. Coskan-Johnson, "What Writer Would Not," 111; Lopenzina, "Good Indian," 750; D. Carlson, *Sovereign Selves*, 137. See also Heflin, "*I Remain Alive*," 41–78.

128. Vizenor, *Manifest Manners*, 54.

129. On the notion of putting together a life, see Spivak, *Critique of Postcolonial Reason*. On the ways narrations of proper resistance can radically limit possibilities for conceptualizing the affects and agency of oppressed people(s), see Cacho, *Social Death*; Holland, *Erotic*

*Life of Racism*; Keeling, *Witch's Flight*; Lazo, "Confederates in the Hispanic Attic"; Million, *Therapeutic Nations*; Ngai, *Ugly Feelings*, 174–208; Quashie, *Sovereignty of Quiet*; Sharpe, *Ghosts of Slavery*.

130. Miranda, *Bad Indians*, xiv; Powell, "Imagining a New Indian," 214.

131. Martínez (*Dakota Philosopher*) observes, "The expulsion and near-annihilation of the Dakota in Minnesota would always symbolize for Eastman just how far white Americans were willing to stray from their Christian and democratic ideals for the sake of robbing Indians of their land and heritage" (127).

132. Eastman, *Indian Boyhood*, 239. When referring to the war elsewhere in the text, Eastman puts *massacre* in scare quotes, indicating his critical relation to that characterization of Dakota violence.

133. Eastman, *From the Deep Woods*, 3, 136.

134. Eastman, *Indian Boyhood*, 242; *From the Deep Woods*, 3.

135. Tatonetti, "Disrupting a Story of Loss," 287.

136. In this way my reading of Eastman's work differs from that of Lucy Maddox (*Citizen Indians*), who argues, "For Eastman as a writer and public figure, representing identity usually meant deliberately turning away from the disturbing details of specific recent histories, tribal or personal, and toward a more optimistic and idealized representation of 'the Indian' as a figure grounded in history but not bound by it, a particular type of universal ideal" (132).

137. In *From the Deep Woods* Eastman actively endorses allotment (164–65), and he describes his time serving as an allotment agent for the Lakota, enrolling them and giving them heteropatriarchal family names (182–84).

138. Piatote, "The Indian/Agent Aporia," 51, 47.

139. Eastman, *From the Deep Woods*, 8, 15, 16, 25. At one point in *From the Deep Woods*, as Eastman matriculates at Dartmouth, he contrasts his efforts at adaptation with the supposed tragic failure of peoples in the Northeast to do so: "thinking of the time when red men lived here in plenty and freedom, it seemed as if I had been destined to come view their graves and bones. No, I said to myself, I have come to continue that which in their last struggle they proposed to take up, in order to save themselves from extinction; . . . it was too late"; "This was my ambition—that the Sioux should accept civilization before it was too late!" (65). On this depiction, see Powell, "Imagining a New Indian," 221–22.

140. Martínez, *Dakota Philosopher*, 142.

141. Eastman, *From the Deep Woods*, 98, 117. On Eastman's response to Wounded Knee, see also Lopenzina, "Good Indian"; Powell, "Imagining a New Indian"; Tatonetti, "Disrupting a Story of Loss," 285–94.

142. Eastman, *From the Deep Woods*, 84. I will raise questions about Eastman's way of understanding Native prophetic movements as merely a response to non-native violence in chapter 4. On the Ghost Dance among the Lakota and the massacre, see Ostler, *Plains Sioux and U.S. Colonialism*.

143. Eastman, *From the Deep Woods*, 92. For a more complex account of prophetic movements, see Eastman, *Soul of the Indian*, 41–42.

144. As others have noted, Eastman had no experience of living on-reservation. See Lopenzina, "Good Indian."

145. Eastman, *Indian Boyhood*, v, 245; *Indian To-day*, 23; *From the Deep Woods*, 132. This reading of Eastman's representation of the reservation offers one way of answering the question posed by Penelope Myrtle Kelsey about his work: "How does one reconcile statements that affirm a type of Vanishing Americanism (symbolic past tense) within an autobiography that praises and affirms traditional Dakota life in the face of cultural genocide (symbolic present tense)?" *Tribal Theory in Native American Literature*, 54.

146. Eastman, *Soul of the Indian*, 6, 23. See Carr, *Inventing the American Primitive*, 197–256; Kent, *African, Native, and Jewish American Literature*, 71–112; Pfister, *Individuality Incorporated*; Schedler, *Border Modernism*, 41–84. On Eastman's use of primitivism, particularly to signify his (and Indians') natural masculinity, see Bayers, "Charles Alexander Eastman's"; Maddox, *Citizen Indians*, 133–40.

147. For example, in *Indian Boyhood* he writes, "To me, as a boy, this wilderness was a paradise. It was a land of plenty. To be sure, we did not have any of the luxuries of civilization, but we had every convenience and opportunity and luxury of Nature"; "the truth is that we lived in blessed ignorance of any life that was better than our own" (184).

148. In *From the Deep Woods* Eastman offers a similar formulation: "I seriously considered the racial attitude toward God, and almost unconsciously reopened the book of my early religious training, asking myself how it was that our simple lives were so imbued with the spirit of worship, while church-going among white and nominally Christian Indians led often to such very small results" (141). He later notes, "I realize that the white man's religion is not responsible for his mistakes" (195).

149. Here my reading differs from Tova Cooper's claim that "Eastman couches his critique of the United States within an assimilationist framework," "offer[ing] a veiled critique" of the United States. "On Autobiography," 1–2. On scholarly efforts to (re)conceptualize the legacy of Indigenous intellectuals' pursuit of citizenship for Native peoples in the early twentieth century, especially the legacy of the Society of American Indians, see Lomawaima, "Mutuality"; Maddox, *Citizen Indians*; Piatote, "Indian/Agent Aporia"; J. Porter, *To Be Indian*; Smithers, "Soul of Unity."

150. Kelsey, *Tribal Theory*, 52; Powell, "Imagining a New Indian," 224.

151. Eastman, *Indian To-day*, 32. In Michael A. Elliott's terms, Eastman can be said to offer "a kind of anti-national nationalism—or perhaps better called an anti-American Americanism—a patriotism that professes loyalty to the United States yet also remains deeply skeptical of the United States because of its history of colonization." *Custerology*, 54.

152. Eastman, *Indian To-day*, 47. This passage expresses a similar perspective to that offered by the Society of American Indians' inaugural executive committee (which included Eastman) in their letter to potential members in 1911: "We earnestly believe the time has come when the American Indian should take the initiative in the struggle for his race betterment.... A great many problems have arisen out of the question of race adaptation to new conditions. Mistaken work will continue unless the Indian expresses himself, for no one knows the heart of the Indian as the Indian himself." Quoted in Maddox, *Citizen Indians*, 9–10.

153. Eastman, *Indian To-day*, 50. However, Eastman notes just afterward that "it seems clear that some of the tribes still need intelligent and honest guardianship. To my mind, this machinery might be adjusted more nearly to the requirements of the present-day Indian."

154. Eastman, *Indian To-day*, 195.
155. Luciano, *Arranging Grief*, 21.

THREE. *The Duration of the Land*

1. For the text of the law, see Kappler, *Indian Affairs*, 3:252–58.
2. For overviews of the allotment program, see Genetin-Pilawa, *Crooked Paths to Allotment*; Hoxie, *Final Promise*; Piatote, *Domestic Subjects*; Rifkin, *When Did Indians*, 181–232.
3. On this process in U.S. Indian policy more generally, see Bruyneel, *Third Space of Sovereignty*; Konkle, *Writing Indian Nations*. I also draw here on Jason Cooke's dissertation in progress on discourses of Indianness in the removal era.
4. In fact, allotment and its accompanying promise of citizenship were often celebrated as an emancipation of Native peoples, one that could be seen as parallel to the prior emancipation of former slaves.
5. My formulation here resonates with Jodi Byrd's use of the "transit of Venus" to think about U.S. settler colonialism; she notes that in the transit, "Venus, the sun, and the earth are all in motion during the astronomical event that is the transit of Venus. Each body pulls gravitationally upon the other to distort possible viewing locations and antagonizes any parallax angle to discern coequal or equivalent, static theories of how U.S. empire functions." *Transit of Empire*, 31.
6. On the effort to coordinate a single system of time in this period, and attendant challenges to the notion of simultaneity, see Galison, *Einstein's Clocks, Poincaré's Maps*; Kern, *Culture of Time*, 10–35, 131–80; West-Pavlov, *Temporalities*, 20–28. On the place of Einsteinian relativity (special and general) in the context of intellectual life in the early to mid-twentieth century, see also Kwinter, *Architectures of Time*; Schleifer, *Modernism and Time*, 149–83.
7. The quoted words are from P. Deloria, *Indians in Unexpected Places*, 231.
8. This account of the reservation as the site of multiple, complex temporalities differs from Charles Eastman's account of it, discussed in chapter 2, as a site of containment and simulacrum of Native pasts.
9. John Joseph Mathews actually was studying at Oxford and traveling in Europe in the early 1920s during the period in which Bergson and Einstein (and their various advocates and detractors) were actively debating the meaning of relativity. See Canales, *Physicist and Philosopher*; Kalter, "Introduction," xlv–l.
10. Bergson, *Matter and Memory*, 24, 30.
11. See Gillan, "Hazards of Osage Fortunes"; Hunter, "Protagonist as a Mixed-Blood"; Keresztesi, *Strangers at Home*, 152–61; Owens, *Other Destinies*, 49–60; Schedler, *Border Modernism*, 41–54, 73–84.
12. Here I would note that in discussing the notion of an Osage spatiotemporal formation, I am not suggesting a relation to the past as opposed to the future but ways of envisioning and experiencing the relation between them in moving toward and engaging the latter. For discussion of the historically adaptive and future-oriented character of Osage social formations and practices, see Bailey, *Osage*, 3–9; Harmon, *Rich Indians*, 171–208; Warrior, *People*, 49–94. Moreover, this reading of the novel does not entail aligning Mathews with "tradition" as against

his "cosmopolitan" investments in non-native contemporary social developments and intellectual movements, especially given his education at Oxford and in Paris and his wide-ranging travel around the world (including to Africa and Mexico). Instead, drawing on conceptions of spacetime offers a way of thinking about how transformations in scientific and philosophical discourses over the previous several decades, of which Mathews likely would have been quite aware, might have influenced his thinking about Osage history and politics. On Mathews's education, travels, and cosmopolitanism, see Foster, "Dividing Canaan," 149–247; Kalter, "Introduction"; Lutenski, "Tribes of Men"; Ruoff, "John Joseph Mathews's"; Warrior, *Tribal Secrets*, 14–26; T. Wilson, "Osage Oxonian." On the travels and broad knowledge of the world of nonelite Osage people, see Revard, *Family Matters, Tribal Affairs*, 27–56.

13. Andersen, "From Difference to Density," 92, 97.

14. Here I differ from Michael Snyder's analysis of queerness in the text in "'He Certainly Didn't Want Anyone to Know That He Was Queer,'" which focuses primarily on the ways it potentially points to homoerotic desire.

15. On stylistic developments with respect to temporality within literary modernism, see Kern, *Modernist Novel*, 101–25. On modernist engagements with history and time, see also Kwinter, *Architectures of Time*; Schleifer, *Modernism and Time*; Stasi, *Modernism*. Here I disagree with Rita Keresztesi's claim that Native authors of the period did not "employ the formal experimentation of Anglo high modernism." *Strangers at Home*, 113. For an alternative perspective on Native modernism(s), and the manipulation of genre and form, see Foster, "Dividing Canaan"; Kent, *Reshaping of Modernism*, 71–112; Piatote, *Domestic Subjects*; Schedler, *Border Modernism*, 41–54.

16. Mathews's *Sundown* will be cited parenthetically throughout. Mathews uses the phrase "the great frenzy" initially in his first novel, *Wah'Kon-Tah* (299–308), a fictionalized version of the papers of former Osage agent Labian J. Miles.

17. While the Osage Nation had been considered part of Indian Territory before 1890, Congress incorporated it into Oklahoma Territory when the latter was created in that year, so that by the time the Curtis Act extended allotment to peoples in Indian Territory, that term no longer encompassed the Osage. See Burns, "*Lu tsa ka Le Ah ke ho*," 202.

18. Kappler, *Indian Affairs*, 3:254–55. Under the 1906 act, Osages held two different kinds of allotted lands: a 160-acre plot known as the "homestead" and two other 160-acre plots—along with a piece of land that resulted from apportioning among all those eligible the remaining land of the reservation after everyone had received their first three plots—that were together referred to as an Osage's "surplus land." In this way there was no excess territory on the reservation after allotment that could then be sold to non-natives, unlike in the terms of the Dawes Act. On that process of allotment, see Burns, "*Lu tsa ka Le Ah ke ho*." Both kinds of Osage allotted lands remained "inalienable" under trust, but the surplus land could be taxed after three years. Moreover, if a member applied for and received a "certificate of competency," he or she could sell the surplus land but not the homestead. While blood quantum may have played an unofficial role in determinations of competency, the latter was not defined statutorily with reference to race, although distinctions in blood quantum would play a large role in later laws with respect to the status of allotted lands and the distribution of funds in ways I will discuss. For an overview of federal law governing Osage land tenure after allotment, see Department of the Interior, "Osage People," 165–92.

19. Kappler, *Indian Affairs*, 3:257, 258. The Office of Indian Affairs had ceased to recognize the prior Osage constitutional government as of 1900. See Burns, *History*, 392–94; Department of the Interior, "Osage People," 11–12; Mathews, *Osages*, 771; Warrior, *People*, 49–94; T. Wilson, *Underground Reservation*, 42–43. On the effort in the early twenty-first century to replace this council with a new constitutional Osage government, see Dennison, *Colonial Entanglement*.

20. Kappler, *Indian Affairs*, 3:255.

21. While many Osages themselves fought against allotment, the preservation of an undivided, collective interest in subsurface rights also can be understood as a function of oil companies' advocacy, since oil production would be easier if negotiations could be conducted with a single entity (the Osage Tribal Council) rather than a multiplicity of different private owners. Oil was discovered on the Osage reservation in 1897. On the negotiation of the initial oil leases and their role in allotment, see Burns, *History*, 400–419; Dennison, *Colonial Entanglement*, 102–8; Mathews, *Osages*, 771–78; T. Wilson, *Underground Reservation*, 74–120.

22. Kihekah serves as a stand-in for the town of Pawhuska on the Osage reservation, where the Indian agency was located and where Mathews himself was born.

23. I will address the full-blood/mixed-blood distinction more fully in the final section.

24. Sometimes scholars have suggested that Chal fails to live up to his name, but one could argue that John's incoherent sense of challenge—with its connotation of a clear site of struggle among opponents—suggests that this way of formulating Osage peoplehood neither engages with the specific dynamics and effects of Indian policy as sketched by the novel (especially its ubiquity) nor addresses modes of emplacement not based on *conquest*—John's way of framing the meaning of challenge (3).

25. Both Mathews and the Windzer family in *Sundown* are part of what might be considered an Osage elite. For the contrast between the situation of the Mathews family and that of less privileged Osages, see Revard, *Family Matters, Tribal Affairs*, 3–26.

26. This episode in the novel parallels the events of 1912 in which the secretary used the authority granted him under the allotment act of 1906 to dismiss the entire council, including Mathews's father. See Department of the Interior, "Osage People," 15–16; T. Wilson, *Underground Reservation*, 115–19. In his depiction of these events, Mathews leaves aside the questions of corruption—specifically the issue of kickbacks and the violation of the Tribal Council's own regulations about getting competitive bids for oil production leases—that may have motivated the secretary's actions.

27. Osages were made citizens by congressional act in 1921. Kappler, *Indian Affairs*, 4:317.

28. For the legislation creating Oklahoma Territory and incorporating the Osage into it, see Kappler, *Indian Affairs*, 3:186–92. However, as was noted previously and will be discussed further, the Osage retained collective ownership of the subsurface rights to the reservation as specified in the 1906 act, and having purchased their land as part of the removal process in 1870, the Osage Nation could also be understood as possessing fee simple ownership of the land. The apparent contradiction between being a "county" in Oklahoma and being a territorial entity via both a government-recognized purchase and the persistence of the mineral estate has generated recent legal questions about whether in the wake of allotment the Osage people could be said to have a reservation. See Dennison, *Colonial Entanglement*, 129–55. On the sale of the Kansas reservation and the purchase of the one now enclosed by Oklahoma,

see Burns, *History*, 292–353; Mathews, *Osages*, 650–731; T. Wilson, *Underground Reservation*, 13–23.

29. For a particularly stark reading the novel as a conflict of values or cultures, see Hunter, "Protagonist as a Mixed-Blood."

30. On these dynamics, see R. Parker, *Invention*, 19–50. However, Parker is less interested in Osage temporality and its relation to U.S. policy than in notions of work and Native masculinity.

31. Kappler, *Indian Affairs*, 3:254–55.

32. In this way the legalities of defining *Indianness* and its relative capacities in the late nineteenth and early twentieth centuries share a good deal with emerging definitions of *disability*. See Samuels, *Fantasies of Identification*.

33. Dennison, *Colonial Entanglement*, 18; Department of the Interior, "Osage People," ix, 19–20, 25, 64–66; T. Wilson, *Underground Reservation*, 151, 158. Moreover, as part of an appropriations act for the Bureau of Indian Affairs in 1918, Congress authorizes the Secretary of the Interior, "where the same would be for the best interest of Osage allottees, to permit the sale of surplus and homestead allotments, wholly or in part, of Osage allottees under such rules and regulations as he may prescribe and upon such terms as he shall approve," and in 1921 restrictions on sales were removed for all lands held by people of "less than one-half Indian blood." Kappler, *Indian Affairs*, 4:165, 317.

34. Kappler, *Indian Affairs*, 3:519, 4:317. The act passed in 1912 regulating Osage inheritance indicates that those Osages deemed "incompetent" under Oklahoma state law "shall in probate matters" "be subject to the jurisdiction of the county courts of the State of Oklahoma," which includes the power to assign guardians to incompetent persons. Kappler, *Indian Affairs*, 3:518. On the operation of the Oklahoma guardian system, particularly its effects on members of the Creek Nation, see Thorne, *World's Richest Indian*. On concerns at the time about Osage wealth and its proper management, see Harmon, *Rich Indians*, 171–208.

35. The situation was so extreme that Congress acted in 1925 to remedy the transfer of Osage wealth to guardians. Kappler, *Indian Affairs*, 4:480–81. However, that action did not prevent continued graft and unethical (and often illegal) takings, including those actively promoted by Oklahoma courts. See Department of the Interior, "Osage People," 50–77. The text also addresses what came to be known as the Osage Reign of Terror, in which a white man named William K. Hale moved to Oklahoma, encouraged male family members to marry Osage women, and began orchestrating the killing of them and their family members so that his family could inherit the headright interests in Osage funds and the mineral estate. See Mathews, *Sundown*, 258, 305–7. For discussion of this part of the novel, see Gillan, "Hazards of Osage Fortunes"; Musiol, "*Sundown* and 'Liquid Modernity.'" On those specific murders, see T. Wilson, *Underground Reservation*, 145–46. On the ways guardianship facilitated a wave of murders far larger than those for which Hale was responsible, see McAuliffe, *Bloodland*.

The 1906 act made "surplus" lands nontaxable for three years, and there was a spike in applications for competency just before the deadline so that people would then be able to sell their lands either to avoid the taxes on all of them or to pay the taxes for the plots of land they retained. In addition, in the wake of the passage in 1921 of the act limiting the quarterly payments to restricted Osages, many sought competency to avoid that curtailment of their income. Department of the Interior, "Osage People," ix. Mathews himself received his certificate of competency in 1921, suggesting that he may have been part of this wave. Kalter, "Introduction," lii.

36. See the discussion of the Osage Reign of Terror in the preceding note.

37. On *Sundown* as chronicling the dynamics of a general "(petro)modernity," including its ways of framing and shaping time, see Musiol, "*Sundown* and 'Liquid Modernity.'"

38. On Mathews's participation in a masculinist cosmopolitanism, see Lutenski, "Tribes of Men."

39. Dennison, *Colonial Entanglement*, 6–8.

40. West-Pavlov, *Temporalities*, 6; Chakrabarty, *Provincializing Europe*, 109, 112. As José Rabasa suggests in his discussion of the implications of the Zapatista movement, "the debates over the recognition of Indian rights to their normative systems would not entail recognition of these rights from within a singular universality, but a transformation of a hegemonic discourse. In fact, this transformation would ultimately question the desires for a new hegemonic discourse that in this case would validate indigenous normative systems under the principles of universality—that is, a theory of natural rights obviously grounded in the modernity of Western discourses." *Without History*, 118.

41. Kappler, *Indian Affairs*, 3:255.

42. On the possibility for capitalist forms of fungibility to support modes of Indigenous continuance and revitalization, see Cattelino, *High Stakes*, 95–124. The work of Osage wealth in enabling modes of occupancy and association at odds with bourgeois ideologies and practices runs against the grain of accounts that understand capitalism as necessarily subsuming other modes of life and generating its own specific experience of time (even if it can be pluralized from within). See Chakrabarty, *Provincializing Europe*; Gaonkar, "On Alternative Modernities"; Schleifer, *Modernism and Time*. On the ethics and politics animating Osages' use of their wealth during the oil boom, see Harmon, *Rich Indians*, 171–208.

43. The novel's emphasis on an animating continuity of occupation, and its operation as a different spatiotemporal formation than that of allotment, runs against the grain of accounts of a modernist "crisis" in which the experience of a rupture in time is primary, or at least the novel reframes such an experience as not so much a function of the onset of a shared modernity as the result of imperial intervention. On modernism as indicative of a sense of a crisis in time, see Esty, *Unseasonable Youth*; Kern, *Culture of Time*; Schleifer, *Modernism and Time*; Stasi, *Modernism*.

44. The land contained within the Osage reservation had been part of Osage hunting grounds from at least as far back as their first contact with Europeans in the late seventeenth century, and from the late eighteenth century onward, if not before, it also served as a residential site for Osage villages. While it was ceded in the treaty of 1825 through which the Osage acquired their reservation in Kansas, several Osage bands either never left or moved back after 1825, and the other Osage bands moved there in the wake of the sale of the Kansas reservation in 1870. See Burns, *History*; Mathews, *Osages*; Rollings, *Unaffected by the Gospel*.

45. Mathews does not conceptualize the physical conditions of the nonhuman environment as themselves unchanging or in some sort of eternal balance. For example, in *Talking to the Moon*, his memoir of a decade of living by himself on a ranch in Osage territory, he observes that the house he built for himself among the blackjacks "was not disturbing a state

that had been constant throughout the years, since there had been no absolute constancy to disturb. No two seasons and no two days had been alike through the years in the flow of earth's life toward some mysterious fulfillment" (2). On the complexities of his account of the environment in that text, see Schweninger, *Listening to the Land*, 75–95.

46. Warrior, *Tribal Secrets*, 82.

47. Million, "Felt Theory," 54, 64.

48. Coulthard, *Red Skin, White Masks*.

49. Vizenor, *Manifest Manners*, 70–72. See also Rifkin, "Shadows of Mashantucket."

50. Bergson, *Time and Free Will*, 216; *Matter and Memory*, 243–44.

51. See Bailey, *Osage*, 61–75; Mathews, *Osages*, 53–64, 75–79; Warrior, *People*, 60, 70–75.

52. On the history of the Peyote religion and the I'n-Lon-Schka, see Bailey, *Osage*, 3–26; Callahan, *Osage Ceremonial Dance*; Foster, "Dividing Canaan," 227–47; Rollings, *Unaffected by the Gospel*, 171–85; Stewart, *Peyote Religion*. At a number of points in his writings, Mathews dismisses Peyotism as something of a Christianized bastardization of prior Osage practices while also presenting it as an extension of prior practices through which to engage with Wah'Kon-Tah, the primal spirit. See Mathews, *Talking to the Moon*, 83–84; *Wah'Kon-Tah*, 301–2, 315. On the historical relation between Christian missionization, the boarding school system, and the spread of Peyotism, see Stewart, *Peyote Religion*, 45–96.

53. In terms of the Peyote religion, those links include the following: Peyote celebrations were built around "fireplaces," which is the same term used for the clans; many of those who had been clan-based religious and political authorities became leading Peyotists; and red, which conventionally signaled fire and the power of the sun, was incorporated into Peyote ritual as the color of both the buildings in which ceremonies were held and the streak painted on the head of the leader of the ritual. See Bailey, *Osage*, 19, 35, 72; Mathews, *Osages*, 31–51, 740–58; *Talking to the Moon*, 239; Rollings, *Unaffected by the Gospel*, 183–85; Stewart, *Peyote Religion*, 110–11.

54. Thanks to Jean Dennison for suggesting this point. On the constitution of 1881, see Burns, *History*, 390–93; Mathews, *Wah'Kon-Tah*, 121–35; T. Wilson, *Underground Reservation*, 24–73. For an analysis of that constitution as embracing "modern" Osage governance, see Warrior, *People*, 49–94.

55. Mathews's depiction of this process resonates with the "blood/land/memory" complex analyzed by Chadwick Allen in later Native writing. Moreover, the notion of "blood memory" that Allen discusses in the work of N. Scott Momaday appears in Mathews's oeuvre as the concept of "racial memory." See C. Allen, *Blood Narrative*, 160–93.

56. On modernist forms of primitivism, particularly as they draw on representations of American Indians, see Carr, *Inventing the American Primitive*, 197–256; Kent, *Reshaping of Modernism*, 71–112; Pfister, *Individuality Incorporated*; Schedler, *Border Modernism*, 41–84.

57. Mathews himself, though, is not immune from portraying Osages in primitivizing ways. For example, in *The Osages* he repeatedly refers to the Osage people as "Neolithic," which while possibly ironic seems to be used in a fairly descriptive fashion. Thus, rather than suggesting the complete absence of such dynamics in his work, I want to suggest the presence of another, less anachronizing strand of thought, in *Sundown* in particular.

58. While critics often describe the novel as a tale of someone caught between two cultures, this framing misapprehends the contours and stakes of Mathews's staging of the tensions

generated by the disjunctions between Osage and state-implemented spatiotemporal formations. Dennison argues that presenting "culture" as the means of defining Indigenous identities can call on Native peoples to disregard the ongoing history of settler colonialism and its profound effects on their socialities and modes of self-representation: "American Indian peoples are forced to overturn a destructive legacy of U.S. policies and reconnect to a culture damaged by the colonial process. As a colonial entanglement, American Indian culture is made to stand for all that is fundamental, pure, and noncolonized." *Colonial Entanglement*, 89. See also Barker, *Native Acts*; Den Ouden, *Beyond Conquest*; Engle, *Elusive Promise*; Niezen, *Origins of Indigenism*; Povinelli, *Cunning of Recognition*. Emerging in the early twentieth century out of a range of ethnological and realist discourses, the notion of culture suggests a kind of wholeness, such that each culture remains coherent and bounded while the person shuttling between or among them occupies some intermediate space leading to either alienation or hybridization. On the emergence of what would be characterized as the "culture concept," see Baker, *Anthropology*; Carr, *Inventing the American Primitive*, 197–206, 229–37; Darnell, *And Along Came Boas*; Elliott, *Culture Concept*; Evans, *Before Cultures*; Hegeman, *Patterns for America*. This concept comes to be mobilized within federal Indian policy in the mid-1930s, under the leadership of Commissioner John Collier, as a means of supposedly recognizing distinctive elements of Native life that had been denigrated and disciplined over the prior half century. See Marden, "Anthropologists"; Patterson, *Social History of Anthropology*, 71–102; Pfister, *Individuality Incorporated*, 185–228; Rifkin, *When Did Indians*, 181–232. Not only does such a framing largely displace the question of U.S. colonial superintendence of Native lands and governance (which Mathews addresses rather directly in ways discussed earlier), but it also tends to envision a more or less stable cultural matrix in which the effects of time are minimized, if not bracketed entirely.

59. In fact, there were numerous efforts to criminalize Peyote possession at various governmental levels from the 1880s through the 1930s. See Maroukis, "Peyote Controversy"; Stewart, *Peyote Religion*, 128–47, 213–37. I should note, though, that participation in the actual Peyote ceremony was limited to men, which, combined with the fact that the expression of the terms of Osage duration comes from either male characters or the narrator, extends the gendered dynamics of Mathews's account of Osage experience discussed in the previous section.

60. On "the road" in Peyotism, see Mathews, *Osages*, 740–58; Stewart, *Peyote Religion*, 60, 90.

61. Miranda, *Bad Indians*, xvi.

62. Of this moment, Keresztesi says, "Mathews expresses a radical and surprisingly forward-looking sentiment: the English settlers of the New World were aliens to the American land, thus their notions of progress and civilization did not match the spatial spirit of the continent they colonized." *Strangers at Home*, 154.

63. In his first novel, *Wah'Kon-Tah*, Mathews explores the ways that non-native perceptions, specifically those of Agent Labian J. Miles, can be transformed through long-term residence in Osage territory and among Osage people.

64. See Bailey, *Osage*, 32, 64; Hunter, "Historical Context," 52; Mathews, *Osages*, 333; Mathews, *Wah'Kon-Tah*, 321; Owens, *Other Destinies*, 51.

65. Goeman, *Mark My Words*, 37, 88. On "abstract space," see Lefebvre, *Production of Space*, 229–91.

66. Snyder, "'He Certainly Didn't Want,'" 28. See also R. Parker, *Invention*, 36–38. Susan Kalter has critiqued Snyder for not engaging with Mathews's various expressions of homophobia. "Introduction," xxiv, xxxvii–xxxviii, 262.

67. Ross, "Beyond the Closet," 168.

68. Rohy, *Anachronism and Its Others*, xiv–xv. For further discussion of the relation between discourses of race and sexuality in the nineteenth and early twentieth centuries, see Bederman, *Manliness and Civilization*; Carter, *Heart of Whiteness*; R. Ferguson, *Aberrations in Black*; Piatote 2013; Rifkin, *When Did Indians*, 143–80; Roscoe, *Changing Ones*; Somerville, *Queering the Color Line*; Stokes, *Color of Sex*. For discussion of the relation between mobility and sexual deviance in modernist writing, see Trask, *Cruising Modernism*. However, Trask does not address the relation between racialization and notions of perversity, which also entailed a conception of stuckness—in both time and place.

69. I borrow the term reprosexual from Dana Luciano, *Arranging Grief*.

70. The novel makes reference at several points to the continuing existence of the clan system as well as polygamy. For examples, see 43, 238–39, 260. On changes in Osage clan relations from the nineteenth to the twentieth century, see Nett, "Historical Changes."

71. The 1906 act specifies that heirs shall be determined "according to the laws of the Territory of Oklahoma." Kappler, *Indian Affairs*, 3:258. On this pattern more broadly, see Rifkin, *When Did Indians*, 143–232.

72. On the politics of the form of the bildungsroman, including its relation to (post)colonial articulations, see Esty, *Unseasonable Youth*; Slaughter, *Human Rights, Inc*. On its role in *Sundown*, see Musiol, "*Sundown* and 'Liquid Modernity'"; R. Parker, *Invention*, 19–50. On the ending of the novel (including Chal's mother's view of him) as indicating Chal's continued lack of direction and failure to achieve a sense of resolution with respect to his struggle to define himself, see Hunter, "Historical Context," 64; Musiol, "*Sundown* and 'Liquid Modernity,'" 368; Schedler, *Border Modernism*, 54, 73. For alternative readings of the ending as more hopeful, see Keresztesi, *Strangers at Home*, 159–60; Owens, *Other Destinies*, 59; Warrior, *Tribal Secrets*, 53, 83. On the narrative of queerness as a failure to achieve proper adulthood, see Halberstam, *In a Queer Time*.

73. On the ways Chal's experience of himself challenges the masculine gender codes of allotment, see Gillan, "Hazards of Osage Fortunes"; R. Parker, *Invention*, 19–50.

74. O'Brien, *Firsting and Lasting*, 105. On the history and problems of understanding indigeneity as a reproductively transmitted blood substance, see Barker, *Native Acts*; Garroutte, *Real Indians*; Kauanui, *Hawaiian Blood*; Rifkin, *When Did Indians*; Sturm, *Becoming Indian*.

75. Kappler, *Indian Affairs*, 4:317, 5:88–89. Laws passed from 1921 to 1938 include various other provisions that specify relations of debt and heirship based on blood quantum. See Kappler, *Indian Affairs*, 4:482, 5:89, 607–8. On the complex uses of "blood" by Osage people to define contemporary Osage identity, see Dennison, *Colonial Entanglement*, 48–74.

76. In addition, while competency per se was not defined in strictly racial terms, certificates of competency from 1906 to 1929 were overwhelmingly issued to persons classified as having less than one-half Indian blood. Department of the Interior, "Osage People," 23, 35.

77. T. Wilson, *Underground Reservation*, 180.

78. Freeman, *Time Binds*, 95, 120.

79. Mathews was well aware of the I'n-Lon-Schka, as suggested by the fact that Alice Callahan thanks him for providing great help while she was writing her book on the topic. *Osage Ceremonial Dance*, xii.

80. For a reading of Chal's dancing as indicating his inability "to reconcile the modernist and Osage aspects of his identity," see Schedler, *Border Modernism*, 46. See also Owens, *Other Destinies*, 58. For discussion of his dancing as expressive of repressed homoeroticism, see Snyder, "'He Certainly Didn't Want,'" 39–40. Given the stereotypical associations of dancing with Native peoples, one might want to deemphasize this aspect of the novel. However, the text also might be read as inhabiting that stereotypical figuration in order to reorient it away from primitivity and to convey a sensuous relation to place to non-native readers. Thanks to Jean Dennison for helping me elaborate this point. On the reconfiguration of stereotype in order to signify indigeneity, see P. Deloria, *Playing Indian*; Mithlo, *"Our Indian Princess"*; Raheja, *Reservation Reelism*; Rifkin, *Erotics of Sovereignty*, 215–66. On the ways stereotype can provide access to unacknowledged forms of affect and sensation, see Keeling, *Witch's Flight*.

81. In contrast, Keresztesi characterizes Chal's dreaming as "constant escape into the world of fantasy." *Strangers at Home*, 160. For other readings that present Chal's dreaming as passivity or lostness, see Hunter, "Protagonist as a Mixed-Blood," 329–32; R. Parker, *Invention*, 26–32; Schedler, *Border Modernism*, 44–45, 160. Once Chal has returned to Osage territory after his service in the military, he recounts "the talk of the strong, practical men who did things" as opposed to those Osages who "dreamed silly things in a mystical dream-world" (280), specifically in the sweat lodge of the village near Kihekah from which Chal recently had emerged. This reference to dreaming, along with others I discuss, indicates a connection between Chal's experiences and those of the Peyotists in the village.

82. In fact, wealth resulting from oil production enables the Osage to defend the Peyote religion against federal efforts to criminalize it. See previous note.

83. On the differences and relations among space as "conceived," "perceived," and "lived," see Lefebvre, *Production of Space*.

84. Mathews, though, was elected to the Osage council in 1934 after he had finished writing *Sundown*, and he served on the council for two four-year terms. See T. Wilson, "Osage Oxonian," 277–79.

FOUR. *Ghost Dancing at Century's End*

1. On Wovoka and the Ghost Dance of 1890, see DeMallie, "Lakota Ghost Dance"; Hittman, *Wovoka*; Kehoe, *Ghost Dance*; Mooney, *Ghost-Dance Religion*; Smoak, *Ghost Dances and Identity*. Some of the disputed points, and ideas that shifted as the Ghost Dance moved through various regions and populations, include whether the Native dead would return to this earth or the reunion would happen in heaven, whether whites would be included in such a reunion, and whether, if the reunion occurred on this earth, whites would be moved to some other space beyond the Americas or eliminated entirely.

2. On the Ghost Dance of 1870 among the Walker Lake Paiutes and its wide dissemination, see Du Bois, *1870 Ghost Dance*; Hittman, "1870 Ghost Dance"; Mooney, *Ghost-Dance Religion*; Smoak, *Ghost Dances and Identity*, 113–51. On prophetic movements in what is now

Washington State, see Ruby and Brown, *Dreamer-Prophets*; Harmon, *Indians in the Making*, 125–30; Kehoe, *Ghost Dance*, 115–22; Christopher Miller, *Prophetic Worlds*; Mooney, *Ghost-Dance Religion*, 708–63; Suttles, *Coast Salish Essays*, 152–98; Trafzer and Beach, "Smohalla." On the Ghost Dance in the Great Basin, see Dobyns and Euler, *Ghost Dance of 1889*.

3. On the circulation of this narrative, see P. Deloria, *Indians in Unexpected Places*, 15–51; DeMallie, "Lakota Ghost Dance"; S. Pratt, "Wounded Knee"; Smoak, *Ghost Dances and Identity*, 1–10; Tatonetti, "Disrupting."

4. I borrow the notion of world making from work in queer studies. In particular, see Berlant and Warner, "Sex in Public."

5. Vizenor, *Manifest Manners*, 3, 8.

6. Goeman, *Mark My Words*, 2.

7. Thanks to Coll Thrush and Mishuana Goeman, whose comments on the chapter led me to think in more detailed ways about the relation between prophecy and the everyday.

8. On Native prophecy and its relation to historiography, see Cruikshank, *Social Life of Stories*, 116–37; Dowd, *Spirited Resistance*; Cary Miller, "Every Dream"; Nabokov, *Forest of Time*, 218–40. On the use of prophecy in contemporary Native filmmaking, see Raheja, *Reservation Reelism*, 145–89.

9. West-Pavlov, *Temporalities*, 12.

10. S. Pratt, "Wounded Knee," 156–57. See also Elliott, *Culture Concept*, 89–123. On the importance of attending to Native ontologies on their own terms, especially when they do not fit Euro-American notions of rationality and empiricism, see Bierwert, *Brushed by Cedar*; Boyd and Thrush, "Introduction"; Cruikshank, *Social Life of Stories*; Fogelson, "Ethnohistory of Events"; Grady, "Ancestors"; Landrum, "Shape-Shifters"; Cary Miller, "Every Dream"; Sarris, *Mabel McKay*; Shorter, *We Will Dance*; Stevenson, *Life beside Itself*; Wilson, *Remember This!* On Euro-Americans' critique of such knowledges as animism while accepting similar relations (albeit in a different form) in their own philosophies and practices, see Bracken, *Magical Criticism*. However, on the ways Indigenous peoples can be called on to perform (constrained and stereotypical iterations of) alterity in order to be recognized by the settler state, see Povinelli, *Cunning of Recognition*.

11. S. Pratt, "Wounded Knee," 162.

12. On the everyday life of prophecy, see Amoss, *Coast Salish Spirit Dancing*; Bierwert, *Brushed by Cedar*; Cruikshank, *Social Life of Stories*, 116–37; Cary Miller, "Every Dream"; Shorter, *We Will Dance*; Thornton, *Being and Place*.

13. For a critique of the use of schizophrenia as metaphor in the novel, see Christie, "Renaissance Man."

14. Many of the readings of the novel focus on its critiques of settler violence and/or of non-native representation. See Burnham, "Sherman Alexie's *Indian Killer*"; T. Chen, "Ethics of Knowledge"; Cox, *Muting White Noise*, 178–99; J. Dean, "Violence of Collection"; Hollrah, "Sherman Alexie's Challenge"; Homans, "Adoption Narratives"; James, "'Indians'"; Krupat, *Red Matters*, 98–121; Tatonetti, "Dancing That Way"; Van Styvendale, "Trans/Historicity of Trauma." I would like to build on these readings by focusing more on the ways that Alexie envisions possibilities for moving beyond the limits of non-native conceptions of the real by offering a reformulated vision for thinking about Native temporalities.

15. Such claims to know or own Indianness can be thought of, in Shannon Sullivan's formulation (in *Revealing Whiteness*), as a form of "ontological expansiveness" associated with whiteness. See also Laminack, "Wounded Whiteness."

16. Cox, *Muting White Noise*, 178.

17. J. Dean, "Violence of Collection," 32–33.

18. Alexie's *Indian Killer* will be cited parenthetically throughout. The novel distinguishes between the use of non-native documents by Native people(s) for the purposes of making claims to non-native institutions (such as for federal acknowledgment) and the kind of individual self-invention that Wilson performs.

19. For a sense of *repertoire* as the daily practices and beliefs that function as an alternative to the *archive*, see Taylor, *Archive*. For a discussion of settler phenomenology, see Rifkin, *Settler Common Sense*. On non-natives drawing on Indianness as a means of (re)fashioning their identities, see Byrd, *Transit of Empire*; P. Deloria, *Playing Indian*; Huhndorf, *Going Native*; Sturm, *Becoming Indian*.

20. When he first meets John, "Wilson was too shocked by John's obvious resemblance to his own hero, Aristotle Little Hawk, to be afraid" (268), suggesting that Wilson understands John through the prism of his own fictional imaginings, which then become the basis for his projection of the supposed *reality* of John's identity as the Indian Killer.

21. I follow the novel's pattern of referring to Wilson and Mather by their last names while using first names for most other characters. On the implications of the novel's critique of Mather for teaching Native literatures and cultures, see T. Chen, "Ethics of Knowledge"; Hollrah, "Sherman Alexie's Challenge."

22. Moreton-Robinson, "Writing Off Treaties," 82, 88.

23. Ahmed, *Queer Phenomenology*, 50.

24. Ahmed, *Queer Phenomenology*, 50.

25. Vizenor, *Manifest Manners*.

26. See Cox, *Muting White Noise*, 178–99; J. Dean, "Violence of Collection"; Van Styvendale, "Trans/Historicity of Trauma."

27. See Cox, *Muting White Noise*, 188–89. On "lasting," see O'Brien, *Firsting and Lasting*.

28. Patrice Hollrah argues that the novel displaces white narrations of Native identity in favor of Native ones, whereas Louis Owens has suggested that the novel, and Alexie's other work, simply reiterates stereotypes of Indianness. Hollrah, "Sherman Alexie's Challenge"; Owens, *Mixedblood Messages*, 74–80. However, I would suggest in response to both that Alexie explores the troubled genealogy and trajectories of Native people's notions of Indianness in ways that parallel and amplify the novel's critique of non-native narration.

29. On the depiction of adoption in the novel, see Homans, "Adoption Narratives"; James, "'Indians'"; Van Styvendale, "Trans/Historicity of Trauma." On the history of the Indian Child Welfare Act (1978) and its efforts to respond to conditions like those faced by John, see Cross and Miller, "Indian Child Welfare Act"; Graham, "Reparations"; Strong, "What Is."

Despite the fact that the Duwamish historically had been acknowledged for various purposes as a tribe by the federal government, the judge who ruled in what has come to be known as the Boldt decision (recognizing the right of Native peoples in Washington to have control over half of the annual salmon run) decided five years later, in 1979, that the Duwamish and four other peoples in the Puget Sound region did not qualify for treaty fishing rights. For the

Duwamish, that determination was based on a supposed break in the tribe's self-governance from 1916 to 1925. See Harmon, *Indians in the Making*, 180–82, 209–12, 241–43; Thrush, *Native Seattle*, 193–99. On the history of legal and political nonrecognition for peoples in western Washington, see F. Porter, "Without Reservation."

30. For a sampling of scholarship on the politics and problems of federal recognition, see Barker, *Native Acts*; Den Ouden and O'Brien, *Recognition*; Klopotek, *Recognition Odysseys*; Lowery, *Lumbee Indians*; M. Miller, *Forgotten Tribes*. On the forced (mis)translation of Pacific Northwest geographies and practices into the terms of settler-state legalities, see Boxberger, "Not So Common"; Bracken, *Potlatch Papers*; Fisher, "Reserved for Whom?"; Harmon, *Indians in the Making*; Rigsby, "Stevens Treaties".

31. We might see the impossibility of finding an avenue to communicate Native experience as most forcefully illustrated by John, especially in his memories and fantasies of Father Duncan. A Spokane Jesuit priest, Duncan baptizes John and provides a model and interlocutor for him until Duncan wanders off into the Arizona desert when John is seven. When standing with John in the Chapel of the North American Martyrs in Seattle looking at "vivid stained glass reproductions of Jesuits being martyred by Indians" (13), Duncan says to John, "You see these windows? You see all of this? It's what is happening inside me right now," and the narrator notes of Duncan's thoughts at that moment, "As a Jesuit, he knew those priests were martyred just like Jesus. As a Spokane Indian, he knew those Jesuits deserved to die for their crimes against Indians" (15). The legacies of settlement, the varied stories circulating around those ongoing histories, and the complex affects generated by them remain trapped within Duncan, locked into a scene of conflict that does not seem to change or evolve. Duncan's apparent suicide suggests his inability to find a livable orientation to the reality produced through the materialization of settler stories, a process represented by the chapel and its images. He provides a touchstone for John throughout the text, such as when John, Marie, and others have fought off an assault by a group of local whites: John "wanted to talk, to finally speak. To tell them about Father Duncan and the desert, the dreams he had of his life on a reservation.... But there was no language in which he could express himself" (377). John's absence of speech here suggests less the silence of nonconscious orientation at play in *Sundown* (as discussed in chapter 3) than a stuckness like Duncan's—an inability to find a way to engage that is not always already stymied by settler narratives' freezing of Native people(s) into a stagnant Indian *realness*. Through John's suicide, in which the narrator suggests he "strode into the desert" like (and perhaps in search of) Father Duncan (413), the narrative casts him as an example of the impossible subjectivity produced for Indians within the dominant settler temporality.

32. The following sketch is drawn from Ackerman, "Kinship"; Amoss, *Coast Salish Spirit Dancing*; Asher, *Beyond the Reservation*; Barsh, "Ethnogenesis and Ethnonationalism"; Blee, *Framing Chief Leschi*; Boxberger, "Not So Common"; Carlson, *Power of Place*, 7–57; Fisher, "Reserved for Whom?"; Friday, "Performing Treaties"; Harmon, *Indians in the Making*; Harmon, "Coast Salish History"; Klingle, *Emerald City*; B. Miller, "Introduction"; B. Miller and Boxberger, "Creating Chiefdoms"; F. Porter, "Without Reservation"; Raibmon, *Authentic Indians*, 74–115; J. Ross, "Spokane"; Ruby and Brown, *Dreamer-Prophets*; Ruby and Brown, *Spokane Indians*; Suttles, *Coast Salish Essays*, 209–32; Thrush, *Native Seattle*.

33. Powell, "X-Blood Files," 91.

34. On Native residence in urban centers, see C. Andersen, "Urban Aboriginality"; Danzinger, *Survival and Regeneration*; Fixico, *Urban Indian Experience*; Goeman, *Mark My Words*; D. Jackson, *Our Elders Lived It*; Lawrence, *"Real" Indians and Others*; Ramirez, *Native Hubs*; Strauss and Valentino, "Retribalization." This scholarship makes clear the need to consider the effects of not only forced dislocation and relocation but also chosen forms of Native movement to urban areas, ongoing relations between people living in urban areas and formally recognized reservations/reserves, and Native attachments to cities owing to their location within Indigenous homelands (such as Seattle) that likely are not recognized as such. As Evelyn Peters and Chris Andersen note, "Most cities are located on sites traditionally used by Indignous peoples.... The creation of Indigenous 'homelands' outside of cities is in itself a colonial invention." *Indigenous in the City*, 7–8.

35. Goeman, *Mark My Words*, 103.

36. Vizenor, *Manifest Manners*.

37. See Weheliye, *Habeas Viscus*.

38. For discussion of *Indian Killer*'s use of the Ghost Dance, especially in comparison to Alexie's earlier work, see Tatonetti, "Dancing That Way." For an account of Alexie as more skeptical of the Ghost Dance in his earlier work, see Farrington, "Ghost Dance". In his later novel *Flight*, Alexie references the Ghost Dance as something of a masochistic white fantasy, with the main character being called on to enact a modern-day version through mass killing by a white man who calls himself Justice, but it remains ambiguous whether that novel seeks to displace the Ghost Dance as a horizon of Native vision and remembrance or to indict certain contemporary ways of narrating the nineteenth-century movement. Jeff Berglund suggests, "Alexie considers *Flight* to be his antidotal response to *Indian Killer*, a novel he has largely disowned since 2001 because of its fundamentalism and dangerously narrow view of tribalism," further arguing, "To end violence, one must get outside it." " 'Imagination,' " xxiv–xxv. In addition, Alexie published a short story called "Ghost Dance" in which the soldiers killed alongside George Armstrong Custer come back to life and, zombielike, go on a spree killing everyone they encounter.

39. On how the claiming of a previously unknown or lost Native identity (accurately or not) by those previously identified as white functions as a form of white privilege, see Sturm, *Becoming Indian*; TallBear, *Native American DNA*, 136–41.

40. See Homans, "Adoption Narratives," 22–23; James, " 'Indians' "; Krupat, *Red Matters*, 98–121; Tatonetti, "Dancing That Way," 18–20; Van Styvendale, "Trans/Historicity of Trauma."

41. Marie attributes this version of the Ghost Dance (as an erasure of whites) to Wovoka, but sources suggest that Wovoka's vision did not entail the return of the Native dead to this world or the elimination of Euramerican presence, instead pointing to a peaceful afterlife in which all people would dwell together and seeking to hasten the entry into that state while also calling on Native people to continue to work for and with whites in the ways they already were doing. See Hittman, *Wovoka*; Kehoe, *Ghost Dance*, 5–8, 27–42; Mooney, *Ghost-Dance Religion*, 764–91. However, later I will address the stakes of seeking to assess the accuracy of Alexie's vision by comparison to the "real" Ghost Dance inaugurated by Wovoka's visions.

42. Bierwert, *Brushed by Cedar*, 4.

43. Alexie in his earlier writings offers the formulation "Survival = Anger × Imagination," adapted from the earlier expression "Poetry = Anger × Imagination." See McFarland, "Sherman Alexie's Polemical Stories."

44. See T. Chen, "Ethics of Knowledge," 164–65; Giles, *Spaces of Violence*, 128–44; James, "'Indians.'"

45. Burnham, "Sherman Alexie's *Indian Killer*," 9.

46. S. Pratt, "Wounded Knee," 156.

47. Given the text's proliferation of criteria for Indianness and its staging of the Native characters' failure to fulfill them, the doubling of the title *Indian Killer* (as the title of both Alexie's novel and Wilson's novel-within-the-novel) can be interpreted as less a substitution—here's what the truth of Indianness actually is—than a displacement of the conventional means by which to determine the boundaries and character of realness.

48. Bierwert, *Brushed by Cedar*, 148.

49. Here I am drawing on and extending a brief gesture that Burnham makes toward the end of her essay "Sherman Alexie's *Indian Killer*."

50. Some critics have suggested that the novel offers a departicularized, generic pan-Indianism, especially in its invocation of Lakota histories. See Cox, *Muting White Noise*, 195; J. Dean, "Violence of Collection," 45–50; Krupat, *Red Matters*, 115–21; Tatonetti, "Dancing That Way," 20–22. Scholars have noted the differences between the outlines of Wovoka's vision (which emphasized a shared heaven and the movement of all people toward it) and the prophecy that arose among Lakotas (the more familiar version, of the return of the Native dead to this world and the cleansing of white presence from Indigenous lands). Some have characterized this distinction in terms of a Siouxian perversion or militarization of what was originally or fundamentally a peaceful vision. See Hittman, *Wovoka*; Kehoe, *Ghost Dance*. For critiques of the dismissal of Lakotas in this account, as well as of the attempt to purify the Ghost Dance by treating Wovoka as a pacific origin and all deviations from it as corrupt, see DeMallie, "Lakota Ghost Dance"; Ostler, *Plains Sioux*; S. Pratt, "Wounded Knee"; Smoak, *Ghost Dances and Identity*.

51. Smoak, *Ghost Dances and Identity*, 191, 199.

52. See Christopher Miller, *Prophetic Worlds*, 23–51; Suttles, *Coast Salish Essays*, 152–98; Walker and Schuster, "Religious Movements," 499–501.

53. Quoted in Mooney, *Ghost-Dance Religion*, 711. Smohalla was from a Sahaptian-speaking people rather than a Salishan one, but I have included him as a "Salish" influence given the incredible significance he and his followers had in the region. On Smohalla, see Christopher Miller, *Prophetic Worlds*, 118–21; Mooney, *Ghost-Dance Religion*, 708–31; Ruby and Brown, *Dreamer-Prophets*, 19–102; Trafzer and Beach, "Smohalla"; Walker and Schuster, "Religious Movements," 501, 505.

54. On Transformer, see Bierwert, *Brushed by Cedar*, 72–111; Blee, *Framing Chief Leschi*, 26–27; Klingle, *Emerald City*, 12–13, 19, 42; Suttles, *Coast Salish Essays*, 3–14, 152–98.

55. Bierwert, *Brushed by Cedar*, 286; quotations are from Thrush, *Native Seattle*, 25. Thrush writes the term as *dookw*, but given the phonetic similarity to Bierwert's transcription *duk* (which also appears in a slightly modified form in Suttles, *Coast Salish Essays*, 185) and the fact that Thrush translates it as "'to change' or 'transform'" (23), *dookw* and *duk* seem to be the same term.

56. Merleau-Ponty, *Phenomenology of Perception*, 125.

57. On the Longhouse religion, see Amoss, *Coast Salish Spirit Dancing*; Bierwert, *Brushed by Cedar*, 160–96; Ruby and Brown, *Dreamer-Prophets*, 29–50; Suttles, *Coast Salish Essays*, 199–208; Walker and Schuster, "Religious Movements," 501–7. The practice of Smohalla's followers has been considered part of the Washat tradition. On the ways "spirit dances" became linked to annual celebrations of treaty promises in the Puget Sound region in the twentieth century, see Friday, "Performing Treaties". While it is not wholly clear when during the year the novel is set, the sections with Mather suggest the beginning of the semester, so either early fall or midwinter. If the latter, that would be the time of year for such dances and the practice of the Longhouse religion. Thanks to Jim Cox for helping me think about the novel's setting in time.

58. There is some disagreement about whether the term *syowen* itself refers to the spirit guide or to the song given by the guide.

59. J. Miller, "Basin Religion and Theology," 73. Also, the prophet dances on the Plateau sometimes specifically appealed to the Transformer as the agent of renewal. Suttles, *Coast Salish Essays*, 153, 159–60, 164.

60. At one point Mather suggests that the "Indian Killer . . . is an inevitable creation of capitalism. A capitalistic society will necessarily create an underclass of powerless workers. . . . Indian Killer is, in fact, a revolutionary construct" (245), and Marie responds to Mather, "I'm not quite the revolutionary construct you had in mind, am I?," further noting that she's "not some demure little Indian woman healer talking spider this . . . babbling about the four directions" (247). In this exchange the novel suggests that modes of realist explanation (the killer is a "construct" created by "capitalism") rely on narrative frames that circulate accreting forms of simulation (the "little Indian woman" invoking "the four directions").

61. Such identification might be thought of as the obverse of Elizabeth Freeman's notion of "temporal drag," a feeling or desire across time that does not so much "reincarnate the lost, nondominant past in the present" in order "to pass it on with a difference" (*Time Binds*, 71) as project pastness onto contemporary phenomena.

62. See Harmon, *Indians in the Making*, 218–41; Thrush, *Native Seattle*, 162–83.

63. On the history of Red Cloud and Crazy Horse, including the latter's murder, see Ostler, *Plains Sioux*. On events in 1973, see Smith and Warrior, *Like a Hurricane*.

64. On the continuing importance of treaties in the Pacific Northwest, see Blee, *Framing Chief Leschi*; Harmon, *Power of Promises*. Even as the text raises questions about the dynamics of a reservation-based sense of Native identity and territoriality, it seems to remain committed to treaties as an index of continuing Native sovereignty and as the most available legal vehicle for insisting on non-native recognition of it.

65. In contrast, Burnham, in "Sherman Alexie's *Indian Killer*," reads the novel as something of a ghost story. Within Coast Salish ontologies, owls "are thought to be manifestations of the dead" (Amoss, *Coast Salish Spirit Dancing*, 74), and in this fashion the proliferation of owls in the novel may signal the presence of those who have passed even as the killer itself is not revenant.

66. On this tradition, see Bergland, *National Uncanny*; Boyd and Thrush, "Introduction."

67. On the speech's production, history, circulation, and reception, see Bierwert, "Remembering Chief Seattle"; Furtwangler, *Answering Chief Seattle*. The speech is reproduced in Furtwangler's book.

68. In Furtwangler, *Answering Chief Seattle*, 17, 15.

69. Thrush, *Native Seattle*, 3–4, 7.

70. Some scholars have sought to recuperate the notion of haunting as a way of indicating those elements of contemporary life that exceed existing modes of explanation and documentation (especially in the social sciences), as well as the continuing collective legacies of (social) death. See Derrida, *Specters of Marx*; A. Gordon, *Ghostly Matters*; Holland, *Raising the Dead*; Lim, *Translating Time*. However, while not rejecting such formulations, I'm seeking to mark the work of the "ghostly" in Seattle Indian stories and Alexie's irony, especially as the novel implicitly distinguishes such spectral tales and the temporality they posit from the materialization of prophecy. However, Bierwert highlights Chief Seattle's own participation in Salish ceremonial practice and the ways his commemoration in the Chief Seattle Days starting in the early twentieth century became the occasion for performing "traditional medicine dances." "Remembering Chief Seattle," 286, 291–95.

71. John is described as "six feet six inches tall and heavily muscled" (23), and accounts of Chief Seattle indicate that he also was a very large man (Furtwangler, *Answering Chief Seattle*, 10, 40). John also resembles Wovoka, who was "nearly six feet in height." Mooney, *Ghost-Dance Religion*, 768–69.

72. See Blee, *Framing Chief Leschi*, 116–24; Klingle, *Emerald City*; Thrush, *Native Seattle*.

73. In Furtwangler, *Answering Chief Seattle*, 16.

74. Although Wovoka did not directly oppose settler expansion, it is notable that the Yerington Paiutes, of whom Wovoka was one, did not have a reservation (unlike those at Walker and Pyramid Lakes) and that Wovoka repeatedly requested access to land on the Walker Lake reservation. Hittman, *Wovoka*, 74–75, 97, 218.

75. On the role of prophets in Native resistance in eastern Washington Territory, see Ruby and Brown, *Dreamer-Prophets*. On the significance of Alexie's use of the name Polatkin, see Hollrah, "Sherman Alexie's Challenge," 161. On the war of 1858, see Ruby and Brown, *Spokane Indians*, 83–140. On Alexie's references to the war of 1858 in his previous work, see McFarland, "Sherman Alexie's Polemical Stories"; Peterson, "'If I Were Jewish.'"

76. Womack, "Theorizing American Indian Experience," 372.

77. Cruikshank, *Social Life of Stories*, 133–34.

78. Das, *Life and Words*, 100, 104.

79. Das, *Life and Words*, 108.

80. This image of pulling out eyes may allude to the role of such eye loss in the making of the sun in Spokane creation stories. Ruby and Brown, *Spokane Indians*, 7.

81. See also Coulthard, *Red Skin, White Masks*, 105–30.

82. The novel at times chronicles John's imagination of his life on the reservation had he been raised by his biological mother (43–48, 287–92). In these daydreams, it appears as a place of comfort and tradition, one from which an untroubled sense of Indian authenticity might arise in ways that other characters' experiences on actual reservations puts into question. In addition, Alexie suggests that when John commits suicide he is going in search of his biological mother, his biological father, and Father Duncan (413), presenting John as ultimately seeking to recapture a past that has been lost. On this dynamic in the novel, see Homans, "Adoption Narratives." I should be clear that I am not suggesting that John's turn to fantasy or his lack of language makes him less authentic or suggests that he has been

"contaminated" by white culture. For such a claim, see Grassian, *Understanding Sherman Alexie*, 104–11; Giles, *Spaces of Violence*, 132–34, 143. On the problems of understanding the reservation as the space of Native authenticity, see Barker, *Native Acts*; Goeman, *Mark My Words*; Ramirez, *Native Hubs*. I will return to this point later in my discussion of *Gardens in the Dunes*.

83. See Berglund, "'Imagination,'" xxiv–xxv; L. Cooper, "Critique of Violent Atonement"; J. Dean, "Violence of Collection," 49–50; Grassian, *Understanding Sherman Alexie*, 104–26; Giles, *Spaces of Violence*, 128–44; Krupat, *Red Matters*, 98–121; Van Styvendale, "Trans/Historicity of Trauma," 220. On the generativity of Native anger, see Berglund, "Facing the Fire"; Carpenter, *Seeing Red*; Coulthard, *Red Skin, White Masks*; Simpson, *Mohawk Interruptus*.

84. Ahmed, *Cultural Politics of Emotion*, 175, 39.

85. Ahmed, *Cultural Politics of Emotion*, 175.

86. On the persistence of Wovoka's influence after Wounded Knee and the expansion of movements inspired by his prophecy to other areas, see Dobyns and Euler, *Ghost Dance of 1889*, 5, 10, 24–26, 33–35, 37; Kehoe, *Ghost Dance*, 27–52; Mooney, *Ghost-Dance Religion*, 653, 927; Hittman, *Wovoka*; Ruuska, "Ghost Dancing." Readers are told that President McKinley has chosen Teddy Roosevelt as a running mate (125), that events in the novel occur in the wake of "war with Spain" (162), and that the characters' travel in Italy comes after the assassination of the king of Italy (276), which cumulatively point to 1900 as the year in which the novel is set.

87. On the relation between notions of origin and racialized modes of authenticating Indianness, see Barker, *Native Acts*, 217–28. On the distinction between racial calculations and more expansive modes of genealogical reckoning among Indigenous peoples, see Kauanui, *Hawaiian Blood*; Rifkin, *When Did Indians*; TallBear, *Native American DNA*.

88. On the fictionality of the Sand Lizard people, see Arnold, "Listening to the Spirits," 163–64, 172–73.

89. Silko's *Gardens in the Dunes* will be cited parenthetically throughout.

90. For examples, see Coltelli, "That Gardens of Memory"; Huhndorf, *Going Native*, 189–98; Regier, "Revolutionary Enunciatory Spaces"; Roppolo, "'We've Got.'"

91. Moore, "Ghost Dancing," 96, 99.

92. Nyong'o, *Amalgamation Waltz*, 176–77, 10.

93. On this effect, see also C. Andersen, *"Métis"*; H. Jackson, *American Blood*, 46–88; Sexton, *Amalgamation Schemes*; Stokes, *Color of Sex*; Young, *Colonial Desire*. One might challenge this sense of a meeting between Natives and Euramericans as producing a third, new option—the hybrid—by instead characterizing that relation as one of *indigenization* or *syncretism*, in which once-alien beliefs, practices, objects, and even persons become part of Native lifeways. This terminological reorientation diffuses the impression of anomaly that attaches to Indigenous adoptions and adaptations, but it still preserves something of the (reproductive) temporality of hybridity, continuing to posit a genealogical unfolding into which at a given historical moment something new is introduced through combination. See Palmer, "Devil in the Details," 277–78.

94. Ahmed, *Queer Phenomenology*, 66. One can see such connections at play in A. M. Regier's discussion of the role of the Ghost Dance in the novel, particularly in the running

comparison of the text's supposed celebration of hybridity with *mestizaje*, itself a racializing matrix through which cultural change is imagined. Regier, "Revolutionary Enunciatory Spaces." See Contreras, *Blood Lines*. Similarly, Shari Huhndorf describes Indigo in the following terms: "As a figure who is racially mixed and who unites past and present, the child is the ultimate crosser of racial and temporal boundaries." *Going Native*, 198.

95. Ahmed, *Queer Phenomenology*, 70, 83.

96. At several points, Silko notes that Jesus is accompanied by his wife, his mother, and his eleven children (30–32, 276–77, 319–20). Moreover, Jesus not only travels with Wovoka (22–24) but "looked like he might be Paiute... with handsome dark skin and black hair and eyes" (220). In addition, he comes from "the mountains beyond Walker Lake, where he was born" (262). Together, these dynamics suggest a connection in the text between deviations from conventional understandings of causality and from normative Christian notions of sociosexual order (in which Jesus himself is celibate). While "claim[ing] to speak for Jesus Christ," those same people who deny Jesus's potential identity as Paiute seek to take away Indigenous children in order to train them for civilization, condemn women's nudity in seeking to institute specific gendered ideals of sexual propriety, and depend for aid on "the authorities [who] punished the reservation Indians for any contact they had with the renegades" living up in the dunes (49).

97. J. Miller, "Basin Religion and Theology," 72–73, 82. On the nonantagonistic coexistence of nonequivalent principles in Native philosophical traditions, see Waters, "Language Matters." On the role of "life seeking like" in Native knowledge systems, see Cajete, *Native Science*.

98. Some critics have suggested that the text substitutes myth for history or fuses them into a hybrid construction. See Cummings, "'Settling' History"; Moore, "Ghost Dancing"; J. Porter, "History in *Gardens*." However, such a formulation seems to me to come close to the kinds of "ontological reduction" discussed earlier and to repeat, albeit in a more supportive key, the portrayal of Ghost Dancing in the Great Basin as a kind of "fantasy pattern." Dobyns and Euler, *Ghost Dance of 1889*, 49. This analysis underplays the significance of the text's account of prophecy as indicating a form of temporality and historicity not reducible to post-Enlightenment notions of linear causality, progress, or succession. On the ways myth might be understood as a mode of history, see Nabokov, *Forest of Time*, 85–104; Shorter, *We Will Dance*; Thornton, *Being and Place*; Walters, *Talking Indian*; Wilson, *Remember This!*

99. There is some question about whether Wovoka presented himself as Jesus or as like him. Certainly, Wovoka was presented by others as "the Christ." See Dobyns and Euler, *Ghost Dance of 1889*, 19, 22; Hittman, *Wovoka*, 7–9, 18, 190–91; Kehoe, *Ghost Dance*, 6; Mooney, *Ghost-Dance Religion*, 780–81, 793–96, 913.

100. This process suggests, in temporal terms, something like Renya Ramirez's conception of the "hub," in which geographically dispersed Native people maintain relations to each other and their homeland(s). See Ramirez, *Native Hubs*.

101. Regier suggests that the novel "implies an ongoing future direction with a wide time scale, as dance gatherings are presumed to take place in the future, possibly even in the present time of the reader." "Revolutionary Enunciatory Spaces," 143.

102. Arnold, "Listening to the Spirits," 167; Ruoff, "Leslie Marmon Silko's," 10–11. Of the Colorado River peoples, Mooney observes, "The agent of the Mohave states officially that

these Indians knew nothing about it, but this must be a mistake, as there is constant communication between the Mohave and the southern Paiute, and, according to Wovoka's statement, Mohave delegates attended the dance in 1890, while the 700 Walapai and Chemehuevi associated with the Mohave are known to have been devoted adherents of the doctrine." *Ghost-Dance Religion*, 805.

103. Shepherd, *We Are*, 54.

104. Shepherd (*We Are*, 45–88) offers a rich and complex portrait of the advantages and problems of the reservation. On the Ghost Dance among the Hualapai, including the fact that it continued in various forms through the 1890s, see Dobyns and Euler, *Ghost Dance of 1889*. For discussions of the operation of the reservation system and the spatial, sexual, and spiritual ideologies that supported it, see Asher, *Beyond the Reservation*; Bruyneel, *Third Space of Sovereignty*, 65–95; Genetin-Pilawa, *Crooked Paths to Allotment*; Hoxie, *Final Promise*; Ostler, *Plains Sioux*; Piatote, *Domestic Subjects*; Rand, *Kiowa Humanity*; Rifkin, *When Did Indians*, 143–232; Rockwell, *Indian Affairs*; Ruuska, "Ghost Dancing"; Simonsen, *Making Home Work*; Trennert, *Alternative to Extinction*. On the ways many of the dynamics and features of Indian policy in the West appeared earlier in New England, see Den Ouden, *Beyond Conquest*; Mandell, *Tribe, Race, History*; O'Brien, *Dispossession by Degrees*.

105. The report is reproduced in Prucha, *Documents*, 128.

106. Goeman, *Mark My Words*, 12.

107. My reading of the novel's critique of reservation-based reckonings of time and origin, and its offering of a vision of sexual freedom as opening up ways of conceptualizing Native identity and historicity, deviates from that by critics who suggest the novel chooses the "mythic" or "spiritual" over the "political." See Murray, "Old Comparisons"; J. Porter, "History in *Gardens*." In addition, as Million (*Therapeutic Nations*, 103–45) has argued, consigning the spiritual to some other realm than the political, such as understanding it as a form of agovernmental "healing," can work as a way of preserving a heteropatriarchal understanding of what can constitute meaningful governance and political expression for Native peoples.

108. We learn that their father was Laguna, had other children before meeting their Chemehuevi mother, and took his children to live in Winslow when she died (334). At points, the novel presents Maytha and Vedna's land as on the Chemehuevi reservation (as when the minister contacts the "Chemehuevi reservation superintendent about unauthorized Indians," Indigo and Sister Salt, living with the Chemehuevi sisters [452]), but the text also indicates that Maytha and Vedna "bought" their land (405). In addition, the Chemehuevi reservation as such did not exist until 1907. Up until that point, those Chemehuevis living in the Chemehuevi Valley (which became the reservation) were formally under the authority of the agent for the Colorado River Indian reservation, but they did not have a federally recognized land base. Also, the allotments that were made on the reservation were eventually vacated owing to the flooding of the Chemehuevi Valley as a result of the damming of the Colorado River, an event that did not occur until the 1930s (even though the novel presents it as a turn-of-the-century event). Most of the Chemehuevis were removed to the Colorado River Indian reservation, at which point they ceased to exist as a separate tribe. Their status as such was not restored until 1970. See *Annual Report* 1887, 1; 1893, 108; 1900, 187; Caylor, "'A Promise Long Deferred,'" 212–13; Chemehuevi Indian Tribe, "History and Culture," n.d.; Roth, "Incorporation," 119–26, 156–76. The novel, then, condenses a good deal of Chemehuevi history

in ways that allow it to present the Chemehuevis as metonymic of the reservation system's effects on Native identity more broadly.

109. I should note that the absence of rigid, exclusive boundaries (geographic or procreative) does not mean the lack of a sense of coherent peoplehood, but it does suggest a different mode of living peoplehood than those institutionalized as tribal identity by the U.S. government.

110. Vizenor, *Manifest Manners*, 8; Miranda, *Bad Indians*, 136.

111. In the 1880s and 1890s, the agents for the Colorado River Indian reservation often characterized the Chemehuevis in Chemehuevi Valley (who were under that agent's authority but outside the boundaries of the reservation per se) as among the most "civilized" in the region. See *Annual Report* 1881, 2; 1885, 1; 1886, 1; 1898, 111; 1900, 187. In contrast to the novel's portrayal, though, Chemehuevis in this period retained a support for mixture with other populations, adopting Euramerican dress and technologies while not necessarily accepting their logics of racial bloodedness and belonging. See Roth, "Incorporation," 90, 112–13, 132–37.

112. Justice, "'Go Away, Water!,'" 156.

113. *Annual Report* 1890, 2; 1898, 112; 1899, 147. Not only did many Mohaves avoid the Colorado River Indian reservation, including those who remained in the vicinity at Fort Mohave when the reservation was created in the 1860s, but those who lived in the vicinity of Needles benefited from the economies and travel made possible by the railroad line running through it. On the role of the railroad in facilitating the spread of the Ghost Dance movement in 1890, see Ruuska, "Ghost Dancing." On Mohave history in this period, see Roth, "Incorporation," 232–72.

114. Merleau-Ponty, *Phenomenology of Perception*, 326.

115. Mormons had been proselytizing in the region since the 1850s (including purchasing Southern Paiute captives from the Utes), and they achieved significant conversions in the northern Great Basin during the 1870s. See Blackhawk, *Violence over the Land*, 226–44; Coates, "Mormons," 90, 99, 105; Roth, "Incorporation," 95, 242. On the ways the U.S. Army's presence in Utah in the 1850s and 1860s to subdue Mormons led to extensive violence against Native peoples, particularly Shoshones, see Blackhawk, *Violence over the Land*, 226–66.

116. On this history, see S. Gordon, *Mormon Question*; Iversen, *Antipolygamy Controversy*; Talbot, *Foreign Kingdom*.

117. S. Gordon, *Mormon Question*, 155.

118. Quoted in S. Gordon, *Mormon Question*, 204. On the relationship between the fear of Mormon polygamy and separatism and late nineteenth-century anxieties about Native sociality and sovereignty, see Rifkin, *When Did Indians*, 163–73.

119. This conception of forms of marriage and governance as interdependent illustrations of relative evolutionary advancement, including the denunciation of polygamous backwardness, was a hallmark of late nineteenth-century ethnology, most prominently in the work of Lewis Henry Morgan. See Ben-Zvi, "Where Did Red Go?"; Rifkin, *When Did Indians*, 163–73; Trautmann, *Lewis Henry Morgan*.

120. On the ways Mormons were blamed by other non-natives for inciting the Ghost Dance and the ways Mormon leaders sought to distance themselves from it, see Coates, "Mormons." On the Book of Mormon as offering a vision of time that challenges conven-

tional teleologies of white ascendance, one in which Native peoples feature centrally, see Hickman, "*Book of Mormon*."

121. As the novel's numerous references to Jesus's role in the Ghost Dance suggest, the issue here is not the introduction of Christianity per se or the idea that Native practices somehow become less authentic when Christian elements become part of them. Rather, the critique is of a particular institutionalized vision of Christian morality. For an intriguing account of the incorporation of Christian religious elements into ritual practice among the Yoeme in the Southwest, see Shorter, *We Will Dance*.

122. The contrast with white norms is made more explicit through Indigo: when she observes Edward's sister Susan committing adultery with her gardener, the narrator notes, "She knew the laws of white people" that "men and women don't touch unless they are husband and wife" (191), and when Indigo sees phallic and vaginal imagery in statutes in Europe and Edward seeks to shield her from them, "Indigo was still surprised at the sights white people didn't want children to see" (302).

123. Luciano, *Arranging Grief*.

124. Many critical accounts of the novel address this passage in some fashion. For examples, see S. Ferguson, "Europe," 44; Moore, "Ghost Dancing," 114–17; Regier, "Revolutionary Enunciatory Spaces," 146–47; Roppolo, "'We've Got,'" 85–86; Ruoff, "Leslie Marmon Silko's," 10. These discussions, though, tend to marshal the passage as evidence of the text's endorsement of hybridity or syncretism in ways that I have sought to complicate.

125. Justice, "'Go Away, Water!,'" 151.

126. Joy Porter suggests that Silko here romanticizes the conditions of sexual exploitation that Native women faced. "History in *Gardens*," 65. While acknowledging the violent conditions that often produced Native women's prostitution and that were enacted through such dynamics, this analysis overlooks the potential for an alternative understanding of the relation between eroticism and exchange, especially given that the novel suggests that Big Candy understands Sister Salt and the Chemehuevi sisters' sex work as an expression of their control over their own bodies rather than something over which he can exert control (217). In addition, all three women choose sex work over the conditions in the reservation-run laundry, which they consider more exploitative. Moreover, agents' reports from the period repeatedly cite the prevalence of prostitution among Native women off-reservation as a sign of their broader perversity, degradedness, and need to be brought into the civilizing fold of direct federal oversight. For examples, see *Annual Report* 1893, 106; 1897, 100, 104; 1898, 112.

127. On syncretism in the Ghost Dance, see J. Miller, "Basin Religion and Theology," 77–82.

128. The phrase "whole perceptual context" is from Merleau-Ponty, *Phenomenology of Perception*, 9. As Justice (2008) suggests, "kinship is best thought of as a verb rather than a noun, because kinship, in most indigenous contexts, is something that's *done* more than something that simply *is*." "'Go Away, Water!,'" 150.

129. On this matrilineal chain, see Li, "Domestic Resistance"; Magoulick, "Landscapes of Miracles"; Miranda, "Gynostemic Revolution"; Roppolo, "'We've Got.'"

130. Miranda, "Gynostemic Revolution," 142.

131. I draw the notion of necropolitics, as well as the notion of death-world, from Achille Mbembe's "Necropolitics." As noted earlier, many different kinds of prophetic visions

emerged out of the movement begun by Wovoka, some more oriented toward non-native death/disappearance than others. While the Ghost Dance movement in the Arizona-California border region arose out of relations with Southern Paiutes, it is not clear how much of Wovoka's original vision defined the terms of the movement there.

132. For examples, see Barilla, "Biological Invasion Discourse"; Miranda, "Gynostemic Revolution"; Ryan, "Nineteenth-Century Garden."

133. As A. LaVonne Brown Ruoff observes in "Leslie Marmon Silko's *Gardens in the Dunes*," "One of the major themes in the novel is Europeans' and Euro-Americans' unending desire to renovate ... in order to introduce something new, which they replace with something even newer" (12–13).

134. Mbembe, "Necropolitics," 11, 40.

135. In the Chemehuevi Valley, where Maytha and Vedna live, the floodplain produced by damming the Colorado River eliminates the "irrigated river bottom land," which was the "best land" for growing crops, as well as wiping out "all the houses and the little church" and removing access to drinking water (431). The people there "made jokes about the rising river, the government's plan to drown all the Indians.... The only good land left to them now was about to be taken away by the backwater of the dam" (433).

136. On settler colonialism as the "logic of elimination," see Wolfe, "Settler Colonialism."

137. Population losses for Colorado River peoples in the period were quite steep, a dynamic one can see in the estimates offered in Indian agents' annual reports. See also Roth, "Incorporation," 129–31, 265–67.

138. Huhndorf, *Going Native*, 191, 196.

139. This relation is also reminiscent of the continuing reactivation of oral tradition in the present as a means of understanding and engaging possibilities for the future. See Basso, *Wisdom Sits in Places*; Bierwert, *Brushed by Cedar*; Carlson, *Power of Place*; Cruikshank, *Social Life of Stories*; Nabokov, *Forest of Time*, 85–149; Shorter, *We Will Dance*; Wilson, *Remember This!*

140. From within a Deleuzian frame, elements from "the past" might be understood as part of an assemblage, virtually present as potential and materialized as actual in particular conjunctures. See DeLanda, *New Philosophy of Society*; Deleuze, *Bergsonism*; Protevi, *Political Affect*.

141. David Scott, *Omens of Adversity*, 5. Scott is addressing the aftermath of failed revolution, specifically in Grenada, and the attendant sense of feeling adrift in time.

142. As others have noted, these sites have a profound effect on Hattie, especially in terms of opening up her capacity for sensory, sensual, and spiritual experience. See Coltelli, "That Gardens of Memory"; Magoulick, "Landscapes of Miracles"; Miranda, "Gynostemic Revolution"; Moore, "Ghost Dancing"; Regier, "Revolutionary Enunciatory Spaces."

143. J. Porter, "History in *Gardens*," 59, 66; Murray, "Old Comparisons," 127. On these issues of transit and translation, see also S. Ferguson, "Europe"; Fitz, *Silko*, 191–232; Magoulick, "Landscapes of Miracles"; Miranda, "Gynostemic Revolution"; Moore, "Ghost Dancing"; Schweninger, "Claiming Europe."

144. For a theorization of such "change in continuity," see Carlson, *Power of Place*.

145. In his discussion of the relation between *Gardens* and contemporary discourses of biological invasion, James Barilla argues, "On the most literal level, Indigo's collecting would ren-

der her an ecological pariah—she introduces exotic species into a niche previously inhabited only by 'sand food,'" adding, "That these exotic species might crowd out the sand food never occurs to Indigo, and it never appears in the teachings of Grandma Fleet." "Biological Invasion Discourse," 174. On the question of preservation and restoration, and the settler conceptions of property and place that often animate it, see Cattelino, "Cultural Politics of Invasive Species."

146. Li, "Domestic Resistance," 19.

147. Dobyns and Euler, *Ghost Dance of 1889*, 18–19.

148. Goeman, *Mark My Words*, 3, 102. The dynamic I am addressing here could also be characterized as *resurgence*. See L. Simpson, *Dancing on Our Turtle's Back*.

149. The phrase "reckon[ing] with an environment" is from Merleau-Ponty, *Phenomenology of Perception*, 483.

150. As Laura Coltelli suggests, "Indigo's cultural and spiritual inspiration is fundamentally buttressed . . . by the native land she always bears within her. Yet this by no means implies sterility of vision or preconceived rejection of the new, but rather a profound knowledge, awareness and experience of 'her' land that is a manner of being and a manner of comprehending life." "That Gardens of Memory," 187. See also Barilla, "Biological Invasion Discourse," 171. However, in an interview with Ellen Arnold, Silko explicitly notes that she wanted to write about a fictionalized people who had been "completely wiped out," that she "wanted them to be gone." Arnold, "Listening to the Spirits," 163, 172. Given the novel's repeated emphasis on the possibilities for potential life, this choice seems somewhat bewildering, except that perhaps the novel offers a(n imagined) history that continues to live on despite that supposed disappearance, and in this way the text provides a model of a way in which the past might return to remake the present.

151. See Povinelli, "Governance of the Prior."

152. As in chapter 1, here I draw on while refiguring Denise Ferreira da Silva's work on affectability and race, *Toward a Global Idea of Race*.

CODA. *Deferring Juridical Time*

1. Jean Dennison, *Colonial Entanglement*, 131; Barker, "For Whom Sovereignty Matters," 3.

2. Dennison, *Colonial Entanglement*, 6–7; Barker, "For Whom Sovereignty Matters," 19.

3. Simpson, *Mohawk Interruptus*, 22, 157. As Dennison suggests, "Culture in this context becomes a burden; American Indian peoples are forced to overturn a destructive legacy of U.S. policies and reconnect to a culture damaged by the colonial process. As a colonial entanglement, American Indian culture is made to stand for all that is fundamental, pure, and noncolonized." *Colonial Entanglement*, 89.

4. Simpson, *Mohawk Interruptus*, 158.

5. Simpson, *Mohawk Interruptus*, 11.

6. Dennison, *Colonial Entanglement*, 8, 154.

7. Turner, *This Is Not*, 90, 81.

8. Turner, *This Is Not*, 106.

9. Million, *Therapeutic Nations*, 50.

10. For an excellent study that addresses this process in the everyday operation of the Hopi judiciary, see Richland, *Arguing with Tradition*.

11. Coulthard, *Red Skin, White Masks*, 46; Turner, *This Is Not*.
12. Dennison, *Colonial Entanglement*, 155.
13. Million, *Therapeutic Nations*, 121.
14. Barker, "For Whom Sovereignty Matters," 21.
15. Simpson, *Mohawk Interruptus*, 175.
16. Million, *Therapeutic Nations*, 50, 116.
17. Dennison, *Colonial Entanglement*, 91, 93.
18. Coulthard, *Red Skin, White Masks*, 65, 13.
19. For the text of the declaration, see United Nations, *Declaration*. On UNDRIP, see Charters and Stavenhagen, *Making the Declaration Work*; Engle, *Elusive Promise*; Rifkin, "(Geo)Politics of Belonging."
20. The quotation is from Million, "Felt Theory," 64.
21. Goeman, *Mark My Words*, 15.
22. Thanks to Pete Coviello for helping me think about these queer commitments.
23. Berlant, *Cruel Optimism*, 1–2.
24. Berlant, *Cruel Optimism*, 259, 262.
25. Freeman, *Time Binds*, xiii.
26. Freeman, *Time Binds*, 65.
27. Ahmed, *Queer Phenomenology*, 66, 107, 178.

BIBLIOGRAPHY

Ackerman, Lillian A. "Kinship, Family and Gender Roles." In *Handbook of North American Indians*. Vol. 12, *Plateau*, edited by Deward E. Walker Jr., 515–24. Washington, DC: Smithsonian Institution, 1998.
Adorno, Theodor W. *Negative Dialectics*. 1966. Translated by E. B. Ashton. Reprint, New York: Continuum, 1987.
Agamben, Giorgio. *State of Exception*. 2003. Translated by Kevin Attell. Chicago: University of Chicago Press, 2005.
Ahmed, Sara. *The Cultural Politics of Emotion*. New York: Routledge, 2004.
———. *Queer Phenomenology: Orientations, Objects, Others*. Durham, NC: Duke University Press, 2006.
Akiwenzie-Damm, Kateri. "Erotic, Indigenous Style." In *(Ad)dressing Our Words: Aboriginal Perspectives on Aboriginal Literatures*, edited by Armand Garnet Ruffo, 143–51. Penticton, BC: Theytus Books, 2001.
Alexie, Sherman. *Flight*. New York: Grove/Atlantic, 2007.
———. "Ghost Dance." In *The Living Dead*, edited by John Joseph Adams, 71–80. 2003. San Francisco: Night Shade Books, 2008.
———. *Indian Killer*. New York: Warner Books, 1996.
Allen, Chadwick. *Blood Narrative: Indigenous Identity in American Indian and Maori Literary and Activist Texts*. Durham, NC: Duke University Press, 2002.
Allen, Thomas M. *A Republic in Time: Temporality and Social Imagination in Nineteenth-Century America*. Chapel Hill: University of North Carolina Press, 2008.
Allewaert, Monique. *Ariel's Ecology: Plantations, Personhood, and Colonialism in the American Tropics*. Minneapolis: University of Minnesota Press, 2013.
Amoss, Pamela. *Coast Salish Spirit Dancing: The Survival of an Ancestral Religion*. Seattle: University of Washington Press, 1978.
Andersen, Chris. "From Difference to Density." *Cultural Studies Review* 15, no. 2 (2009): 80–100.
———. *"Métis": Race, Recognition, and the Struggle for Indigenous Peoplehood*. Vancouver: UBC Press, 2014.
———. "Urban Aboriginality as a Distinctive Identity, in Twelve Parts." In *Indigenous in the City: Contemporary Identities and Cultural Innovation*, edited by Evelyn Peters and Chris Andersen, 46–68. Vancouver: UBC Press, 2013.

Andersen, Holly. "The Development of the 'Specious Present' and James's Views on Temporal Experience." In *Subjective Time: The Philosophy, Psychology, and Neuroscience of Temporality*, edited by Valtteri Arstila and Dan Lloyd, 25–42. Cambridge, MA: MIT Press, 2014.

Anderson, Ben. "Modulating the Excess of Affect: Morale in a State of 'Total War.'" In *The Affect Theory Reader*, edited by Melissa Gregg and Gregory J. Seigworth, 161–85. Durham, NC: Duke University Press, 2010.

Anderson, Gary Clayton. *The Indian Southwest, 1580–1830: Ethnogenesis and Reinvention*. Norman: University of Oklahoma Press, 1999.

———. *Kinsmen of Another Kind: Dakota-White Relations in the Upper Mississippi Valley, 1650–1852*. 1984. 2nd ed. St. Paul: Minnesota Historical Society Press, 1997.

———. *Little Crow: Spokesman for the Sioux*. St. Paul: Minnesota Historical Society Press, 1986.

Anderson, Gary Clayton, and Alan R. Woolworth, eds. *Through Dakota Eyes: Narrative Accounts of the Minnesota Indian War of 1862*. St. Paul: Minnesota Historical Society Press, 1988.

*Annual Report of Office of Indian Affairs*. 31st Cong., 2nd sess., 1850, S. Exec. Doc. 1/3.

*Annual Report of Office of Indian Affairs*. 32nd Cong., 1st sess., 1851, S. Exec. Doc. 1/12.

*Annual Report of Office of Indian Affairs*. 32nd Cong., 2nd sess., 1852, S. Exec. Doc. 1/5.

*Annual Report of Office of Indian Affairs*. 33rd Cong., 2nd sess., 1854, S. Exec. Doc. 1/7.

*Annual Report of Office of Indian Affairs*. 34th Cong., 1st sess., 1855, S. Exec. Doc. 1/7.

*Annual Report of Office of Indian Affairs*. 34th Cong., 3rd sess., 1856, S. Exec. Doc. 5/6.

*Annual Report of the Commissioner of Indian Affairs*. 35th Cong., 2nd sess., 1858, S. Exec. Doc. 1/9.

*Annual Report of the Commissioner of Indian Affairs*. 36th Cong., 1st sess., 1859, S. Exec. Doc. 2/5.

*Annual Report of the Commissioner of Indian Affairs*. 36th Cong., 2nd sess., 1860, S. Exec. Doc. 1/4.

*Annual Report of the Commissioner of Indian Affairs*. 37th Cong., 2nd sess., 1861, S. Exec. Doc. 1/5.

*Annual Report of the Commissioner of Indian Affairs*. 41st Cong., 2nd sess., 1869, H.R. Exec. Doc. 1/11.

*Annual Report of the Commissioner of Indian Affairs*. 41st Cong., 3rd sess., 1870, H.R. Exec. Doc. 1/11.

*Annual Report of the Commissioner of Indian Affairs*. Washington, DC: Government Printing Office, 1881.

*Annual Report of the Commissioner of Indian Affairs*. Washington, DC: Government Printing Office, 1885.

*Annual Report of the Commissioner of Indian Affairs*. Washington, DC: Government Printing Office, 1886.

*Annual Report of the Commissioner of Indian Affairs*. Washington, DC: Government Printing Office, 1887.

*Annual Report of the Commissioner of Indian Affairs*. Washington, DC: Government Printing Office, 1890.

*Annual Report of the Commissioner of Indian Affairs*. Washington, DC: Government Printing Office, 1893.

*Annual Report of the Commissioner of Indian Affairs.* Washington, DC: Government Printing Office, 1897.

*Annual Report of the Commissioner of Indian Affairs.* Washington, DC: Government Printing Office, 1898.

*Annual Report of the Commissioner of Indian Affairs.* Washington, DC: Government Printing Office, 1899.

*Annual Report of the Commissioner of Indian Affairs.* Washington, DC: Government Printing Office, 1900.

Arista, Valtteri, and Dan Lloyd. "Subjective Time: From Past to Future." In *Subjective Time: The Philosophy, Psychology, and Neuroscience of Temporality*, edited by Valtteri Arstila and Dan Lloyd, 309–22. Cambridge, MA: MIT Press, 2014.

Armstrong, William H. *Warrior in Two Camps: Ely S. Parker, Union General and Seneca Chief.* Syracuse, NY: Syracuse University Press, 1989.

Arnold, Ellen L. "Listening to the Spirits: An Interview with Leslie Marmon Silko." In *Conversations with Leslie Marmon Silko*, edited by Ellen L. Arnold, 162–95. Jackson: University Press of Mississippi, 2000.

Asher, Brad. *Beyond the Reservation: Indians, Settlers, and the Law in Washington Territory, 1853–1889.* Norman: University of Oklahoma Press, 1999.

Bailey, Garrick A. *The Osage and the Invisible World: From the Works of Francis La Flesche.* Norman: University of Oklahoma Press, 1995.

Baker, Lee D. *Anthropology and the Racial Politics of Culture.* Durham, NC: Duke University Press, 2010.

Barilla, James. "Biological Invasion Discourse and Leslie Marmon Silko's *Gardens in the Dunes*." In *Reading Leslie Marmon Silko: Critical Perspectives through "Gardens in the Dunes,"* edited by Laura Coltelli, 165–76. Pisa, Italy: Pisa University Press, 2007.

Barker, Joanne. "For Whom Sovereignty Matters." In *Sovereignty Matters: Locations of Contestation and Possibility in Indigenous Struggles for Self-Determination*, edited by Joanne Barker, 1–32. Lincoln: University of Nebraska Press, 2005.

———. *Native Acts: Law, Recognition, and Cultural Authenticity.* Durham, NC: Duke University Press, 2011.

Barsh, Russel Lawrence. "Ethnogenesis and Ethnonationalism from Competing Treaty Claims." In *The Power of Promises: Rethinking Indian Treaties in the Pacific Northwest*, edited by Alexandra Harmon, 215–43. Seattle: University of Washington Press, 2008.

Basso, Keith. *Wisdom Sits in Places: Language and Landscape among the Western Apache.* Albuquerque: University of New Mexico Press, 1996.

Bayers, Peter L. "Charles Alexander Eastman's *From the Deep Woods to Civilization* and the Shaping of Native Manhood." *Studies in American Indian Literatures* 20, no. 3 (2008): 52–73.

Bederman, Gail. *Manliness and Civilization: A Cultural History of Gender and Race in the United States, 1880–1917.* Chicago: University of Chicago Press, 1995.

Bennett, Jane. *The Enchantment of Modern Life: Attachments, Crossings, and Ethics.* Princeton, NJ: Princeton University Press, 2001.

Ben-Zvi, Yael. "Where Did Red Go? Lewis Henry Morgan's Evolutionary Inheritance and U.S. Racial Imagination." *CR: The New Centennial Review* 7, no. 2 (2007): 201–29.

Berg, Scott W. *38 Nooses: Lincoln, Little Crow, and the Beginning of the Frontier's End*. New York: Pantheon Books, 2012.

Bergland, Renée L. *The National Uncanny: Indian Ghosts and American Subjects*. Hanover: University Press of New England, 2000.

Berglund, Jeff. "Facing the Fire: American Indian Literature and the Pedagogy of Anger." *American Indian Quarterly* 27, nos. 1–2 (2003): 80–90.

———. "'Imagination Turns Every Word into a Bottle Rocket': An Introduction to Sherman Alexie." In *Sherman Alexie: A Collection of Critical Essays*, edited by Jeff Berglund and Jan Roush, xi–xxxix. Salt Lake City: University of Utah Press, 2010.

Bergson, Henri. *Matter and Memory*. 1908. Translated by Nancy Margaret Paul and W. Scott Palmer. Masfield Centre, CT: Martino, 2011.

———. *Time and Free Will: An Essay on the Immediate Data of Consciousness*. 1889. Translated by F. L. Pogson. Mineola, NY: Dover, 2001.

Berlant, Lauren. *Cruel Optimism*. Durham, NC: Duke University Press, 2011.

Berlant, Lauren, and Michael Warner. "Sex in Public." *Critical Inquiry* 24, no. 2 (1998): 547–66.

Bierwert, Crisca. *Brushed by Cedar, Living by the River: Coast Salish Figures of Power*. Tucson: University of Arizona Press, 1999.

———. "Remembering Chief Seattle: Reversing Cultural Studies of a Vanishing Native American." *American Indian Quarterly* 22, no. 3 (1998): 280–304.

Blackhawk, Ned. *Violence over the Land: Indians and Empires in the Early American West*. Cambridge, MA: Harvard University Press, 2006.

Blaeser, Kimberly. "Wild Rice Rights: Gerald Vizenor and an Affiliation of Story." In *Centering Anishinaabeg Studies: Understanding the World through Stories*, edited by Jill Doerfler, Niigaanwewidam James Sinclair, and Heidi Kiiwetinepinesiik Stark, 237–58. East Lansing: Michigan State University Press, 2013.

Blee, Lisa. *Framing Chief Leschi: Narratives and the Politics of Historical Justice*. Chapel Hill: University of North Carolina Press, 2013.

Blight, David W. *American Oracle: The Civil War in the Civil Rights Era*. Cambridge, MA: Harvard University Press, 2011.

———. *Race and Reunion: The Civil War in American History*. Cambridge, MA: Harvard University Press, 2001.

Blu, Karen I. "'Where Do You Stay At?' Homeplace and Community among the Lumbee." In *Senses of Place*, edited by Steven Feld and Keith H. Basso, 197–228. Santa Fe: SAR Press, 1996.

Board of Commissioners (appointed by the legislature of Minnesota), ed. *Minnesota in the Civil and Indian Wars, 1861–1865*. Vol. 2, *Official Reports and Correspondence*. St. Paul: Pioneer Press, 1893.

Boxberger, Daniel L. "The Not So Common." In *Be of Good Mind: Essays on the Coast Salish*, edited by Bruce Ganville Miller, 55–81. Vancouver: UBC Press, 2007.

Boyd, Colleen E., and Coll Thrush. "Introduction: Bringing Ghosts to Ground." In *Phantom Past, Indigenous Presence: Native Ghosts in North American Culture and History*, edited by Colleen E. Boyd and Coll Thrush, vii–xl. Lincoln: University of Nebraska Press, 2011.

Bracken, Christopher. *Magical Criticism: The Recourse of Savage Philosophy*. Chicago: University of Chicago Press, 2007.

———. *The Potlatch Papers: A Colonial Case History*. Chicago: University of Chicago Press, 1997.

Brant, Beth. *Writing as Witness: Essay and Talk*. Toronto: Women's Press, 1994.

Bruyneel, Kevin. *The Third Space of Sovereignty: The Postcolonial Politics of U.S.-Indigenous Relations*. Minneapolis: University of Minnesota Press, 1997.

Burnham, Michelle. "Sherman Alexie's *Indian Killer* as Indigenous Gothic." In *Phantom Past, Indigenous Presence: Native Ghosts in North American Culture and History*, edited by Colleen E. Boyd and Coll Thrush, 3–25. Lincoln: University of Nebraska Press, 2011.

Burns, Louis F. *A History of the Osage People*. Tuscaloosa: University of Alabama Press, 2004.

———. "Lu tsa ka Le Ah ke ho, 'Can't Go Beyond': Allotting the Osage Reservation, 1906–1909." *Chronicles of Oklahoma* 72, no. 2 (1994): 200–211.

Byrd, Jodi. *The Transit of Empire: Indigenous Critiques of Colonialism*. Minneapolis: University of Minnesota Press, 2011.

Cacho, Lisa Marie. *Social Death: Racialized Rightlessness and the Criminalization of the Unprotected*. New York: New York University Press, 2012.

Cajete, Gregory. *Native Science: Natural Laws of Interdependence*. Santa Fe: Clear Light, 2000.

Callahan, Alice Anne. *The Osage Ceremonial Dance I'n-Lon-Schka*. Norman: University of Oklahoma Press, 1990.

Calloway, Colin G. *The Scratch of a Pen: 1763 and the Transformation of North America*. New York: Oxford University Press, 2006.

Canales, Jimena. *The Physicist and the Philosopher: Einstein, Bergson, and the Debate That Changed Our Understanding of Time*. Princeton, NJ: Princeton University Press, 2015.

Canku, Clifford, and Michael Simon, eds. *The Dakota Prisoner of War Letters: Dakota Kaśkapi Okicize Wowapi*. St. Paul: Minnesota Historical Society Press, 2013.

Carlson, David J. *Sovereign Selves: American Indian Autobiography and the Law*. Urbana: University of Illinois Press, 2006.

Carlson, Keith Thor. *The Power of Place, the Problem of Time: Aboriginal Identity and Historical Consciousness in the Cauldron of Colonialism*. Toronto: University of Toronto Press, 2010.

Carpenter, Cari M. *Seeing Red: Anger, Sentimentality, and American Indians*. Columbus: Ohio State University Press, 2008.

Carr, Helen. *Inventing the American Primitive: Politics, Gender and the Representation of Native American Literary Traditions, 1789–1936*. New York: New York University Press, 1996.

Carter, Julian B. *The Heart of Whiteness: Normal Sexuality and Race in America, 1880–1940*. Durham, NC: Duke University Press, 2007.

Casey, Edward S. "How to Get from Space to Place in a Fairly Short Stretch of Time: Phenomenological Prolegomena." In *Senses of Place*, edited by Steven Feld and Keith H. Basso, 13–52. Santa Fe: SAR Press, 1996.

Cattelino, Jessica. "The Cultural Politics of Invasive Species in the Florida Everglades." Paper presented at the Native American and Indigenous Studies Association meeting, Austin, TX, May 30, 2014.

———. *High Stakes: Florida Seminole Gaming and Sovereignty*. Durham, NC: Duke University Press, 2008.

Caylor, Ann. "'A Promise Long Deferred': Federal Reclamation on the Colorado River Indian Reservation." *Pacific Historical Review* 69, no. 2 (2000): 193–215.

Cervenak, Sarah Jane. *Wandering: Philosophical Performances of Racial and Sexual Freedom*. Durham, NC: Duke University Press, 2014.

Chakrabarty, Dipesh. *Provincializing Europe: Postcolonial Thought and Historical Difference*. Princeton, NJ: Princeton University Press, 2000.

Chan, Stephanie. "'Lincoln': What the Critics Are Saying." *Hollywood Reporter*, November 15, 2012.

Charters, Clare, and Rodolfo Stavenhagen, eds. *Making the Declaration Work: The United Nations Declaration on the Rights of Indigenous Peoples*. Copenhagen: IWGIA, 2009.

Cheah, Pheng. *Inhuman Conditions: On Cosmopolitanism and Human Rights*. Cambridge, MA: Harvard University Press, 2006.

Chemehuevi Indian Tribe. "History and Culture." N.d. http://www.chemehuevi.net/history-culture/.

Chen, Mel Y. *Animacies: Biopolitics, Racial Mattering, and Queer Affect*. Durham, NC: Duke University Press, 2012.

Chen, Tina. "Towards an Ethics of Knowledge." *MELUS* 30, no. 2 (2005): 157–73.

Cherokee Nation v. Georgia, 30 U.S. 1 (1831).

Cheyfitz, Eric. "The Navajo-Hopi Land Dispute: A Brief History." *Interventions* 2, no. 2 (2000): 248–75.

Chomsky, Carol. "The United States-Dakota War Trials: A Study in Military Injustice." *Stanford Law Review* 43 (1990): 13–96.

Chow, Rey. *The Protestant Ethnic and the Spirit of Capitalism*. New York: Columbia University Press, 2002.

Christie, Stuart. "Renaissance Man: The Tribal 'Schizophrenic' in Sherman Alexie's *Indian Killer*." *American Indian Culture and Research Journal* 25, no. 4 (2001): 1–19.

Clifford, James. *Returns: Becoming Indigenous in the Twenty-First Century*. Cambridge, MA: Harvard University Press, 2013.

Coates, Lawrence G. "The Mormons and the Ghost Dance." *Dialogue: A Journal of Mormon Thought* 18, no. 1 (1985): 85–111.

Cohen, Cathy J. "Punks, Bulldaggers, and Welfare Queens: The Radical Potential of Queer Politics?" *GLQ: A Journal of Lesbian and Gay Studies* 3, no. 4 (1997): 437–65.

Collins, John. *Global Palestine*. New York: Columbia University Press, 2011.

Coltelli, Laura. "That Gardens of Memory between the Old and the New World: 'All Who Are Lost Will Be Found.'" In *Reading Leslie Marmon Silko: Critical Perspectives through "Gardens in the Dunes,"* edited by Laura Coltelli, 185–205. Pisa, Italy: Pisa University Press, 2007.

Conable, Mary H. "A Steady Enemy: The Ogden Land Company and the Seneca Indians." PhD diss., University of Rochester, 1994.

Conn, Steven. *History's Shadow: Native Americans and Historical Consciousness in the Nineteenth Century*. Chicago: University of Chicago Press, 2004.

Contreras, Sheila Marie. *Blood Lines: Myth, Indigenism, and Chicana/o Literature*. Austin: University of Texas Press, 2008.

Cooper, Lydia R. "The Critique of Violent Atonement in Sherman Alexie's *Indian Killer* and David Treur's *The Hiawatha*." *Studies in American Indian Literatures* 22, no. 4 (2010): 32–57.

Cooper, Tova. "On Autobiography, Boy Scouts and Citizenship: Revisiting Charles Eastman's *Deep Woods*." *Arizona Quarterly* 65, no. 4 (2009): 1–35.

Cordova, V. F. *How It Is: The Native Philosophy of V. F. Cordova*. Edited by Kathleen Dean Moore, Kurt Peters, Ted Jojola, and Amber Lacy. Tucson: University of Arizona Press, 2007.

Coskan-Johnson, Gale P. "What Writer Would Not Be an Indian for a While? Charles Alexander Eastman, Critical Memory, and Audience." *Studies in American Indian Literature* 18, no. 2 (2006): 105–31.

Coulthard, Glen Sean. *Red Skin, White Masks: Rejecting the Colonial Politics of Recognition*. Minneapolis: University of Minnesota Press, 2014.

Cox, James H. *Muting White Noise: Native American and European American Novel Traditions*. Norman: University of Oklahoma Press, 2006.

Crenshaw, Kimberlé. "Race, Reform, and Retrenchment: Transformation and Legitimation in Anti-discrimination Law." In *Critical Race Theory: The Key Writings That Formed the Movement*, edited by Kimberlé Crenshaw, Neil Gotanda, Gary Peller, and Kendall Thomas, 103–22. New York: New Press, 1995.

Cross, Terry L., and Robert J. Miller. "The Indian Child Welfare Act of 1978 and Its Impact on Tribal Sovereignty and Governance." In *Facing the Future: The Indian Child Welfare Act at 30*, edited by Matthew L. M. Fletcher, Wenone T. Singel, and Kathryn E. Fort, 13–28. East Lansing: Michigan State University Press, 2008.

Cruikshank, Julie. *The Social Life of Stories: Narrative and Knowledge in the Yukon Territory*. Lincoln: University of Nebraska Press, 2000.

Cummings, Denise K. "'Settling' History: Understanding Leslie Marmon Silko's *Ceremony*, *Storyteller*, *Almanac of the Dead*, and *Gardens in the Dunes*." *Studies in American Indian Literatures* 12, no. 4 (2000): 65–90.

Dainton, Barry. *Time and Space*. 2nd ed. Montreal: McGill-Queen's University Press, 2010.

———. "Time, Passage, and Immediate Experience." In *The Oxford Handbook of Philosophy of Time*, edited by Craig Callender, 382–419. New York: Oxford University Press, 2011.

Danzinger, Edmund Jefferson, Jr. *Survival and Regeneration: Detroit's American Indian Community*. Detroit: Wayne State University Press, 1991.

Darnell, Regna. *And Along Came Boas: Continuity and Revolution in Americanist Anthropology*. Philadelphia: John Benjamins, 1998.

Das, Veena. *Life and Words: Violence and the Descent into the Ordinary*. Berkeley: University of California Press, 2007.

Dayan, Colin. *The Law Is a White Dog: How Legal Rituals Make and Unmake Persons*. Princeton, NJ: Princeton University Press, 2011.

Dean, Janet. "The Violence of Collection: *Indian Killer*'s Archives." *Studies in American Indian Literatures* 20, no. 3 (2008): 29–51.

Dean, Tim. "Bareback Time." In *Queer Times, Queer Becomings*, edited by E. L. McCallum and Mikko Tuhkanen, 75–100. Albany: SUNY Press, 2011.

DeLanda, Manuel. *A New Philosophy of Society: Assemblage Theory and Social Complexity*. New York: Continuum, 2006.

Deleuze, Gilles. *Bergsonism*. 1966. Translated by Hugh Tomlinson and Barbara Habberjam. New York: Zone Books, 1991.

Deloria, Philip J. *Indians in Unexpected Places*. Lawrence: University of Kansas Press, 2004.

——. *Playing Indian*. New Haven, CT: Yale University Press, 1998.

Deloria, Vine, Jr. *God Is Red: A Native View of Religion*. 1973. 3rd ed. Golden, CO: Fulcrum, 2003.

DeMallie, Raymond J. "The Lakota Ghost Dance: An Ethnohistorical Account." *Pacific Historical Review* 51, no. 4 (1982): 385–405.

Denetdale, Jennifer Nez. "Chairmen, Presidents, and Princesses: The Navajo Nation, Gender, and the Politics of Tradition." *Wicazo Sa Review* 21, no. 1 (2006): 9–28.

——. *Reclaiming Diné History: The Legacies of Navajo Chief Manuelito and Juanita*. Tucson: University of Arizona Press, 2007.

Dennison, Jean. *Colonial Entanglement: Constituting a Twenty-First-Century Osage Nation*. Chapel Hill: University of North Carolina Press, 2012.

Den Ouden, Amy E. *Beyond Conquest: Native Peoples and the Struggle for History in New England*. Lincoln: University of Nebraska Press, 2005.

Den Ouden, Amy E., and Jean M. O'Brien, eds. *Recognition, Sovereignty Struggles, and Indigenous Rights in the United States: A Sourcebook*. Chapel Hill: University of North Carolina Press, 2013.

Densmore, Christopher. *Red Jacket: Iroquois Diplomat and Orator*. Syracuse, NY: Syracuse University Press, 1999.

Department of the Interior. "The Osage People and Their Trust Property." A Field Report of the Bureau of Indian Affairs, Anadarko Area Office, Osage Agency, 1953.

Derrida, Jacques. *Specters of Marx: The State of the Debt, the Work of Mourning, and the New International*. Translated by Peggy Kamuf. New York: Routledge, 1994.

Dinshaw, Carolyn, Lee Edelman, Roderick A. Ferguson, Carla Freccero, Elizabeth Freeman, Judith (Jack) Hablerstam, Annamarie Jagose, Christopher Nealon, and Nguyen Tan Hoang. "Theorizing Queer Temporalities: A Roundtable Discussion." *GLQ: A Journal of Lesbian and Gay Studies* 13, nos. 2–3 (2007): 177–95.

Dobyns, Henry F., and Robert C. Euler. *The Ghost Dance of 1889: Among the Pai Indians of Northwestern Arizona*. Prescott, AZ: Prescott College Press, 1967.

Doerfler, Jill, Niigaanwewidam James Sinclair, and Heidi Kiiwetinepinesiik Stark. "*Bagijige*: Making an Offering." In *Centering Anishinaabeg Studies: Understanding the World through Stories*, edited by Jill Doerfler, Niigaanwewidam James Sinclair, and Heidi Kiiwetinepinesiik Stark, xv–xxvii. East Lansing: Michigan State University Press, 2013.

——, eds. *Centering Anishinaabeg Studies: Understanding the World through Stories*. East Lansing: Michigan State University Press, 2013.

Dowd, Gregory Evans. *A Spirited Resistance: The North American Indian Struggle for Unity, 1745–1815*. Baltimore: Johns Hopkins University Press, 1992.

Dred Scott v. Sandford. 60 U.S. 393 (1857).

Driskill, Qwo-Li. "Doubleweaving Two-Spirit Critiques: Building Alliances between Native and Queer Studies." *GLQ: A Journal of Lesbian and Gay Studies* 16, nos. 1–2 (2010): 69–92.

Driskill, Qwo-Li, Chris Finley, Brian Joseph Gilley, and Scott Lauria Morgensen, eds. *Queer Indigenous Studies: Critical Interventions in Theory, Politics, and Literature*. Tucson: University of Arizona Press, 2011.

Du Bois, Cora. *The 1870 Ghost Dance*. 1938. Reprint, Lincoln: University of Nebraska Press, 2007.

Duggan, Lisa. "The New Homonormativity: The Sexual Politics of Neoliberalism." In *Materializing Democracy: Toward a Revitalized Cultural Politics*, edited by Russ Castronovo and Dana D. Nelson, 175–94. Durham, NC: Duke University Press, 2002.

Eastman, Charles Alexander. *From the Deep Woods to Civilization*. 1916. Reprint, Lincoln: University of Nebraska Press, 1977.

———. *Indian Boyhood*. 1902. Reprint, New York: Dover, 1971.

———. *The Indian To-day: The Past and Future of the First American*. 1915. Reprint, Minneapolis: Filiquarian, 2014.

———. *The Soul of the Indian*. 1911. Reprint, Mineola, NY: Dover, 2003.

Elliott, Michael A. *The Culture Concept: Writing and Difference in the Age of Realism*. Minneapolis: University of Minnesota Press, 2002.

———. *Custerology: The Enduring Legacy of the Indian Wars and George Armstrong Custer*. Chicago: University of Chicago Press, 2007.

Engle, Karen. *The Elusive Promise of Indigenous Development: Rights, Culture, Strategy*. Durham, NC: Duke University Press, 2010.

Escobar, Arturo. *Territories of Difference: Place, Movements, Life, Redes*. Durham, NC: Duke University Press, 2008.

Esty, Jed. *Unseasonable Youth: Modernism, Colonialism, and the Fiction of Development*. New York: Oxford University Press, 2012.

Evans, Brad. *Before Cultures: The Ethnographic Imagination in American Literature, 1865–1920*. Chicago: University of Chicago Press, 2005.

Fabian, Johannes. *Time and the Other: How Anthropology Makes Its Object*. 1983. Reprint, New York: Columbia University Press, 2002.

Farrington, Tom. "The Ghost Dance and the Politics of Exclusion in Sherman Alexie's 'Distances.'" *Journal of American Studies* 47, no. 2 (2013): 521–40.

Feld, Steven, and Keith H. Basso, eds. *Senses of Place*. Santa Fe: SAR Press, 1996.

Fellows v. Blacksmith. 60 U.S. 366 (1857).

Fenton, William N. *The Great Law of the Longhouse: A Political History of the Iroquois Confederacy*. Norman: University of Oklahoma Press, 1998.

Ferguson, Roderick. *Aberrations in Black: Toward a Queer of Color Critique*. Minneapolis: University of Minnesota Press, 2004.

———. *The Reorder of Things: The University and Its Pedagogies of Minority Difference*. Minneapolis: University of Minnesota Press, 2012.

Ferguson, Suzanne. "Europe and the Quest for Home in James Welch's *The Heartsong of Changing Elk* and Leslie Marmon Silko's *Gardens in the Dunes*." *Studies in American Indian Literatures* 18, no. 2 (2006): 34–53.

Finkelman, Paul. "Lincoln and the Preconditions for Emancipation: The Moral Grandeur of a Bill of Lading." In *Lincoln's Proclamation: Emancipation Reconsidered*, edited by William A. Blair and Karen Fisher Younger, 13–44. Chapel Hill: University of North Carolina Press, 2009.

Fisher, Andrew H. "Reserved for Whom? Defending and Defining Treaty Rights on the Columbia River, 1880–1920." In *The Power of Promises: Rethinking Indian Treaties in the Pacific Northwest*, edited by Alexandra Harmon, 186–214. Seattle: University of Washington Press, 2008.

Fitz, Brewster E. *Silko: Writing Storyteller and Medicine Woman*. Norman: University of Oklahoma Press, 2004.

Fixico, Donald L. *The American Indian Mind in a Linear World: American Indian Studies and Traditional Knowledge*. New York: Routledge, 2003.

———. *The Urban Indian Experience in America*. Albuquerque: University of New Mexico Press, 2000.

Fogelson, Raymond D. "The Ethnohistory of Events and Nonevents." *Ethnohistory* 36, no. 2 (1989): 133–47.

Foster, Tol. "Dividing Canaan: Oklahoma Writers and the Multicultural Frontier." PhD diss., University of Wisconsin–Madison, 2006.

Freccero, Carla. *Queer/Early/Modern*. Durham, NC: Duke University Press, 2006.

Freeman, Elizabeth. *Time Binds: Queer Temporalities, Queer Histories*. Durham, NC: Duke University Press, 2010.

Friday, Chris. "Performing Treaties: The Culture and Politics of Treaty Remembrance and Celebration." In *The Power of Promises: Rethinking Indian Treaties in the Pacific Northwest*, edited by Alexandra Harmon, 157–85. Seattle: University of Washington Press, 2008.

Fritzsche, Peter. *Stranded in the Present: Modern Time and the Melancholy of History*. Cambridge, MA: Harvard University Press, 2004.

Furtwangler, Albert. *Answering Chief Seattle*. Seattle: University of Washington Press, 1997.

Galison, Peter. *Einstein's Clocks, Poincaré's Maps: Empires of Time*. New York: W. W. Norton, 2003.

Gallagher, Shaun. *How the Body Shapes the Mind*. New York: Oxford University Press, 2005.

———. "Time in Action." In *The Oxford Handbook of Philosophy of Time*, edited by Craig Callender, 420–38. New York: Oxford University Press, 2011.

Gallagher, Shaun, and Dan Zahavi. "Primal Impression and Enactive Perception." In *Subjective Time: The Philosophy, Psychology, and Neuroscience of Temporality*, edited by Valtteri Arstila and Dan Lloyd, 83–100. Cambridge, MA: MIT Press, 2014.

Ganter, Granville, ed. *The Collected Speeches of Sagoyewatha, or Red Jacket*. Syracuse, NY: Syracuse University Press, 2006.

Gaonkar, Dilip Parameshwar. "On Alternative Modernities." In *Alternative Modernities*, edited by Dilip Parameshwar Gaonkar, 1–23. Durham, NC: Duke University Press, 2001.

Garroutte, Eva Marie. *Real Indians: Identity and the Survival of Native America*. Berkeley: University of California Press, 2003.

Garroutte, Eva Marie, and Kathleen Delores Westcott. "The Story Is a Living Being: Companionship with Stories in Anishinaabeg Studies." In *Centering Anishinaabeg Studies: Understanding the World through Stories*, edited by Jill Doerfler, Niigaanwewidam James Sinclair, and Heidi Kiiwetinepinesiik Stark, 61–80. East Lansing: Michigan State University Press, 2013.

Gell, Alfred. *The Anthropology of Time: Cultural Constructions of Temporal Maps and Images*. Washington, DC: Berg, 1992.

Genetin-Pilawa, C. Joseph. *Crooked Paths to Allotment: The Fight over Federal Indian Policy after the Civil War*. Chapel Hill: University of North Carolina Press, 2012.

———. "Ely S. Parker and the Paradox of Reconstruction Politics in Indian Country." Unpublished manuscript, n.d.

Giles, James R. *The Spaces of Violence*. Tuscaloosa: University of Alabama Press, 2006.

Gillan, Jennifer. "The Hazards of Osage Fortunes: Gender and the Rhetoric of Compensation in Federal Policy and American Indian Fiction." *Arizona Quarterly* 54, no. 3 (1998): 1–25.

Gilman, Rhoda R. *Henry Hastings Sibley: Divided Heart*. St. Paul: Minnesota Historical Society Press, 2004.

Goeman, Mishuana. *Mark My Words: Native Women Mapping Our Nations*. Minneapolis: University of Minnesota Press, 2013.

Goldberg, Jonathan, and Madhavi Menon. "Queering History." *PMLA* 120, no. 5 (2005): 1608–17.

Gonzales, Christian Michael. "Cultural Colonizers: Persistence and Empire in the Indian Antiremoval Movement, 1815–1859." PhD diss., University of California, San Diego, 2010.

Goodwin, Doris Kearns. "Foreword." In *Lincoln: The Screenplay*, by Tony Kushner, ix–x. New York: Theatre Communications Group, 2012.

Gordon, Avery F. *Ghostly Matters: Haunting and the Sociological Imagination*. Minneapolis: University of Minnesota Press, 1997.

Gordon, Sarah Barringer. *The Mormon Question: Polygamy and Constitutional Conflict in Nineteenth-Century America*. Chapel Hill: University of North Carolina Press, 2002.

Gould, Janice. "Disobedience (in Language) in Texts by Lesbian Native Americans." *ARIEL: A Review of International English Literature* 25, no. 1 (1994): 32–44.

Grady, C. Jill. "Ancestors, Ethnohistorical Practice, and the Authentication of Native Place and Past." In *Phantom Past, Indigenous Presence: Native Ghosts in North American Culture and History*, edited by Colleen E. Boyd and Coll Thrush, 280–300. Lincoln: University of Nebraska Press, 2011.

Graham, Lorie M. "Reparations, Self-Determination, and the Seventh Generation." In *Facing the Future: The Indian Child Welfare Act at 30*, edited by Matthew L. M. Fletcher, Wenone T. Singel, and Kathryn E. Fort, 50–110. East Lansing: Michigan State University Press, 2008.

Grande, Sandy. *Red Pedagogy: Native American Social and Political Thought*. New York: Rowman and Littlefield, 2004.

Grassian, Daniel. *Understanding Sherman Alexie*. Columbia: University of South Carolina Press, 2005.

Graymont, Barbara. "New York State Indian Policy after the Revolution." *New York History* 57, no. 4 (1976): 438–74.

Greene, Brian. *The Fabric of the Cosmos: Space, Time, and the Texture of Reality*. New York: Vintage Books, 2004.

Grosz, Elizabeth. *Time Travels: Feminism, Nature, Power*. Durham, NC: Duke University Press, 2005.

Guelzo, Allen C. *Fateful Lightning: A New History of the Civil War and Reconstruction*. New York: Oxford University Press, 2012.

Guerlac, Suzanne. *Thinking in Time: An Introduction to Henri Bergson*. Ithaca, NY: Cornell University Press, 2006.

Guha, Ranajit. "On Some Aspects of the Historiography of Colonial India." In *Selected Subaltern Studies*, edited by Ranajit Guha and Gayatri Chakravorty Spivak, 37–44. New York: Oxford University Press, 1988.

Halberstam, J. Jack. *In a Queer Time and Place: Transgender Bodies, Subcultural Lives*. New York: New York University Press, 2005.

Hampson, Rick. "Lincoln's Bold Move Still Sparks Debate; On the 150th Anniversary of the Emancipation Proclamation, Historians and Civil Rights Activists Question President's Motives for Signing It." *USA Today*, December 26, 2012.

Hanchard, Michael. "Afro-Modernity: Temporality, Politics, and the African Diaspora." *Public Culture* 11, no. 1 (1999): 245–68.

Harmon, Alexandra. "Coast Salish History." In *Be of Good Mind: Essays on the Coast Salish*, edited by Bruce Ganville Miller, 30–54. Vancouver: UBC Press, 2007.

———. *Indians in the Making: Ethnic Relations and Indian Identities around Puget Sound*. Berkeley: University of California Press, 1998.

———, ed. *The Power of Promises: Rethinking Indian Treaties in the Pacific Northwest*. Seattle: University of Washington Press, 2008.

———. *Rich Indians: Native People and the Problem of Wealth in American History*. Chapel Hill: University of North Carolina Press, 2010.

Harris, Cheryl I. "Whiteness as Property." *Harvard Law Review* 106, no. 8 (1993): 1707–91.

Hartman, Saidiya V. *Scenes of Subjection: Terror, Slavery, and Self-Making in Nineteenth-Century America*. New York: Oxford University Press, 1997.

Hartman, Saidiya V., and Frank B. Wilderson III. "The Position of the Unthought: An Interview with Saidiya V. Hartman Conducted by Frank B. Wilderson, III." *Qui Parle* 13, no. 2 (2003): 183–201.

Hauptman, Laurence M. *Conspiracy of Interests: Iroquois Dispossession and the Rise of New York State*. Syracuse, NY: Syracuse University Press, 1999.

———. *The Iroquois in the Civil War: From Battlefield to Reservation*. Syracuse, NY: Syracuse University Press, 1993.

———. *The Tonawanda Senecas' Heroic Battle against Removal*. Albany: SUNY Press, 2011.

Heflin, Ruth J. *"I Remain Alive": The Sioux Literary Renaissance*. Syracuse, NY: Syracuse University Press, 2000.

Hegeman, Susan. *Patterns for America: Modernism and the Concept of Culture*. Princeton, NJ: Princeton University Press, 1999.

Herbert, Maeve. "Explaining the Sioux Military Commission of 1862." *Columbia Human Rights Law Review* 40 (2009): 743–98.

Hickman, Jared. "*The Book of Mormon* as Amerindian Apocalypse." *American Literature* 86, no. 3 (2014): 429–61.

Hirsch, Marianne. *The Generation of Postmemory: Writing and Visual Culture after the Holocaust*. New York: Columbia University Press, 2012.

Hittman, Michael. "The 1870 Ghost Dance at the Walker River Reservation: A Reconstruction." *Ethnohistory* 20, no. 3 (1973): 247–78.

———. *Wovoka and the Ghost Dance*. Rev. ed. Lincoln: University of Nebraska Press, 1997.

Hogan, Jackie. *Lincoln, Inc.: Selling the Sixteenth President in Contemporary America*. New York: Rowman and Littlefield, 2011.

Holland, Sharon. *The Erotic Life of Racism*. Durham, NC: Duke University Press, 2012.

———. *Raising the Dead: Readings of Death and (Black) Subjectivity*. Durham, NC: Duke University Press, 2000.

Hollrah, Patrice. "Sherman Alexie's Challenge to the Academy's Teaching of Native American Literature, Non-Native Writers, and Critics." In *Sherman Alexie: A Collection of Critical Essays*, edited by Jeff Berglund and Jan Roush, 159–85. Salt Lake City: University of Utah Press, 2010.

Holzer, Harold. *Emancipating Lincoln: The Proclamation in Text, Context, and Memory*. Cambridge, MA: Harvard University Press, 2012.

Homans, Margaret. "Adoption Narratives, Trauma, and Origins." *Narrative* 14, no. 1 (2006): 4–26.

Hoxie, Frederick E. *A Final Promise: The Campaign to Assimilate the Indians, 1880–1920*. 1984. Reprint, Cambridge: Cambridge University Press, 1992.

Huhndorf, Shari M. *Going Native: Indians in the American Cultural Imagination*. Ithaca, NY: Cornell University Press, 2001.

Hunter, Carol. "The Historical Context in John Joseph Mathews' *Sundown*." *MELUS* 9, no. 1 (1982): 61–72.

———. "The Protagonist as a Mixed-Blood in John Joseph Mathews' Novel: *Sundown*." *American Indian Quarterly* 6, nos. 3–4 (1982): 319–37.

Hyman, Colette A. *Dakota Women's Work: Creativity, Culture, and Exile*. St. Paul: Minnesota Historical Society Press, 2012.

Ingold, Tim. *Being Alive: Essays on Movement, Knowledge, and Description*. New York: Routledge, 2011.

Ismael, Jenann. "Temporal Experience." In *The Oxford Handbook of Philosophy of Time*, edited by Craig Callender, 460–82. New York: Oxford University Press, 2011.

Iversen, Joan Smyth. *The Antipolygamy Controversy in U.S. Women's Movements, 1880–1925: A Debate on the American Home*. New York: Garland, 1997.

Jackson, Deborah Davis. *Our Elders Lived It: American Indian Identity in the City*. DeKalb: Northern Illinois University Press, 2002.

Jackson, Holly. *American Blood: The Ends of the Family in American Literature, 1850–1900*. New York: Oxford University Press, 2013.

Jackson, Shona. *Creole Indigeneity: Between Myth and Nation in the Caribbean*. Minneapolis: University of Minnesota Press, 2012.

James, Meredith. "'Indians Do Not Live in Cities, They Only Reside There': Captivity and the Urban Wilderness in *Indian Killer*." In *Sherman Alexie: A Collection of Critical Essays*, edited by Jeff Berglund and Jan Roush, 171–85. Salt Lake City: University of Utah Press, 2010.

Jones, Dorothy V. *License for Empire: Colonialism by Treaty in Early America*. Chicago: University of Chicago Press, 1982.

Justice, Daniel Heath. "'Go Away, Water!' Kinship Criticism and the Decolonization Imperative." In *Reasoning Together: The Native Critics Collective*, edited by Craig S. Womack, Daniel Heath Justice, and Christopher B. Teuton, 147–68. Norman: University of Oklahoma Press, 2008.

———. "Notes toward a Theory of Anomaly." *GLQ: A Journal of Lesbian and Gay Studies* 16, nos. 1–2 (2010): 207–42.

Kalter, Susan. "Introduction." In *Twenty Thousand Mornings: An Autobiography*, by John Joseph Mathews, edited by Susan Kalter, xvii–lii. Norman: University of Oklahoma Press, 2012.

Kappler, Charles J. *Indian Affairs: Laws and Treaties*. Vols. 2–5. Washington, DC: Government Printing Office, 1929.

Kauanui, J. Kēhaulani. *Hawaiian Blood: Colonialism and the Politics of Sovereignty and Indigeneity*. Durham, NC: Duke University Press, 2008.

———. "A Sorry State: Apology Politics and Legal Fictions in the Court of the Conqueror." In *Formations of United States Colonialism*, edited by Alyosha Goldstein, 110–36. Durham, NC: Duke University Press, 2014.

Kazanjian, David. *The Colonizing Trick: National Culture and Imperial Citizenship in Early America*. Minneapolis: University of Minnesota Press, 2003.

Keeling, Kara. *The Witch's Flight: The Cinematic, the Black Femme, and the Image of Common Sense*. Durham, NC: Duke University Press, 2007.

Kehoe, Alice Beck. *The Ghost Dance: Ethnohistory and Revitalization*. 2nd ed. Long Grove, IL: Waveland, 2006.

Kelman, Ari. *A Misplaced Massacre: Struggling over the Memory of Sand Creek*. Cambridge, MA: Harvard University Press, 2013.

Kelsey, Penelope Myrtle. *Tribal Theory in Native American Literature: Dakota and Haudenosaunee Writing and Indigenous Worldviews*. Lincoln: University of Nebraska Press, 2008.

Kent, Alicia A. *African, Native, and Jewish American Literature and the Reshaping of Modernism*. New York: Palgrave, 2007.

Keresztesi, Rita. *Strangers at Home: American Ethnic Modernism between the World Wars*. Lincoln: University of Nebraska Press, 2005.

Kern, Stephen. *The Culture of Time and Space, 1880–1918*. 1983. Reprint, Cambridge, MA: Harvard University Press, 2003.

———. *The Modernist Novel: An Introduction*. New York: Cambridge University Press, 2011.

King, Thomas. *The Truth about Stories: A Native Narrative*. Minneapolis: University of Minnesota Press, 2003.

Klingle, Matthew. *Emerald City: An Environmental History of Seattle*. New Haven, CT: Yale University Press, 2007.

Klopotek, Brian. *Recognition Odysseys: Indigeneity, Race, and Federal Tribal Recognition Policy in Three Louisiana Indian Communities*. Durham, NC: Duke University Press, 2011.

Kohn, Eduardo. *How Forests Think: Toward an Anthropology beyond the Human*. Berkeley: University of California Press, 2013.

Konkle, Maureen. *Writing Indian Nations: Native Intellectuals and the Politics of Historiography, 1827–1863*. Chapel Hill: University of North Carolina Press, 2004.

Krupat, Arnold. *Red Matters: Native American Studies*. Philadelphia: University of Pennsylvania Press, 2002.

Kushner, Tony. *Lincoln: The Screenplay*. New York: Theatre Communications Group, 2012.

Kwinter, Sanford. *Architectures of Time: Toward a Theory of the Event in Modernist Culture*. Cambridge, MA: MIT Press, 2001.

Laminack, Zachary S. "Wounded Whiteness: Masculinity, Sincerity, and Settlement in Contemporary U.S. Fiction." PhD diss., University of North Carolina at Greensboro, 2015.

Landrum, Cynthia. "Shape-Shifters, Ghosts, and Residual Power: An Examination of North Plains Spiritual Beliefs, Location, Objects, and Spiritual Colonialism." In *Phantom Past, Indigenous Presence: Native Ghosts in North American Culture and History*, edited by Colleen E. Boyd and Coll Thrush, 255–79. Lincoln: University of Nebraska Press, 2011.

Landsberg, Alison. *Prosthetic Memory: The Transformation of American Remembrance in the Age of Mass Culture*. New York: Columbia University Press, 2004.

Latour, Bruno. *Reassembling the Social: An Introduction to Actor-Network Theory*. New York: Oxford University Press, 2005.

———. "A Relativistic Account of Einstein's Relativity." *Social Studies of Science* 18, no. 1 (1988): 3–44.

———. *We Have Never Been Modern*. 1991. Translated by Catherine Porter. Cambridge, MA: Harvard University Press, 1993.

Laverty, Philip. "The Ohlone/Costanona-Esselen Nation of Monterey, California: Dispossession, Federal Neglect, and the Bitter Irony of the Federal Acknowledgment Process." *Wicazo Sa Review* 18, no. 2 (2003): 41–77.

Lawrence, Bonita. *"Real" Indians and Others: Mixed-Blood Urban Native Peoples and Indigenous Nationhood*. Lincoln: University of Nebraska Press, 2004.

Lazo, Rodrigo. "Confederates in the Hispanic Attic: The Archive against Itself." In *Unsettled States: Nineteenth-Century American Literary Studies*, edited by Dana Luciano and Ivy G. Wilson, 31–54. New York: New York University Press, 2014.

Lee, Emily S., ed. *Living Alterities: Phenomenology, Embodiment, and Race*. Albany: SUNY Press, 2014.

Lee, James Kyung-Jin. *Urban Triage: Race and the Fictions of Multiculturalism*. Minneapolis: University of Minnesota Press, 2004.

Lefebvre, Henri. *The Production of Space*. 1974. Translated by Donald Nicholson-Smith. Malden, MA: Blackwell, 1991.

"Letter from the Secretary of War, Addressed to Mr. Schenck, Chairman of the Committee on Military Affairs, Transmitting a Report by Colonel Parker on Indian Affairs." 39th Cong., 2nd sess., 1866–67, H.R. Mis. Doc. 37.

Li, Stephanie. "Domestic Resistance: Gardening, Mothering, and Storytelling in Leslie Marmon Silko's *Gardens in the Dunes*." *Studies in American Indian Literatures* 21, no. 1 (2009): 18–37.

Lim, Bliss Cua. *Translating Time: Cinema, the Fantastic, and Temporal Critique*. Durham, NC: Duke University Press, 2009.

Lomawaima, K. Tsianina. "The Mutuality of Citizenship and Sovereignty: The Society of American Indians and the Battle to Inherit America." *American Indian Quarterly* 37, no. 3 (2013): 331–51.

Lopenzina, Drew. "'Good Indian': Charles Eastman and the Warrior as Civil Servant." *American Indian Quarterly* 27, nos. 3–4 (2003): 727–57.

Lowery, Malinda Maynor. *Lumbee Indians in the Jim Crow South: Race, Identity, and the Making of a Nation*. Chapel Hill: University of North Carolina Press, 2010.

Luciano, Dana. *Arranging Grief: Sacred Time and the Body in Nineteenth-Century America*. New York: New York University Press, 2007.

Lutenski, Emily. "Tribes of Men: John Joseph Mathews and Indian Internationalism." *Studies in American Indian Literatures* 24, no. 2 (2012): 39–64.
Lyons, Scott Richard. *X-Marks: Native Signatures of Assent*. Minneapolis: University of Minnesota Press, 2010.
Maddox, Lucy. *Citizen Indians: Native American Intellectuals, Race, and Reform*. Ithaca, NY: Cornell University Press, 2005.
Magoulick, Mary. "Landscapes of Miracles and Matriarchy in Silko's *Gardens in the Dunes*." In *Reading Leslie Marmon Silko: Critical Perspectives through "Gardens in the Dunes,"* edited by Laura Coltelli, 21–36. Pisa, Italy: Pisa University Press, 2007.
Manalansan, Martin F., IV. "Race, Violence, and Neoliberal Spatial Politics in the Global City." *Social Text* 84–85, nos. 3–4 (2005): 141–55.
Mandell, Daniel R. *Tribe, Race, History: Native Americans in Southern New England, 1780–1880*. Baltimore: Johns Hopkins University Press, 2008.
Marden, David L. "Anthropologists and Federal Indian Policy Prior to 1940." *The Indian Historian* 5 (winter 1972): 19–26.
Maroukis, Thomas C. "The Peyote Controversy and the Demise of the Society of American Indians." *Studies in American Indian Literatures* 25, no. 2 (2013): 159–80.
Marratto, Scott L. *The Intercorporeal Self: Merleau-Ponty on Subjectivity*. Albany: SUNY Press, 2012.
Martínez, David. *Dakota Philosopher: Charles Eastman and American Indian Thought*. Minneapolis: Minnesota Historical Society Press, 2009.
———. "Remembering the Thirty-Eight: Abraham Lincoln, the Dakota, and the U.S. War on Barbarism." *Wicazo Sa Review* 28, no. 2 (2013): 5–29.
Massey, Doreen. *For Space*. London: Sage, 2005.
Massumi, Brian. *Parables for the Virtual: Movement, Affect, Sensation*. Durham, NC: Duke University Press, 2002.
Mathews, John Joseph. *The Osages: Children of the Middle Waters*. Norman: University of Oklahoma Press, 1961.
———. *Sundown*. 1934. Reprint, Norman: University of Oklahoma Press, 1988.
———. *Talking to the Moon: Wildlife Adventures on the Plains and Prairies of Oklahoma*. 1945. Reprint, Norman: University of Oklahoma Press, 1981.
———. *Wah'Kon-Tah: The Osage and the White Man's Road*. Norman: University of Oklahoma Press, 1932.
Maudlin, Tim. *Philosophy of Physics: Space and Time*. Princeton, NJ: Princeton University Press, 2012.
Mbembe, Achille. "Necropolitics." Translated by Libby Meintjes. *Public Culture* 15, no. 1 (2003): 11–40.
McAuliffe, Dennis, Jr. *Bloodland: A Family Story of Oil, Greed, and Murder on the Osage Reservation*. San Francisco: Council Oak Books, 1999.
McCurry, Stephanie. "War, Gender, and Emancipation in the Civil War South." In *Lincoln's Proclamation: Emancipation Reconsidered*, edited by William A. Blair and Karen Fisher Younger, 120–50. Chapel Hill: University of North Carolina Press, 2009.
McFarland, Ron. "Sherman Alexie's Polemical Stories." *Studies in American Indian Literatures* 9, no. 4 (1997): 27–38.

Mehta, Uday Singh. *Liberalism and Empire: A Study in Nineteenth-Century British Thought.* Chicago: University of Chicago Press, 1999.
Melamed, Jodi. *Represent and Destroy: Rationalizing Violence in the New Racial Capitalism.* Minneapolis: University of Minnesota Press, 2011.
Merleau-Ponty, Maurice. *Phenomenology of Perception.* 1945. Translated by Colin Smith. New York: Routledge, 2002.
Mermin, N. David. *It's about Time: Understanding Einstein's Relativity.* Princeton, NJ: Princeton University Press, 2005.
"Message from the President of the United States in Answer to Resolution of the House of the 18th December Last, Respecting the Cause of the Recent Outbreaks of the Indian Tribes in the Northwest." 37th Cong., 3rd sess., 1862–63, H.R. Exec. Doc. 68.
"Message of the President of the United States in Answer to a Resolution of the Senate of the 5th Instant in Relation to the Indian Barbarities in Minnesota." 37th Cong., 3rd sess., 1862–63, S. Exec. Doc. 7.
Meyer, Roy W. *History of the Santee Sioux: United States Indian Policy on Trial.* Rev. ed. Lincoln: University of Nebraska Press, 1993.
Mignolo, Walter D. *Local Histories/Global Designs: Coloniality, Subaltern Knowledges, and Border Thinking.* Princeton, NJ: Princeton University Press, 2000.
Miller, Bruce Ganville. "Introduction." In *Be of Good Mind: Essays on the Coast Salish*, edited by Bruce Ganville Miller, 1–29. Vancouver: UBC Press, 2007.
———. *Oral History on Trial: Recognizing Aboriginal Narratives in the Courts.* Vancouver: UBC Press, 2011.
Miller, Bruce Ganville, and Daniel L. Boxberger. "Creating Chiefdoms: The Puget Sound Case." *Ethnohistory* 41, no. 2 (1994): 267–93.
Miller, Cary. "Every Dream Is a Prophecy: Rethinking Revitalization—Dreams, Prophets, and Routinized Cultural Evolution." In *Centering Anishinaabeg Studies: Understanding the World through Stories*, edited by Jill Doerfler, Niigaanwewidam James Sinclair, and Heidi Kiiwetinepinesiik Stark, 119–32. East Lansing: Michigan State University Press, 2013.
Miller, Christopher L. *Prophetic Worlds: Indians and Whites on the Columbia Plateau.* 1985. Reprint, Seattle: University of Washington Press, 2003.
Miller, Jay. "Basin Religion and Theology: A Comparative Study of Power (*Puha*)." *Journal of California and Great Basin Anthropology* 5, nos. 1–2 (1983): 66–86.
Miller, Mark Edwin. *Forgotten Tribes: Unrecognized Indians and the Federal Acknowledgment Process.* Lincoln: University of Nebraska Press, 2004.
Million, Dian. "Felt Theory: An Indigenous Feminist Approach to Affect and History." *Wicazo Sa Review* 24, no. 2 (2009): 53–76.
———. *Therapeutic Nations: Healing in an Age of Indigenous Human Rights.* Tucson: University of Arizona Press, 2013.
———. "There Is a River in Me: Theory from Life." In *Theorizing Native Studies*, edited by Audra Simpson and Andrea Smith, 31–42. Durham, NC: Duke University Press, 2014.
Miranda, Deborah A. *Bad Indians: A Tribal Memoir.* Berkeley, CA: Heyday, 2013.
———. "Dildos, Hummingbirds, and Driving Her Crazy." *Frontiers* 23, no. 2 (2002): 135–49.
———. "Extermination of the *Joyas*: Gendercide in Spanish California." *GLQ: A Journal of Lesbian and Gay Studies* 16, nos. 1–2 (2010): 253–84.

———. "A Gynostemic Revolution: Some Thoughts about Orchids, *Gardens in the Dunes*, and Indigenous Feminism at Work." In *Reading Leslie Marmon Silko: Critical Perspectives through "Gardens in the Dunes*," edited by Laura Coltelli, 133–48. Pisa, Italy: Pisa University Press, 2007.

Mitchell, Melanie. *Complexity: A Guided Tour*. New York: Oxford University Press, 2009.

Mithlo, Nancy Marie. *"Our Indian Princess": Subverting the Stereotypes*. Santa Fe: SAR Press, 2008.

Mooney, James. *The Ghost-Dance Religion and the Sioux Outbreak of 1890*. 1896. Reprint, Lincoln: University of Nebraska Press, 1991.

Moore, David L. "Ghost Dancing through History in Silko's *Gardens in the Dunes* and *Almanac of the Dead*." In *Reading Leslie Marmon Silko: Critical Perspectives through "Gardens in the Dunes*," edited by Laura Coltelli, 91–118. Pisa, Italy: Pisa University Press, 2007.

Moreton-Robinson, Aileen. "Writing Off Treaties: White Possession in the United States Critical Whiteness Studies Literature." In *Transnational Whiteness Matters*, edited by Aileen Moreton-Robinson, Maryrose Casey, and Fiona Nicoll, 81–96. New York: Rowman and Littlefield, 2008.

Morgensen, Scott L. *Spaces between Us: Queer Settler Colonialism and Indigenous Decolonization*. Minneapolis: University of Minnesota Press, 2011.

———. "White Settlers and Indigenous Solidarity: Confronting White Supremacy, Answering Decolonial Alliances." *Decolonization: Indigeneity, Education, and Society*, May 26, 2014. http://decolonization.wordpress.com/2014/05/26/white-settlers-and-indigenous-solidarity-confronting-white-supremacy-answering-decolonial-alliances.

Muñoz, José Esteban. *Cruising Utopia: The Then and There of Queer Futurity*. New York: New York University Press, 2009.

Murray, David. "Old Comparisons, New Syncretisms and *Gardens in the Dunes*." In *Reading Leslie Marmon Silko: Critical Perspectives through "Gardens in the Dunes*," edited by Laura Coltelli, 119–30. Pisa, Italy: Pisa University Press, 2007.

Musiol, Hanna. "*Sundown* and 'Liquid Modernity' in Pawhuska, Oklahoma." *Journal of American Studies* 46, no. 2 (2012): 357–73.

Nabokov, Peter. *A Forest of Time: American Indian Ways of History*. New York: Cambridge University Press, 2002.

Nealon, Christopher. *Foundlings: Lesbian and Gay Historical Emotion before Stonewall*. Durham, NC: Duke University Press, 2001.

Neely, Mark E., Jr. *Lincoln and the Triumph of the Nation: Constitutional Conflict in the American Civil War*. Chapel Hill: University of North Carolina Press, 2011.

Nett, Betty. "Historical Changes in the Osage Kinship System." *Southwestern Journal of Anthropology* 8, no. 2 (1952): 164–81.

Ngai, Sianne. *Ugly Feelings*. Cambridge, MA: Harvard University Press, 2005.

Nguyen, Mimi Thi. *The Gift of Freedom: War, Debt, and Other Refugee Passages*. Durham, NC: Duke University Press, 2012.

Nichols, David A. *Lincoln and the Indians: Civil War Policy and Politics*. 1978. Reprint, St. Paul: Minnesota Historical Society Press, 2012.

Nichols, Robert. "Contract and Usurpation: Enfranchisement and Racial Governance in Settler-Colonial Contexts." In *Theorizing Native Studies*, edited by Audra Simpson and Andrea Smith, 99–121. Durham, NC: Duke University Press, 2014.

Niezen, Ronald. *The Origins of Indigenism: Human Rights and the Politics of Identity*. Berkeley: University of California Press, 2003.

Nixon, Rob. *Slow Violence and the Environmentalism of the Poor*. Cambridge, MA: Harvard University Press, 2011.

Nunpa, Chris Mato. "Dakota Commemorative March: Thoughts and Reactions." *American Indian Quarterly* 28, nos. 1–2 (2004): 216–37.

Nyong'o, Tavia. *The Amalgamation Waltz: Race, Performance, and the Rules of Memory*. Minneapolis: University of Minnesota Press, 2009.

O'Brien, Jean. *Dispossession by Degrees: Indian Land and Identity in Natick, Massachusetts, 1650–1790*. New York: Cambridge University Press, 1997.

———. *Firsting and Lasting: Writing Indians Out of Existence in New England*. Minneapolis: University of Minnesota Press, 2010.

Ortiz, Alfonso. *The Tewa World: Space, Time, Being, and Becoming in a Pueblo Society*. Chicago: University of Chicago Press, 1969.

Osborne, Peter. *The Politics of Time: Modernity and Avant-Garde*. New York: Verso, 1995.

Ostler, Jeffrey. *The Plains Sioux and U.S. Colonialism from Lewis and Clark to Wounded Knee*. New York: Cambridge University Press, 2004.

Owens, Louis. *Mixedblood Messages: Literature, Film, Family, Place*. Norman: University of Oklahoma Press, 1998.

———. *Other Destinies: Understanding the American Indian Novel*. Norman: University of Oklahoma Press, 1992.

Palmer, Vera B. "The Devil in the Details: Controverting an American Indian Conversion Narrative." In *Theorizing Native Studies*, edited by Audra Simpson and Andrea Smith, 266–96. Durham, NC: Duke University Press, 2014.

Parker, Arthur Caswell. *The History of the Seneca Indians*. 1926. Reprint, Long Island, NY: Ira J. Friedman, 1967.

———. *The Life of General Ely S. Parker: Last Grand Sachem of the Iroquois and General Grant's Military Secretary*. Buffalo, NY: Baker, Jones, Hausauer, 1919.

Parker, Robert Dale. *The Invention of Native American Literature*. Ithaca, NY: Cornell University Press, 2003.

Patterson, Thomas C. *A Social History of Anthropology in the United States*. New York: Berg, 2001.

Peacock, John. "An Account of the Dakota-US War of 1862 as Sacred Text: Why My Dakota Elders Value Spiritual Closure over Scholarly 'Balance.'" *American Indian Culture and Research Journal* 27, no. 3 (2013): 185–206.

Peters, Evelyn, and Chris Andersen, eds. *Indigenous in the City: Contemporary Identities and Cultural Innovation*. Vancouver: UBC Press, 2013.

Peterson, Nancy J. "'If I Were Jewish, How Would I Mourn the Dead?' Holocaust and Genocide in the Work of Sherman Alexie." *MELUS* 35, no. 3 (2010): 63–84.

Pexa, Christopher J. "Transgressive Adoptions: Dakota Prisoners' Resistances to State Domination Following the 1862 U.S.-Dakota War." *Wicazo Sa Review* 30, no. 1 (2015): 29–56.

Pfister, Joel. *Individuality Incorporated: Indians and the Multicultural Modern*. Durham, NC: Duke University Press, 2004.

Phillips, Ian. "The Temporal Structure of Experience." In *Subjective Time: The Philosophy, Psychology, and Neuroscience of Temporality*, edited by Valtteri Arstila and Dan Lloyd, 139–58. Cambridge, MA: MIT Press, 2014.

Piatote, Beth H. *Domestic Subjects: Gender, Citizenship, and Law in Native American Literature*. New Haven, CT: Yale University Press, 2013.

———. "The Indian/Agent Aporia." *Studies in American Indian Literatures* 25, no. 2 (2013): 45–62.

Porter, Frank W., III. "Without Reservation: Federal Indian Policy and the Landless Tribes of Washington." In *State and Reservation: New Perspectives on Federal Indian Policy*, edited by George Pierre Castile and Robert L. Bee, 110–35. Tucson: University of Arizona Press, 1992.

Porter, Joy. "History in *Gardens in the Dunes*." In *Reading Leslie Marmon Silko: Critical Perspectives through "Gardens in the Dunes,"* edited by Laura Coltelli, 57–71. Pisa, Italy: Pisa University Press, 2007.

———. *To Be Indian: The Life of Iroquois-Seneca Arthur Caswell Parker*. Norman: University of Oklahoma Press, 2001.

Povinelli, Elizabeth A. *The Cunning of Recognition: Indigenous Alterities and the Making of Australian Multiculturalism*. Durham, NC: Duke University Press, 2002.

———. *Economies of Abandonment: Social Belonging and Endurance in Late Liberalism*. Durham, NC: Duke University Press, 2011.

———. "The Governance of the Prior." *Interventions* 13, no. 1 (2011): 13–30.

———. *Labor's Lot: The Power, History, and Culture of Aboriginal Action*. Chicago: University of Chicago Press, 1993.

Powell, Malea. "Imagining a New Indian: Listening to the Rhetoric of Survivance in Charles Eastman's *From the Deep Woods to Civilization*." *Para-doxa: Studies in World Literary Genres* 15 (2001): 211–26.

———. "The X-Blood Files: Whose Story? Whose Indian?" In *Native Authenticity: Transnational Perspectives on Native American Literary Studies*, edited by Deborah L. Madsen, 87–102. Albany: SUNY Press, 2010.

Pratt, Lloyd. *Archives of American Time: Literature and Modernity in the Nineteenth Century*. Philadelphia: University of Pennsylvania Press, 2010.

Pratt, Scott L. "Wounded Knee and the Prospect of Pluralism." *Journal of Speculative Philosophy* 19, no. 2 (2005): 150–66.

Protevi, John. *Political Affect: Connecting the Social and the Somatic*. Minneapolis: University of Minnesota Press, 2009.

Prucha, Francis Paul. *American Indian Treaties: The History of a Political Anomaly*. Berkeley: University of California Press, 1994.

———. *Documents of United States Indian Policy*. 3rd ed. Lincoln: University of Nebraska Press, 2000.

Puar, Jasbir K. *Terrorist Assemblages: Homonationalism in Queer Times*. Durham, NC: Duke University Press, 2007.

Quashie, Kevin. *The Sovereignty of Quiet: Beyond Resistance in Black Culture*. New Brunswick, NJ: Rutgers University Press, 2012.

Rabasa, José. *Without History: Subaltern Studies, the Zapatista Insurgency, and the Specter of History*. Pittsburgh: University of Pittsburgh Press, 2010.

Raheja, Michelle H. "'I Leave It with the People of the United States to Say': Autobiographical Disruption in the Personal Narratives of Black Hawk and Ely S. Parker." *American Indian Culture and Research Journal* 30, no. 1 (2006): 87–108.

———. *Reservation Reelism: Redfacing, Visual Sovereignty, and Representations of Native Americans in Film*. Lincoln: University of Nebraska Press, 2010.

Raibmon, Paige. *Authentic Indians: Episodes of Encounter from the Late-Nineteenth-Century Northwest Coast*. Durham, NC: Duke University Press, 2005.

Ramirez, Renya K. *Native Hubs: Culture, Community, and Belonging in Silicon Valley and Beyond*. Durham, NC: Duke University Press, 2007.

Rand, Jacki Thompson. *Kiowa Humanity and the Invasion of the State*. Lincoln: University of Nebraska Press, 2008.

Rappaport, Joanne. *The Politics of Memory: Native Historical Interpretation in the Colombian Andes*. Durham, NC: Duke University Press, 1998.

Reddy, Chandan. *Freedom with Violence: Race, Sexuality, and the US State*. Durham, NC: Duke University Press, 2011.

Regier, A. M. "Revolutionary Enunciatory Spaces: Ghost Dancing, Transatlantic Travel, and Modernist Arson in *Gardens in the Dunes*." *Modern Fiction Studies* 51, no. 1 (2005): 134–57.

Reilly, Alexander. "Sovereign Apologies." In *Sovereignty: Frontiers of Possibility*, edited by Julie Evans, Ann Genovese, Alexander Reilly, and Patrick Wolfe, 196–219. Honolulu: University of Hawai'i Press, 2013.

"Report of the Committee on Indian Affairs, to Whom Were Referred Sundry Petitions and Memorials from Citizens of New York, and Others, Praying That the Tonawanda Band of the Seneca Tribe of Indians May Be Exempted from the Operation of the Treaty of the 20th May, 1842." 29th Cong., 2nd sess., S. Doc. 156.

Revard, Carter. *Family Matters, Tribal Affairs*. Tucson: University of Arizona Press, 1998.

Rhea, Steven. "As Lincoln, Daniel Day-Lewis Stands Very Tall." *Philadelphia Inquirer*, November 16, 2012.

Richland, Justin. *Arguing with Tradition: The Language of Law in Hopi Tribal Court*. Chicago: University of Chicago Press, 2008.

Rifkin, Mark. *The Erotics of Sovereignty: Queer Native Writing in the Era of Self-Determination*. Minneapolis: University of Minnesota Press, 2012.

———. *Manifesting America: The Imperial Construction of U.S. National Space*. New York: Oxford University Press, 2009.

———. "On the (Geo)Politics of Belonging: Agamben and the U.N. Declaration on the Rights of Indigenous Peoples." *Settler Colonial Studies* 6, no. 4 (2016).

———. *Settler Common Sense: Queerness and Everyday Colonialism in the American Renaissance*. Minneapolis: University of Minnesota Press, 2014.

———. "Shadows of Mashantucket: William Apess and the Representation of Pequot Place." *American Literature* 84, no. 4 (2012): 691–714.

———. *When Did Indians Become Straight? Kinship, the History of Sexuality, and Native Sovereignty*. New York: Oxford University Press, 2011.

Rigsby, Bruce. "The Stevens Treaties, Indian Claims Commission Docket 264, and the Ancient One Known as Kennewick Man." In *The Power of Promises: Rethinking Indian Treaties in the Pacific Northwest*, edited by Alexandra Harmon, 244–76. Seattle: University of Washington Press, 2008.

Rockwell, Stephen J. *Indian Affairs and the Administrative State in the Nineteenth Century*. New York: Cambridge University Press, 2010.

Rohy, Valerie. *Anachronism and Its Others: Sexuality, Race, Temporality*. Albany: SUNY Press, 2009.

Rollings, Willard Hughes. *Unaffected by the Gospel: Osage Resistance to the Christian Invasion, 1673–1906: A Cultural Victory*. Albuquerque: University of New Mexico Press, 2004.

Roppolo, Kimberly. "'We've Got to Get Ourselves Back to the Garden': Indigenous Views of the Life-Death Cycle as Resistance in Leslie Marmon Silko's *Gardens in the Dunes*." In *Reading Leslie Marmon Silko: Critical Perspectives through "Gardens in the Dunes,"* edited by Laura Coltelli, 73–89. Pisa, Italy: Pisa University Press, 2007.

Roscoe, Will. *Changing Ones: Third and Fourth Genders in Native North America*. New York: St. Martin's Griffin, 1998.

Ross, John Alan. "Spokane." In *Handbook of North American Indians*. Vol. 12, *Plateau*, edited by Deward E. Walker Jr., 271–82. Washington, DC: Smithsonian Institution, 1998.

Ross, Marlon B. "Beyond the Closet as Raceless Paradigm." In *Black Queer Studies: A Critical Anthology*, edited by E. Patrick Johnson and Mae G. Henderson, 161–89. Durham, NC: Duke University Press, 2005.

Roth, George Edwin. "Incorporation and Changes in Ethnic Structure: The Chemehuevi Indians." PhD diss., Northwestern University, 1976.

Ruby, Robert H., and John A. Brown. *Dreamer-Prophets of the Columbian Plateau: Smohalla and Skolaskin*. Norman: University of Oklahoma Press, 1989.

———. *The Spokane Indians: Children of the Sun*. 2nd ed. Norman: University of Oklahoma Press, 2006.

Ruoff, A. LaVonne Brown. "John Joseph Mathews's *Talking to the Moon*: Literary and Osage Contexts." In *Multicultural Autobiography: American Lives*, edited by James Robert Payne, 1–31. Knoxville: University of Tennessee Press, 1992.

———. "Leslie Marmon Silko's *Gardens in the Dunes*: Contact Zones and Cross Currents." In *Reading Leslie Marmon Silko: Critical Perspectives through "Gardens in the Dunes,"* edited by Laura Coltelli, 7–20. Pisa, Italy: Pisa University Press, 2007.

Ruuska, Alex. "Ghost Dancing and the Iron Horse: Surviving through Tradition and Technology." *Technology and Culture* 52, no. 3 (2011): 574–97.

Ryan, Terre. "The Nineteenth-Century Garden: Imperialism, Subsistence, and Subversion in Leslie Marmon Silko's *Gardens in the Dunes*." *Studies in American Indian Literatures* 19, no. 3 (2007): 115–32.

Samuels, Ellen. *Fantasies of Identification: Disability, Gender, Race*. New York: New York University Press, 2014.

Sarris, Greg. *Keeping Slug Woman Alive: A Holistic Approach to American Indian Texts*. Berkeley: University of California Press, 1993.

———. *Mabel McKay: Weaving the Dream*. Berkeley: University of California Press, 1994.

Savitt, Steven. "Time in the Special Theory of Relativity." In *The Oxford Handbook of Philosophy of Time*, edited by Craig Callender, 546–70. New York: Oxford University Press, 2011.

Schedler, Christopher. *Border Modernism: Intercultural Readings in American Literary Modernism*. New York: Routledge, 2002.

Schleifer, Ronald. *Modernism and Time: The Logic of Abundance in Literature, Science, and Culture, 1880–1930.* New York: Cambridge University Press, 2000.

Schwartz, Barry. *Abraham Lincoln in the Post-heroic Era: History and Memory in Late-Twentieth-Century America.* Chicago: University of Chicago Press, 2008.

Schweninger, Lee. "Claiming Europe: Native American Literary Responses to the Old World." *American Indian Culture and Research Journal* 27, no. 2 (2003): 61–76.

———. *Listening to the Land: Native American Literary Responses to the Landscape.* Athens: University of Georgia Press, 2008.

Scott, A. O. "A President Engaged in a Great Civil War." *New York Times,* November 9, 2012.

Scott, Darieck. *Extravagant Abjection: Blackness, Power, and Sexuality in the African American Literary Imagination.* New York: New York University Press, 2010.

Scott, David. *Omens of Adversity: Tragedy, Time, Memory, Justice.* Durham, NC: Duke University Press, 2014.

Sedgwick, Eve Kosofsky. *Touching Feeling: Affect, Pedagogy, Performativity.* Durham, NC: Duke University Press, 2003.

Seigworth, Gregory J., and Melissa Gregg. "An Inventory of Shimmers." In *The Affect Theory Reader,* edited by Melissa Gregg and Gregory J. Seigworth, 1–25. Durham, NC: Duke University Press, 2010.

Senier, Siobhan, and Clare Barker. "Introduction." *Journal of Literary and Cultural Disability Studies* 7, no. 2 (2013): 123–40.

Seremetakis, C. Nadia. "The Memory of the Senses, Part I: Marks of the Transitory." In *The Senses Still: Perception and Memory as Material Culture in Modernity,* edited by C. Nadia Seremetakis, 1–18. Chicago: University of Chicago Press, 1994.

Sexton, Jared. *Amalgamation Schemes: Antiblackness and the Critique of Multiracialism.* Minneapolis: University of Minnesota Press, 2008.

Sharpe, Jenny. *Ghosts of Slavery: A Literary Archaeology of Black Women's Lives.* Minneapolis: University of Minnesota Press, 2002.

Shepherd, Jeffrey P. *We Are an Indian Nation: A History of the Hualapai People.* Tucson: University of Arizona Press, 2010.

Shorter, David Delgado. *We Will Dance Our Truth: Yaqui History in Yoeme Performances.* Lincoln: University of Nebraska Press, 2009.

Silko, Leslie Marmon. *Gardens in the Dunes.* New York: Simon and Schuster, 1999.

Silva, Denise Ferreira da. *Toward a Global Idea of Race.* Minneapolis: University of Minnesota Press, 2007.

Simonsen, Jane E. *Making Home Work: Domesticity and Native American Assimilation in the American West, 1860–1919.* Chapel Hill: University of North Carolina Press, 2006.

Simpson, Audra. *Mohawk Interruptus: Political Life across the Borders of Settler States.* Durham, NC: Duke University Press, 2014.

Simpson, Leanne. *Dancing on Our Turtle's Back: Stories of Nishnaabeg Re-Creation, Resurgence, and a New Emergence.* Winnipeg: Arbeiter Ring Publishing, 2011.

Slaughter, Joseph R. *Human Rights, Inc.: The World Novel, Narrative Form, and International Law.* New York: Fordham University Press, 2007.

Smith, Paul Chaat. *Everything You Know about Indians Is Wrong.* Minneapolis: University of Minnesota Press, 2009.

Smith, Paul Chaat, and Robert Warrior. *Like a Hurricane: The Indian Movement from Alcatraz to Wounded Knee.* New York: New Press, 1997.

Smithers, Gregory D. "The Soul of Unity: The Quarterly Journal of the Society of American Indians, 1913–1915." *American Indian Quarterly* 37, no. 3 (2013): 263–89.

Smoak, Gregory E. *Ghost Dances and Identity: Prophetic Religion and American Indian Ethnogenesis in the Nineteenth Century.* Berkeley: University of California Press, 2006.

Smolin, Lee. *Three Roads to Quantum Gravity.* New York: Basic Books, 2001.

———. *Time Reborn: From the Crisis in Physics to the Future of the Universe.* Boston: Houghton Mifflin Harcourt, 2013.

Snyder, Michael. "'He Certainly Didn't Want Anyone to Know That He Was Queer': Chal Windzer's Sexuality in John Joseph Mathews's *Sundown*." *Studies in American Indian Literatures* 20, no. 1 (2008): 27–54.

Society of Friends. *The Case of the Seneca Indians in the State of New York: Illustrated by Facts.* Philadelphia: Merrihew and Thompson, 1840.

Somerville, Siobhan B. *Queering the Color Line: Race and the Invention of Homosexuality in American Culture.* Durham, NC: Duke University Press, 2000.

Spivak, Gayatri Chakravorty. *A Critique of Postcolonial Reason: Toward a History of the Vanishing Present.* Cambridge, MA: Harvard University Press, 1999.

Stanley, Amy Dru. *From Bondage to Contract: Wage Labor, Marriage, and the Market in the Age of Slave Emancipation.* New York: Cambridge University Press, 1998.

Stasi, Paul. *Modernism, Imperialism, and the Historical Sense.* New York: Cambridge University Press, 2012.

Stevenson, Lisa. *Life beside Itself: Imagining Care in the Canadian Artic.* Berkeley: University of California Press, 2014.

Stewart, Omer C. *Peyote Religion: A History.* Norman: University of Oklahoma Press, 1987.

Stokes, Mason. *The Color of Sex: Whiteness, Heterosexuality, and the Fictions of White Supremacy.* Durham, NC: Duke University Press, 2001.

Strauss, Terry, and Debra Valentino. "Retribalization in Urban Indian Communities." *American Indian Culture and Research Journal* 22, no. 4 (1998): 103–15.

Strong, Pauline Turner. "What Is an Indian Family? The Indian Child Welfare Act and the Renascence of Tribal Sovereignty." *American Studies* 46, nos. 3–4 (2005): 205–31.

Sturm, Circe. *Becoming Indian: The Struggle over Cherokee Identity in the Twenty-First Century.* Santa Fe: SAR Press, 2010.

———. *Blood Politics: Race, Culture, and Identity in the Cherokee Nation of Oklahoma.* Berkeley: University of California Press, 2002.

Sullivan, Shannon. *Revealing Whiteness: The Unconscious Habits of Racial Privilege.* Bloomington: Indiana University Press, 2006.

Suttles, Wayne. *Coast Salish Essays.* Seattle: University of Washington Press, 1987.

Sweet, John Hope. *Bodies Politic: Negotiating Race in the American North, 1730–1830.* Philadelphia: University of Pennsylvania Press, 2006.

Talbot, Christine. *A Foreign Kingdom: Mormons and Polygamy in American Political Culture, 1852–1890.* Urbana: University of Illinois Press, 2013.

TallBear, Kimberly. *Native American DNA: Tribal Belonging and the False Promise of Genetic Science.* Minneapolis: University of Minnesota Press, 2013.

Tatonetti, Lisa. "Dancing That Way, Things Began to Change: The Ghost Dance as Pantribal Metaphor in Sherman Alexie's Writings." In *Sherman Alexie: A Collection of Critical Essays*, edited by Jeff Berglund and Jan Roush, 1–24. Salt Lake City: University of Utah Press, 2010.

———. "Disrupting a Story of Loss: Charles Eastman and Nicholas Black Elk Narrate Survivance." *Western American Literature* 39, no. 3 (2004): 279–311.

———. *The Queerness of Native American Literature*. Minneapolis: University of Minnesota Press, 2014.

Taylor, Diana. *The Archive and the Repertoire: Performing Cultural Memory in the Americas*. Durham, NC: Duke University Press, 2003.

Thorne, Tanis C. *The World's Richest Indian: The Scandal over Jackson Barnett's Oil Fortune*. New York: Oxford University Press, 2003.

Thornton, Thomas F. *Being and Place among the Tlingit*. Seattle: University of Washington Press, 2008.

Thrush, Coll. *Native Seattle: Histories from the Crossing-Over Place*. Seattle: University of Washington Press, 2007.

Tiro, Karim M. *The People of the Standing Stone: The Oneida Nation from the Revolution through the Era of Removal*. Amherst: University of Massachusetts Press, 2011.

Trafzer, Clifford E., and Margery Ann Beach. "Smohalla, the Washani, and Religion as a Factor in Northwestern Indian History." *American Indian Quarterly* 9, no. 3 (1985): 309–24.

Trask, Michael. *Cruising Modernism: Class and Sexuality in American Literature and Social Thought*. Ithaca, NY: Cornell University Press, 2003.

Trautmann, Thomas R. *Lewis Henry Morgan and the Invention of Kinship*. Berkeley: University of California Press, 1987.

Trennert, Robert A., Jr. *Alternative to Extinction: Federal Indian Policy and the Beginnings of the Reservation System, 1846–51*. Philadelphia: Temple University Press, 1975.

Trigg, Dylan. *The Memory of Place: A Phenomenology of the Uncanny*. Athens: Ohio University Press, 2012.

Turner, Dale. *This Is Not a Peace Pipe: Towards a Critical Indigenous Philosophy*. Toronto: University of Toronto Press, 2006.

United Nations. *United Nations Declaration on the Rights of Indigenous Peoples*. New York: United Nations, 2008. http://www.un.org/esa/socdev/unpfii/documents/DRIPS_en.pdf.

Valone, Stephen J. "William Seward, Whig Politics, and the Compromised Indian Removal Policy in New York State, 1838–1843." *New York History* 82, no. 2 (2001): 107–34.

Van Styvendale, Nancy. "The Trans/Historicity of Trauma in Jeannette Armstrong's *Slash* and Sherman Alexie's *Indian Killer*." *Studies in the Novel* 40, nos. 1–2 (2008): 203–23.

Vizenor, Gerald. *Fugitive Poses: Native American Indian Scenes of Absence and Presence*. Lincoln: University of Nebraska Press, 1998.

———. *Manifest Manners: Postindian Warriors of Survivance*. Hanover, CT: Wesleyan University Press, 1994.

Wagner, Bryan. *Disturbing the Peace: Black Culture and the Police Power after Slavery*. Cambridge, MA: Harvard University Press, 2009.

Wakefield, Sarah F. *Six Weeks in the Sioux Tepees: A Narrative of Indian Captivity*. Edited by June Namias. Norman: University of Oklahoma Press, 1997.
Wakeham, Pauline. *Taxidermic Signs: Reconstructing Aboriginality*. Minneapolis: University of Minnesota Press, 2008.
Walker, Deward E., and Helen H. Schuster. "Religious Movements." In *Handbook of North American Indians*. Vol. 12, *Plateau*, edited by Deward E. Walker Jr., 499–514. Washington, DC: Smithsonian Institution, 1998.
Walters, Anna Lee. *Talking Indian: Reflections on Survival and Writing*. Ithaca, NY: Firebrand Books, 1992.
Warren, Kenneth W. *What Was African American Literature?* Cambridge, MA: Harvard University Press, 2011.
Warrior, Robert. *The People and the Word: Reading Native Nonfiction*. Minneapolis: University of Minnesota Press, 2005.
——. *Tribal Secrets: Recovering American Indian Intellectual Traditions*. Minneapolis: University of Minnesota Press, 1995.
Waters, Anne. "Language Matters: Nondiscrete Nonbinary Dualism." In *American Indian Thought: Philosophical Essays*, edited by Anne Waters, 97–115. Malden, MA: Blackwell, 2004.
Waziyatawin. *What Does Justice Look Like? The Struggle for Liberation in Dakota Homeland*. St. Paul: Living Justice Press, 2008.
Weheliye, Alexander G. *Habeas Viscus: Racializing Assemblages, Biopolitics, and Black Feminist Theories of the Human*. Durham, NC: Duke University Press, 2014.
Weigman, Robyn. *Object Lessons*. Durham, NC: Duke University Press, 2012.
West-Pavlov, Russell. *Temporalities*. New York: Routledge, 2013.
White, Ed. *The Backcountry and the City: Colonization and Conflict in Early America*. Minneapolis: University of Minnesota Press, 2005.
Wilderson, Frank B., III. *Red, White, and Black: Cinema and the Structures of U.S. Antagonisms*. Durham, NC: Duke University Press, 2010.
Wilkinson, Norman B. "Robert Morris and the Treaty of Big Tree." In *The Rape of Indian Lands*, edited by Paul Wallace Gates, 257–78. New York: Arno, 1979.
Williams, Robert A. *Linking Arms Together: American Indian Treaty Visions of Law and Peace, 1600–1800*. New York: Oxford University Press, 1997.
Wilson, Terry P. "Osage Oxonian: The Heritage of John Joseph Mathews." *Chronicles of Oklahoma* 59, no. 3 (1981): 264–93.
——. *The Underground Reservation: Osage Oil*. Lincoln: University of Nebraska Press, 1985.
Wilson, Waziyatawin Angela. "Decolonizing the 1862 Death Marches." *American Indian Quarterly* 28, nos. 1–2 (2004): 185–215.
——. "Introduction: Manipi Hena Owas'in Wicunkiksuyapi (We Remember All Those Who Walked)." *American Indian Quarterly* 28, nos. 1–2 (2004): 151–69.
——. *Remember This! Dakota Decolonization and the Eli Taylor Narratives*. Lincoln: University of Nebraska Press, 2005.
Wingerd, Mary Lethert. *North Country: The Making of Minnesota*. Minneapolis: University of Minnesota Press, 2010.
Witt, John Fabian. *Lincoln's Code: The Laws of War in American History*. New York: Free Press, 2012.

Wolfe, Patrick. "Settler Colonialism and the Elimination of the Native." *Journal of Genocide Research* 8, no. 4 (2000): 387–409.
Womack, Craig. "Theorizing American Indian Experience." In *Reasoning Together: The Native Critics Collective*, edited by Craig S. Womack, Daniel Heath Justice, and Christopher B. Teuton, 353–410. Norman: University of Oklahoma Press, 2008.
Wright, Michelle W. *Physics of Blackness: Beyond the Middle Passage Epistemology*. Minneapolis: University of Minnesota Press, 2015.
Wynter, Sylvia. "1492: A New World View." In *Race, Discourse, and the Origin of the Americas: A New World View*, edited by Vera Lawrence Hyatt and Rex Nettleford, 5–57. Washington, DC: Smithsonian Institution Press, 1995.
Yirush, Craig. *Settlers, Liberty, and Empire: The Roots of Early American Political Theory, 1675–1775*. New York: Cambridge University Press, 2011.
Young, Robert J. C. *Colonial Desire: Hybridity in Theory, Culture, and Race*. New York: Routledge, 1995.

INDEX

abandonment, 75–76
aboriginality, 6
activism, 1, 74, 79, 87, 105, 137, 149–50, 180, 190
Agamben, Giorgio, 50
agriculture, 95
Ahmed, Sara, 2, 11, 51, 67, 73; on anger, 155; on Native figures, 60; on objects, 137; on straightness, 158–59, 191
Alexie, Sherman, xii–xiii; on Ghost Dance, 130, 132–35, 141–43, 149, 152, 229n38, 230n50; on prophecy, 133, 145–46, 177. *See also Indian Killer*
Allegany reservation, 76
Allen, Thomas M., 43
allotment: conceptualization of, 96; General Allotment Act, 164; of Osage Nation, 99–107, 113, 219n21, 221n44; policy of, 95; in *Sundown*, 99–107, 113; with time, 97–98
*Alternative Modernities* (Gaonkar), 10
anachronism, 7, 16
Andersen, Chris, 98
anger: Ahmed on, 155; in *Indian Killer*, 154–56; self-determination by, 155
*Arranging Grief* (Luciano), 37, 39, 93, 165
authenticity, 85, 205n153; in *Gardens in the Dunes*, 162; in *Indian Killer*, 140
automobile, 9–10

*Bad Indians* (Miranda, D.), 86, 193n6
Barilla, James, 238n145
Barker, Joanne, 6, 180, 184
"Basin Religion and Theology" (Miller), 159

Battle of Steptoe Butte, 153
becoming. *See* process of becoming
being-in-time, vii–viii, 1, 3–5, 26, 29, 57, 182; articulation of, 26; conditions of, 9–10, 13; forms of, 41; modes of, 4, 142; orientation of, 16–17; with policy, 88; of Seneca people, 80; settler time and, 9; as temporal sovereignty, 45; tradition in, 29–30
Bergson, Henri, 22, 97, 112
Berlant, Lauren, 190
between two worlds notion, 193n5
Bierwert, Crisca, 143
Big Foot, 129
bird (fictional character), 149–50
*The Birth of a Nation*, 57
Blaeser, Kimberley, 45
Bouck, William C., 77
Brant, Beth, 42
*Brushed by Cedar* (Bierwert), 143, 145
Bruyneel, Kevin, 5
Bureau of Indian Affairs, 149, 220n33
Byrd, Jodi, 42, 51, 63, 217n5

California, 18, 26, 28, 31–32, 35; Ghost Dance in, 129, 133–34, 168–69, 239n131
Camp McClellan, 62
capitalism, 231n60
Carlotta Lott (fictional character), 138, 140
Cattaraugus reservation, 76
Chakrabarty, Dipesh, 12, 92; on Europe time, 19
Chal Windzer (fictional character), 98–99; background of, 101; desire of, 123–25; dreaming of, 125–26, 225n81; on government,

Chal Windzer (fictional character) (*continued*) 102; name of, 218n24; queerness of, 119–21; sense of stasis, 100. *See also Sundown*

Chemehuevi reservation, 161, 170, 235n108, 235n111

Chiricahua Apache, 9

Christianity, 90, 237n121

chronobiopolitics, 37, 120–21

chronogeopolitics, 38, 42, 45, 53, 72–74, 82, 120–21, 163

citizenship, 84–85, 88, 103–4, 164, 181

civilization, xii, 52, 76, 82, 161–62; distinction of, 88–90, 92, 95, 99; of Osage people, 106, 111, 114; progress toward, 121–23, 163

Civil War, x–xi, 39, 207n33; continued envisioning of, 50; death in, 55–56; democratic form from, 54–55; as emancipation horizon, 54–59; exceptionalization in, 59–60; in *Lincoln*, 54; purpose in, 56–57; slavery as aim in, 57; violence in, 93. *See also Lincoln*

Coast Salish peoples, 133, 143; Transformer figure of, 146–47

coevalness, viii, 197n56, 198n60; inaccuracy of, 21

colonialism, xv, 2, 4, 6, 8–9, 10, 32, 50, 52, 127–28; dynamics of, 92–94, 130; as ongoing, 152, 180; with sovereignty, 43

Colorado River, 238n135

Coltelli, Laura, 239n150

Columbia Plateau peoples, 132–33, 138–39, 145–46, 152–53

commercial rareness, 168–69

competency, 103–5, 122

Confederacy, x, 56

conformity, xi

continuity, 15, 25, 31, 39–41, 44, 50; change and, ix–x, 3, 13, 46, 64, 157, 174, 189; of duration, 23–24, 84, 131; in *Lincoln*, 92–93; meaning of, 33, 67; narratives of, 34–36; of natural time, 19; of Osage people, 97–98, 109, 113, 114–15; from stories, 44

"Contract and Usurpation" (Nichols), 59

Cordova, V. F., 2, 17, 31; on reality, 25

Coulthard, Glen, 14, 69, 111, 183, 187

Cox, Jacob D., 160–61

Cox, James, 135

Crow Creek reservation, 62

cruel optimism, 190

Cruikshank, Julie, 34, 153

*The Cultural Politics of Emotion* (Ahmed), 155

culture, 3, 5, 6, 18, 180–83, 200n81, 202n105; loss of, 15, 27, 86; of material, 7–8, 26; Miranda, D., on, 26–27; of Osage Nation, 122, 128, 185–86, 222n58; stories in, 86

Curtis Act, 100

Dainton, Barry, 20–21

Dakota people, 53, 206n15, 208n43; dispossession of, 68–69; ethnic cleansing of, 68–69; futurity of, 87; Indianness of, 67; land of, 71; Lincoln, A., on, 62; oral tradition of, 72; recognition of, 68; removal of, 66–67; treaties on, 60–62, 64–65, 70–71, 210n63; violence of, 64–65, 72, 210n69. *See also* Eastman, Charles Alexander

*Dakota Philosopher* (Martínez), 85

Dakota War, x–xi, 208n39, 211n78; Eastman on, 86–87; Galbraith on, 61–62, 67, 70; as interruption, 73; Little Crow in, 61–62; outcomes of, 71–72; Parker in, 52; as shared past, 70; sovereignty in, 64; as time experience, 72; violence in, 60–66, 210n62

dance: in *Gardens in the Dunes*, 157–59, 168; I'n-Lon-Schka, 112; of Messiah, 159; in *Sundown*, 123–24, 225n80. *See also* Ghost Dance

Das, Veena, 9, 27–28; on past, 154

Dawes Act, 100

Dean, Janet, 135

death: in Civil War, 55–56; in *Indian Killer*, 105, 150–52, 232n82

death-worlds, 168–78, 237n131

*Deep Woods* (Eastman), 87, 88, 89; development in, 91–92

Deloria, Philip, 7, 8, 12, 70

democracy: from Civil War, 54–55; idea of, 56

Dennison, Jean, 107, 180, 184

development, 4, 19, 24, 30, 37, 51–52, 67, 78, 80, 89, 91–92, 130, 163; of land, 95, 120; personal, 122–23; trajectory of, 99, 168, 173

Dinshaw, Carolyn, 38

dispossession, Native land, 31; of Dakota people, 68–69; in *Indian Killer*, 139–40; as inevitable, 84; race in, 60; from treaties, 53

"Disrupting a Story of Loss" (Tatonetti), 87

Dobyns, Henry F., 174

dreaming, 125–26, 225n81

Dred Scott case, 81

duration, xiii, 3, 22, 26, 46, 155, 186, 189, 190–91; characteristics of, 97; continuity of, 23–24, 84, 131; Eastman on, 86; experiences of, 3, 23, 29, 36, 41, 186; in Indianness, 189; of land, 95–99, 108–9, 112, 113–16, 119, 124–25, 128; in natural time, 23; queer times and storied landscapes with, 33–47
Duwamish peoples, 151, 227n29

Eastman, Charles Alexander, xi, 215n139, 216n145; on advancement, 88; background on, 53; on Dakota War, 86–87; on development, 91–92; on duration, 86; on future, 90; on national participation, 91; on peoplehood, 54; recognition of, 92; on reservations, 89; stories of, 84–85; on temporal sovereignty, 88–89
*Economies of Abandonment* (Povinelli), 75
Einstein, Albert, 20
emancipation, 50; Civil War as horizon for, 54–59. *See also* Lincoln
Emancipation Proclamation, 57
emancipation sublime, 51, 55
erotohistoriography, 123
Esselen people, 15. *See also Bad Indians*
ethnic cleansing, 68–69
Euclid, 55
Euler, Robert C., 174
Euro-American historicism, 4, 12, 19, 21, 34, 92, 144, 147, 194n8; time in, 39–40
Europe, 42, 157, 171–72; colonialism in, 10; racial hierarchy in, 10–11; on time, 19
*Everything You Know about Indians Is Wrong* (Smith), vii
exceptionalization, 67; in Civil War, 59–60

Fabian, Johannes, viii, 18, 197n55
family, 159, 162–63, 165–67; of Messiah, 170–72; as nuclear, 35, 37, 95–96, 99, 120–21, 164, 166, 168, 181; queerness with, 37; stories from, 36
Ferguson, Roderick, 57
*Firsting and Lasting* (O'Brien), 5, 67
flesh, 58
Fort Snelling, 62
"For Whom Sovereignty Matters" (Barker), 180
frame of reference, ix; change in, 32–33; as homogenizing, 32; limitations to, 96–97;

for non-natives, 25; normativities of, 188–89; of Osage Nation, 100; Parker on territoriality as, 83; of settlers, 192; as shared, 43; stories for, 35, 46; in *Sundown*, 106, 110–11; for Tonawanda people, 78–79
Freccero, Carla, 41
freedom, 207n25; terminology of, 56; through Union, 55; with violence, 56
Freeman, Elizabeth, 37, 123, 190, 231n61
"From Difference to Density" (Andersen), 98
*From the Deep Woods to Civilization* (Eastman), 87
futurity: commercial rareness and, 169; of Dakota people, 87; Eastman on, 90; in *Gardens in the Dunes*, 168–70; in Ghost Dance, 168; in necropolitics, 169–70; of Osage Nation, xi; prophecy and, xii; seeds as, 173–74; as shared, 197n59; from State, 93; in *Sundown*, 115–16; treaties in, 75

Galbraith, Thomas J.: on Dakota War, 61–62, 67, 70; on treaties, 64
Galison, Peter, 20
Gaonkar, Dilip Parameshwar, 10
*Gardens in the Dunes* (Silko), xii, 186; authenticity in, 162; biological invasion in, 238n145; commercial rareness in, 168–69; dance in, 157–59, 168; fertility in, 172; futurity in, 168–70; Ghost Dance in, 130, 132–34, 156, 166–68; hybridization in, 158; identity in, 166–67; land connection in, 166–68; legal space in, 163; Messiah in, 170–71, 234n96; moral depravity in, 162–63; Mormonism in, 163–64 236n115; myth in, 234n98; Needles in, 162–63; overview of, 157; past in, 171–72; process of becoming in, 172; prophecy in, 156–57, 177; reservations in, 160–63; seeds in, 173–75; serpents in, 172–73; settler time displacement in, 176–77; sexuality in, 164–66, 237n126; violence in, 170–71; Wovoka in, 159
gender, 42, 95, 107, 134, 185, 188, 203n122
General Allotment Act, 164. *See also* Dawes Act
general relativity, 20–21; reinterpretation of, 24
Geronimo, 9
Gettysburg Address, 55

Ghost Dance, xii, 89; cross-time affiliations of, 173; 1890 failure of, 153–54; emergence of, 129, 225n1; existence from, 167–68; as futility figure, 130; futurity in, 168; in *Gardens in the Dunes*, 130, 132–34, 156, 166–68; of Hualapai people, 174–75, 235n104; in *Indian Killer*, 130, 132–35, 141–43, 145, 149, 152, 229n38, 230n50; narratives of, 130; persistence of, 233n86; rejuvenation from, 68; self-determination through, 130–31; temporalities in, 176–77; violence with, 129

"Ghost Dancing through History in Silko's *Gardens in the Dunes* and *Almanac of the Dead*" (Moore), 158

"'Go Away, Water!'" (Justice), 162

Goeman, Mishuana, 45, 46, 118, 130, 140, 161; on women's stories, 175, 189

Goldberg, Jonathan, 38

Goodwin, Doris Kearns, 54

government, xi, 1, 15, 47, 52–54, 60, 64, 66, 71, 75; imposed heteronormativity from, 38; sovereignty acknowledgment from, 6; in *Sundown*, 102. *See also* Parker, Ely S.; sovereignty

Grandma Fleet (fictional character), 157, 164

Grant, Ulysses S., x, 49

Greene, Brian, 22

grounded normativity, 187

"A Gynostemic Revolution" (Miranda, D.), 166

Halberstam, J. Jack, 37

Harrington, John P., 44

Hattie (fictional character), 166

Haudenosaunee peoples, 79. *See also* Seneca people

haunting, 232n70. *See also* Ghost Dance

healing, 175

hermeneutics, 4, 47; temporal sovereignty as, 185, 189

heteronormativity: government imposed, 38; queerness with, 37. *See also* straightness

history: actors in, 11–12; bodies in, 196n44; Euro-American, 21, 39–40, 194n8; Osage Nation in, xi; Parker in, 73–74; portrayal in, vii; straightness in, 39; in *Sundown*, 103. *See also* Euro-American historicism

home, 87, 98, 101, 109, 111, 113, 117, 138, 175–77, 187

Hualapai people, 160; Ghost Dance of, 174–75, 235n104

Hualapai reservation, 160

Huhndorf, Shari, 171

hybridization, 193n5, 204n147; in *Gardens in the Dunes*, 158; of modernity, 10

identity, 11–12, 15, 26–27, 32, 53, 60, 85–86, 135, 140, 145; cultural, 166–67; in *Gardens in the Dunes*, 166–67; in *Indian Killer*, 138–39; of Osage people, 107, 114, 122–23; from place, 176; in settler time, 177–78

"Imagining a New Indian" (Powell), 86

*Indian Boyhood* (Eastman), 87, 89

Indian Homestead Acts of 1875 and 1884, 139

*Indian Killer* (Alexie), xii, 186; activism in, 149–50; anger in, 154–56; authenticity in, 140; capitalism in, 231n60; closure in, 143–44; death in, 105, 150–52, 232n82; dispossession in, 139–40; Ghost Dance in, 130, 132–35, 141–43, 145, 149, 152, 229n38, 230n50; imagination in, 153; killer in, 144–47, 150, 152–56; nonfiction novel within, 135–36, 147; non-natives on Indianness in, 133, 135, 138, 149; non-native stories in, 137–38; overview of, 134–35; past in, 150–51; prophecy in, 145–46, 177; Puget Sound region in, 138–39; settler time displacement in, 176–77; shaped reality in, 143; time experience in, 147–49; tribal identity in, 138–39; warriors in, 152–53; Wounded Knee massacre in, 141

Indianness, 145–49, 154, 176–77; of Dakota people, 67; defining, 156–57; duration modes in, 189; in *Gardens in the Dunes*, 160, 162, 166, 168, 170; in *Indian Killer*, 133, 135, 138, 140–41; in *Lincoln*, 73; non-natives on, 133, 135, 138, 149; as past, 149; queerness and, 119–20; in *Sundown*, 103

*Indians in Unexpected Places* (Deloria), 7

*The Indian To-day* (Eastman), 89

Indigo (fictional character), 157, 237n122, 239n150; travel of, 172. *See also Gardens in the Dunes*

I'n-Lon-Schka dance, 112–13, 124, 127–28, 222n52

Jack Wilson (fictional character), 135–36, 137

Jenkins' Ferry battlefield, 55

John Smith (fictional character), 134, 140, 142–43, 228n31. *See also Indian Killer*
Johnson, Andrew, 62
John Windzer (fictional character), 101–2; death of, 105, 232n82. *See also Sundown*
Jones, Robinson, 61
juridical time, 181; deferring, 179–92; queerness and, 190–91; temporalities with, 187–88; temporal sovereignty with, 192
Justice, Daniel, 162

Keckley, Elizabeth, 55
Kushner, Tony, 49, 54

Lakota people, 145, 150. *See also* Ghost Dance; Wounded Knee Massacre
land, 202n102; of Dakota people, 71; development of, 95, 120; duration of, 95–99, 108–9, 112, 113–16, 119, 124–25, 128; in *Gardens in the Dunes*, 166–68; of Osage Nation, 218n18, 220n35; Puget Sound region recognition of, 139; Sand Lizard tribe connection to, 166; seeds and, 173–75; of Seneca people, 76; in *Sundown*, 116. *See also* dispossession, Native land; treaties
Lee, Robert E., 49
*Liberalism and Empire* (Mehta), 13
Lim, Bliss Cua, 16, 197n55
*Lincoln* (film), x, 49; as Civil War storying, 54; confidence in, 206n9; emancipation sublime in, 51, 55; Indianness in, 73; national continuity in, 92–93; in settler temporalities, 58; slavery in, 55; time in, 50; in tradition, 50. *See also* Parker, Ely S.
Lincoln, Abraham, 55; on Dakota people, 62; on slavery, 56; on Thirteenth Amendment, 57
Lincoln, Mary Todd, 55
literary modernism, 218n15
Little Crow, 208n38; in Dakota War, 61–62; intention of, 62
long house religion, 231n57
Lopenzina, Drew, 85
Luciano, Dana, 37, 39, 93, 165
Lyons, Scott Richard, 12–13, 213n100

*Manifest Manners* (Vizenor), 84, 111–12, 130, 214n124
Marie Polatkin (fictional character), 136, 141–42
market economy, 95, 100, 169, 173, 179

*Mark My Words* (Goeman), 140
Martínez, David, 85, 88
Mather, Clarence (fictional character), 136, 137, 141–42
Mathews, John Joseph, xi–xii, 97–99; allotment in *Sundown* by, 99–107, 113; *Sundown* by, 107–28. *See also* Osage Nation
*Matter and Memory* (Bergson), 23, 97
Mbembe, Achille, 169
Mdewakanton people, 61, 69, 71
Mehta, Uday Singh, 13
memory, ix, 202n111; prosthetic, 205n5
*The Memory of Place* (Trigg), 28
Menon, Madhavi, 38
Merleau-Ponty, Maurice, 27–28, 35, 201n89–201n90
Mermin, N. David, 21
Messiah dance, 168. *See also* Ghost Dance
Mignolo, Walter, 10
Miller, Jay, 159
Million, Dian, 35, 53, 183–84
Minneconjous people, 129
Miranda, Deborah, 3, 6, 14–15, 166, 193n6; on California, 31–32; on culture, 26–27; on stories, 34, 36, 86
Miranda, Tom, 14–15
missions, 27
mixed-blood Native peoples, 193n5; in *Sundown*, 122–23
modernity, 194n24, 196n47; critique of, 179; elements of, 8–10; hybridization of, 10; of Native peoples, 5–7, 15; of non-natives, 7; as normative, 13; presentness and, 9; as shared, 13; in *Sundown*, 127; of West, 10
modernization, 4, 195n34
*Mohawk Interruptus* (Simpson), 14
momentum, 32
Moore, David L., 158
Moreton-Robinson, Aileen, 60, 136
Mormonism: assault on, 164, 236n120; in *Gardens in the Dunes*, 163–64, 236n115
*Muting White Noise* (Cox, James), 135

*Native Acts* (Barker), 6
Native peoples. *See specific topics*
natural time: continuity of, 19; cultural construction with, 20; duration in, 23; as present, 18–19
Nealon, Christopher, 204n138

necropolitics, 237n131; futurity in, 169–70
*New York Times*, 54
Nichols, Robert, 59
non-natives, 4–5, 9, 13–14, 21, 23, 41–42, 46, 66–68, 73; aggression of, 79; frame of reference for, 25; on Indianness, 133, 135, 138, 149; modernity of, 7; Native peoples relationship with, vii, 7, 12, 25; Native stories from, 137–38; in present, 1
Nyong'o, Tavia, 158

O'Brien, Jean, 5, 9, 14, 67, 122
Office of Indian Affairs, 68. *See also* Bureau of Indian Affairs
Ogden Land Company, 74–75; Tonawanda struggle with, 78
Oklahoma, 102, 219n28
Olivia Smith (fictional character), 135
oral tradition, 34; of Dakota people, 72. *See also* stories
orientation, 194n11; of being-in-time, 17; in everyday, 191–92; of settlement, 4; as temporal, 2–3; to territories, 3
Osage Allotment Act, 95
Osage Nation, 217n12; allotment of, 99–107, 113, 219n21, 221n44; bloodedness of, 122; civilizedness of, 105; competency notion of, 103–5, 122; constitutionalism of, 107–8, 184; country moving notion of, 112; culture of, 185–86, 222n58; density of experience in, 128; frame of reference of, 100; futurity of, xi; in history, xi; I'n-Lon-Schka dance of, 112; land of, 218n18, 220n35; in market economy, 100; mineral estate of, 108–9; oil of, 106, 108, 109; in Oklahoma, 102; "ought" experience of, 104; peoplehood of, 98, 110–15, 121–22, 125, 128; Peyote religion of, 113, 127; settlers with, 115, 117; sociospatiality of, 97; sovereignty of, 110. *See also Sundown*

Palmer, Vera B., 64
Parker, Ely S., x–xi, 212n94; annual report of, 214n117; as commissioner of Indian Affairs, 80–81; in Dakota War, 52; in history, 73–74; as Native flesh, 58; in peacetime, 73–84; policy of, 214n120; as Seneca, 52; silence of, 49–50, 59; temporalities from, 94; on territoriality frame of reference, 83; on treaties, 81–82, 84

past, vii, 6, 86, 92–94, 101, 106, 114, 117, 123, 131; Dakota War as shared, 70; Das on, 154; distinction of, 38–39; in *Gardens in the Dunes*, 171–72; in *Indian Killer*, 150–51; Indianness as, 149; as shared, 7, 9, 70. *See also* present
peace: Parker in, 73–84; from treaties, 82
Peacock, John, 72
peoplehood, 13, 15–16, 26–27, 30–31, 36, 75, 84; dilemmas of, 53; Eastman on, 54; of Osage Nation, 98, 110–15, 121–22, 125, 128; queerness in, 189–90; sense and expression of, 180–90; in *Sundown*, 112–17; threats to, 84
people of color: alliance with, 42–43; sexuality of, xii, 42
Peyote religion, 112, 127, 222n52; legality of, 223n59
*Phenomenology of Perception* (Merleau-Ponty), 27, 201n89–201n90
Piatote, Beth, 88
Pine Ridge reservation, 53, 129
plurality: of processes of becoming, 17; of temporalities, ix; of time, 16
policy, x, xii, 38, 45–46, 52–53, 60, 66, 71–72, 76, 88, 99; being-in-time with, 88; of Parker, 214n120. *See also* allotment; treaties
political nationhood, 194n24
politics: of recognition, 14; refusal strategy in, 183
polygamy, 158, 164, 224n70, 236n118
Pope, John, 61–62, 64, 66, 70
Porter, Joy, 237n126
Povinelli, Elizabeth, 53, 75
Powell, Malea, 86, 139–40
Pratt, Lloyd, 43
Pratt, Scott L., 131
present, 200n77; modernity and, 9; natural time as, 18–19; non-natives in, 1; participation in, 18; story influence on, 36, 44; tradition and, 30
process of becoming, 3, 10, 17, 25, 30, 34, 35, 40, 76, 108, 123, 131; in *Gardens in the Dunes*, 172; in Ghost Dance, 143, 145; momentum of, 32; plurality of, 17; temporality of, 136
prophecy: Alexie on, 133, 145–46, 177; as everyday, xiii; futurity and, xii; in *Gardens in the Dunes*, 156–57, 177; in *Indian Killer*, 145–46, 177; Silko on, 133; in temporali-

ties, xii; traditions of, 146; of Wovoka, 130, 152–53, 159, 229n41. *See also* Ghost Dance

Puget Sound region: in *Indian Killer*, 138–39; Native land recognition in, 139

Quakers, 76

quasi-event, 52–53, 69; of invasion, 77–78

"Queering History" (Goldberg and Menon), 38

queerness: with family, 37; with heteronormativity, 37; Indianness and, 119–20; juridical time and, 190–91; in peoplehood, 189–90; scholarship on, 38; sexuality of, 120; in *Sundown*, 99, 118–28; terminology of, xii; in time, 37, 40–41, 159

*Queer Phenomenology* (Ahmed), 2, 67, 73, 137, 158–59

Rabasa, José, 11

race, 91–93, 99, 114, 116, 118, 158; chronobiopolitics of, 120–21; in dispossession, 60. *See also* Indianness

racial hierarchy, 10–11

racialization, 42, 51, 195n36

racism, 59

recognition, 5–6, 8, 30, 34, 67, 92, 100, 136, 138, 179, 196n35; politics of, 14; as temporal, 4–16, 39, 86

Reddy, Chandan, 56

Reggie (fictional character), 149–50

relativity, 200n79; general, 20–21, 24; special, 20–21, 198n69; of time, 20–21

*Remember This!* (Wilson, W. A.), 35

removal: of Dakota people, 66–67; of Seneca people, 79; struggle against, 80

*The Reorder of Things* (Ferguson), 57

reservations: alternatives to, 160; Eastman on, 89; in *Gardens in the Dunes*, 160–63; moral depravity outside, 162–63

Revolution, the, x

Rohy, Valerie, 38, 120

Ross, Marlon, 42, 120

Running Elk (fictional character), 103

Sand Lizard tribe: home of, 175–76; land connection of, 166; sexuality in, 164–66. *See also Gardens in the Dunes*

Savitt, Steven, 21

Scott, A. O., 54, 57

Scott, Darieck, 200n84

Scott, David, 74

Seattle, 140, 145, 149, 151. *See also Indian Killer*

Second Battle of Bull Run, 61

Sedgwick, Eve Kosofsky, 47

Seeathl (Chief Seattle), 151–52

seeds, 173–75

self-determination, 1, 3–4, 9, 14, 30, 50, 58, 85, 91, 95, 178, 187; as aberrance, 53; by anger, 155; dissmisal of, 133; as normative, 33; temporalities with, 188; through Ghost Dance, 130–31. *See also* sovereignty

Seneca people: being-in-time of, 80; land of, 76; Parker as part of, 52; removal of, 79; struggles of, 84

settlers: frame of reference of, 192; temporalities of, 58; violence of, 9, 201n101

settler simulation, 140

settler time, vii, 15, 26, 33, 44–47, 66, 84, 91, 130–31, 150; being-in-time and, 9; displacement of, 176–77; identity in, 177–78; temporality of, viii, 58

sexuality, 37, 42, 120, 204n138; connections from, 165–66; in *Gardens in the Dunes*, 164–66, 237n126; of people of color, xii, 42; in Sand Lizard tribe, 164–66. *See also* queerness; straightness

Shepherd, Jeffrey P., 160

Sibley, Henry H., 61–62, 64, 66

Silko, Leslie Marmon, xii–xiii; on Ghost Dance, 130, 132–34, 156, 166–68; on prophecy, 133. *See also Gardens in the Dunes*

Simpson, Audra, 14, 31, 181, 184

simultaneity, 28; of time, 20–22; as time physical property, 22

Sioux people, 63, 88–89. *See also* Dakota people

Sisseton people, 71

Sister Salt (fictional character), 157, 165

slavery, 51; as Civil War aim, 57; Lincoln, A., on, 56; in *Lincoln*, 55

Smith, Paul Chaat, vii, 30

Smoak, Gregory, 145

Smohalla, 146

Smolin, Lee, 24, 200n78

Snyder, Michael, 120

*The Social Life of Stories* (Cruikshank), 34

Society of Friends, 76

*The Soul of the Indian* (Eastman), 90

sovereignty: in Dakota War, 64; disappearance and, 5; government recognition of, 50; as normative, 33; of Osage Nation, 110; political, 180–82; possession of, 181; of state, 50; as state recognized, 182–83; variation in, 184–85; violence with state, 50–51. *See also* temporal sovereignty
spacetime, 96, 196n48, 198n68, 198n70–198n73
special relativity, 20–21, 198n69
speed of light, 21–22, 198n68
Spillers, Hortense, 58
Spokane reservation, 145, 150, 228n31. *See also Indian Killer*
Stark, Heidi Kiiwetinepinesiik, 46
state, 2, 6, 14, 27, 30, 34, 38, 41–42, 49, 59, 66, 69, 98; futurity from, 93; sovereignty of, 50; sovereignty recognition of, 182–83; violence organized by, 57–59; violence with sovereignty of, 50–51
Stephens, Alexander, 56
stories, 202n114; connection from, 35, 44; continuity from, 44; in culture, 86; of dislocation, 85; of Eastman, 84–85; from family, 36; for frame of reference, 35, 46; in land question, 45; Miranda, D., on, 34, 36, 86; from non-natives, 137–38; present influence of, 36, 44; from women, 175, 189
straightness: Ahmed on, 158–59, 191; in history, 39; in *Sundown*, 120–22; of time, 37, 42, 120–22, 158–59
*Sundown* (Mathews), xi, 97; allotment in, 99–107, 113; cheating in, 105; competency in, 103–5; dance in, 123–24, 225n80; frame of reference in, 106, 110–11; futurity in, 115–16; gender in, 107; government in, 102; harmony in, 110; history in, 103; Indianness in, 103; land in, 116; mineral estate in, 108–9; mixed-blood struggle in, 122–23; modernity in, 127; modes of time in, 98; oil in, 108; "ought" in, 104; peoplehood in, 112–17; political economy of, 101–2; queerness in, 99, 118–28; red in, 117–18; straightness of time in, 120–22; style of, 99; temporalities in, 105–6; Time in, 98; violence in, 104–5
Sun-on-His-Wings (fictional character), 103

Tammany, 84
Tatonetti, Lisa, 87

technology, 8
temporal drag, 40–41, 231n61
temporalities, 33, 35–36, 47, 52, 58, 84, 130–31; in Ghost Dance, 176–77; with juridical time, 187–88; of Native peoples, viii–ix; normative framework of, 26; of orientation, 2–3; from Parker, 94; plurality of, ix; of process of becoming, 136; prophecy in, xii; with self-determination, 188; of settlers, 58; of settler time, viii; sharedness of, 8; in *Sundown*, 105–6; of time, 2
temporal multiplicity, 15–16, 198n61
temporal sovereignty, x, 26, 52, 182; being-in-time as, 45; denial of, 2; Eastman on, 88–89; emergence of, 179; as hermeneutic, 185, 189; in juridical time, 192; time and, 5, 186–87
*Therapeutic Nations* (Million), 53, 184
"There Is a River in Me" (Million), 35
Thirteenth Amendment, 49; function of, 56; Lincoln, A., on, 57; ratification of, 51
Thrush, Coll, 151
time, 196n40, 200n78; allotment with, 97–98; as contingent, ix; Dakota War as experience in, 72; distinctions of, 5; enumeration of, 23; in Euro-American historicism, 39–40; Europe on, 19; experience of, 22, 24; in *Indian Killer*, 147–49; legal space and, 186; in *Lincoln*, 50; plurality of, 16; queerness in, 37, 40–41, 159; relationship with, vii–viii; relativity of, 20–21; simultaneity of, 20–22; space and, 43; straightness of, 37, 42, 120–22, 158–59; in *Sundown*, 98; temporalities of, 2; temporal sovereignty and, 5, 186–87; unity of, 18. *See also* being-in-time; juridical time; natural time; settler time; spacetime
*Time and Free Will* (Bergson), 22, 112
*Time and Space* (Dainton), 20–21
*Time Reborn* (Smolin), 24
Tonawanda people, 212n94, 213n110; frame of reference for, 78–79; harassment of, 79; Ogden Land Company struggle with, 78; treaties for, 77–79
Trade and Intercourse Acts, 70
tradition, 4; in being-in-time, 29–30; of knowledge, 30; *Lincoln* in, 50; new ideas with, 90; oral, 34, 72; perceptual, 28–29; present and, 30; of prophecy, 146
trajectory, 24

276 • INDEX

transformer figure, 146–47
*Translating Time* (Lim), 16
translation, 92; of experience, 25; reverse process of, 46
treaties: coercion in, 75; consent in, 78–79; on Dakota people, 60–62, 64–65, 70–71, 210n63; dispossession from, 53; in futurity, 75; Galbraith on, 64; for land transfers, 80–81; as legalized land theft, 60; Parker on, 81–82, 84; peace from, 82; for Tonawanda people, 77–79
Treaty of Big Tree, 75
Treaty of Buffalo Creek, 49, 74, 77, 79
Treaty of Canandaigua, 79
*Tribal Secrets* (Warrior), 110
Trigg, Dylan, 28
Tucker, John Randolph, 164
Turner, Dale, 4, 26, 75, 183; on knowledge, 30

UNDRIP. *See* United Nations Declaration on the Rights of Indigenous Peoples
union: freedom, 55; racial equality of, 57
United Nations Declaration on the Rights of Indigenous Peoples (UNDRIP), 187, 188
University of Oklahoma, 102
urban areas, 140, 150, 205n153, 229n34

Mrs. Van Wagnen (fictional character), 164
violence: in Civil War, 93; of Dakota people, 64–65, 72, 210n69; in Dakota War, 60–66, 210n62; freedom with, 56; in *Gardens in the Dunes*, 170–71; with Ghost Dance, 129; of missions, 26; of settlers, 9, 201n101; as state-organized, 58–59; with state sovereignty, 50–51; in *Sundown*, 104–5
Vizenor, Gerald, 84–85, 111–12, 130

Wahpekute people, 69, 71
Wahpeton people, 71
Walker Lake Paiute tribe, 129
Walters, Anna Lee, viii
Wanapum people, 146
Warrior, Robert, 110
warriors, 152–53; Mdewakanton as, 61
Watching Eagle (fictional character), 109, 114
Weheliye, Alexander G., 58
West, the, 10
West-Pavlov, Russell, 16, 107, 130
whiteness, 68
Wilkinson, Morton S., 66
Wilson, Waziyatawin Angela, 35, 68, 203n118
Winnebago people, 66
Womack, Craig, 153
World War I, 106
Wounded Knee massacre, xii, 53, 88; events of, 129; in *Indian Killer*, 141. *See also* Ghost Dance
Wovoka, Jack, 129; in *Gardens in the Dunes*, 159; prophecy of, 130, 152–53, 159, 229n41. *See also* Ghost Dance
Wright, Silas, 77–79
"Writing Off Treaties" (Moreton-Robinson), 136
Wynter, Sylvia, 10, 58

"The X-Blood Files" (Powell), 139–40
*X-Marks* (Lyons), 12, 213n100

Yellow Medicine, 61
Yerington Paiute tribe, 129

Zapatista movement, 221n40

www.ingramcontent.com/pod-product-compliance
Lightning Source LLC
Chambersburg PA
CBHW050209240426
43671CB00013B/2273